STATISTICAL
INFERENCE
BASED ON RANKS

STATISTICAL INFERENCE BASED ON RANKS

THOMAS P. HETTMANSPERGER

Pennsylvania State University

JOHN WILEY & SONS

New York • Chichester • Brisbane • Toronto • Singapore

Library of Congress Cataloging in Publication Data:

Hettmansperger, Thomas P., 1939–
    Statistical inference based on ranks.

    (Wiley series in probability and mathematical
statistics. Probability and mathematical statistics.
ISSN 0271-6232)
    Includes index.
    1. Experimental design.   2. Nonparametric statistics.
I. Title.   II. Series.

QA279.H48   1984   519.5   83-23519
ISBN 0-471-88474-X

Printed in the United States of America

10 9 8 7 6 5 4 3 2 1

To Ann

"Woe, Woe, bitter are the consequences of ill considered inferences."

From *A Modern Madrigal Opera: The Fable of Chicken Little.* Libretto by Alicia Carpenter and music by Gregg Smith.

Preface

The history of nonparametric methods based on ranks is rather brief, extending back roughly forty years. I would like to mention a few of the major contributions. The list is subjective and by no means exhaustive. The field continues to be an active area of research and much is owed to a few key contributions.

The systematic development and assessment of the nonparametric methods studied in this book began with the work of F. Wilcoxon in 1945 and H. B. Mann and D. R. Whitney in 1947. For the next decade, nonparametric tests for location were studied using Pitman's asymptotic efficiency to assess their local power properties. In a series of papers, J. L. Hodges and E. L. Lehmann discovered the surprising result that rank tests suffer negligible efficiency loss when compared to the t-test at the normal model and may be much more efficient at heavy-tailed models. At about this time, nonparametric tests began to gain some acceptance from data analysts and found their way into the final chapter of elementary textbooks. The first book on applied nonparametric statistics was published by S. Siegel in 1956. It was a great success, especially among behavioral scientists, and, in fact, it ranked second on the list of most-cited mathematics and statistics books, 1961–1972, with 1824 citations. Since 1970, new books on nonparametrics have appeared at the rate of roughly one per year.

In the 1960s, Hodges and Lehmann derived point estimates and confidence intervals for location parameters from rank test statistics. They further showed that the estimation methods inherit their efficiency properties from the parent test statistics. It was also found that these estimates are robust according to the new criteria proposed by J. W. Tukey, P. J. Huber, and F. Hampel for assessing stability of estimates. During this same period, J. Hajek developed a new and powerful approach to the asymptotic distribution theory needed for the construction of general rank score test statistics.

Aligned rank tests for analysis of designed experiments were introduced in the early 1960s by Hodges and Lehmann and extensively developed by

M. L. Puri and P. K. Sen. J. N. Adichie proposed and studied rank tests and the corresponding estimates for simple regression models.

In the 1970s the previous work was consolidated and extended to rank-based tests and estimates in the linear model. Much of the asymptotic distribution theory needed for the linear model derives from basic results due to J. Jureckova published just prior to 1970. Based on her work, it is possible to develop unified approaches based on ranks to the analysis of complex data sets. Hopefully, in the 1980s we will see the computer implementation and more widespread use of these efficient and robust statistical methods.

The major goal of this book is to develop a coherent and unified set of statistical methods (based on ranks) for carrying out inferences in various experimental situations. The book begins with the simple one-sample location model and progresses through the two-sample location model, the one- and two-way layouts to the general linear model. A final chapter develops methods for the multivariate location models. In all cases, testing and estimation are developed together as an interconnected set of methods for each model.

The basic tools and results from mathematical statistics are introduced as they are needed. The tools fall into two groups: tools to assess the statistical properties of the procedures, and tools to assess the stability properties. In the former case, the major tools are asymptotic relative efficiency and asymptotic local power. In the latter case, the main tools are the influence curve and the tolerance (breakdown). The stability criteria are central to the modern theory of robust statistical methods. The statistical efficiency properties are described for all the methods introduced in the book. The robustness properties are developed extensively in the one-sample location model and discussed briefly for the simple regression model. The goal is to help the student develop a working knowledge of both efficiency and robustness.

The text is organized around statistical models because this is the context in which statistical inference and data analysis are carried out. By acquiring a firm understanding of the methods and their properties in the simple models, the reader will be prepared to deal with the methods in the general linear model. We provide a rigorous development of methods based on rank sums. These methods include the Wilcoxon signed rank statistic, the Mann–Whitney–Wilcoxon statistic, the Kruskal–Wallis statistic, the Friedman statistic, and rank tests based on residuals in the linear model. The more general sums of rank scores are discussed and integrated into the discussion with references to the sources of their rigorous development. We have concentrated on rank sums for two reasons: they are the most commonly used by researchers, and their properties can be explored with the least amount of mathematical sophistication.

The linear model, which includes multiple regression and analysis of variance designs, is not generally treated systematically in texts on nonparametric statistics, applied or theoretical. This is a serious omission since most data analysis is carried out in the context of the linear model. Part of the reason that serious researchers have not used nonparametric methods more extensively is the lack of their systematic development for the linear model. The present text provides a development of these methods. Furthermore, in the near future, there will be statistical software available to implement these methods. The Minitab statistical computing system, which already includes the major nonparametric methods for the simple designs, will include a rank regression command that will provide both rank tests and estimates. Hence, the procedures developed in the text will be fully operational and can be used by researchers for analysis in complex data sets.

The book contains many exercises and problems. Major results in exercises are explicitly presented. Thus, the equations are available to the reader who does not want to take the time to derive them. An appendix of important results (without proofs) from the main body of mathematical statistics is provided. All major procedures are illustrated on data sets.

The first three chapters cover the one- and two-sample location models. Finite sample and asymptotic distribution theory is developed. Tests, point estimates, and confidence intervals are derived. Their properties are explored through asymptotic efficiency, influence curves, and tolerance (breakdown). This material can be covered in a one-semester course at the first- or second-year graduate level. The prerequisites are an introductory course in mathematical statistics and a course in advanced calculus.

Most of the robustness material is located at the end of sections. If more statistical inference is desired, then by skipping the robustness material, topics in the one- and two-way layout designs (Chapter 4) can be covered or the multivariate versions of the one- and two-sample univariate tests (Chapter 6) can be covered. The rank methods in the linear model (Chapter 5) can be developed in a follow-up seminar. The material in Chapter 5 requires a deeper background in statistics. The reader should have prior knowledge of the linear model in matrix notation.

The book has grown out of lectures given at Penn State over the last 15 years. My interest in the subject derives from discussions and work at the University of Iowa with Bob Hogg and Tim Robertson. I thank them for many hours of stimulating conversations. Graduate students whom I have worked with over the years provided the continual challenges needed to sort out many of the ideas. I would like to especially acknowledge my appreciation to Joe McKean and Jay Aubuchon.

I would also like to thank Bill Harkness, head of the Statistics Department at Penn State, for continual support. The Office of Naval Research, which sponsored part of the research that appears in Chapter 5, is gratefully acknowledged. Bea Shube, at John Wiley, has been most helpful throughout the work on the book. Finally, thanks to the typists Jane Uhrin, Peggy Lynch, Bonnie Cain, Barbara Itinger, and especially Bonnie Henninger, who struggled mightily to transcribe my notes.

THOMAS P. HETTMANSPERGER

State College, Pennsylvania
January 1984

Contents

STATISTICAL
INFERENCE
BASED ON RANKS

CHAPTER 1

The One-Sample Location Model with an Arbitrary, Continuous Distribution

1.1. INTRODUCTION

We begin by considering observations drawn from a single population about which we wish to make a minimal number of assumptions. We will only suppose that the underlying distribution is continuous and that we wish to make statistical inferences about the location of the population.

Hence we need to define a measure of location for an arbitrary, continuous distribution. Since we make no shape assumptions on the distribution, the two most common measures, the mean and the median, will not necessarily coincide. In the case of symmetry, they do coincide and naturally locate the center of the underlying population.

The median has two advantages over the mean in the general setting. The first is that the median always exists, being approximately that point which divides the population distribution in half. The mean, on the other hand, need not exist, as in the case of a Cauchy distribution. Second, the median is very resistant to slight perturbations in the underlying distribution. Hence, if there are outliers or gross errors present in the population, they will have little influence on the median but may produce extreme changes in the population mean. Later in the chapter, we consider in greater detail the stability properties of the mean and median. For now, we will simply use the population median θ as the measure of location.

In this chapter we introduce a test, confidence interval, and point estimate of θ and investigate some of the properties of these procedures. Many of the ideas presented here are used throughout the text and the simple location model provides a nice context for their discussion. This chapter contains discussions of asymptotic approximations for the significance level and power of a test, the consistency of a test, and the derivation of estimation procedures from a test statistic.

Suppose X is a random variable with an arbitrary, continuous distribution function $F(x) = P(X \leqslant x)$, we will refer interchangably to the median of X or the median of F. The median is defined as a point θ such that

$$P(X \leqslant \theta) = P(X \geqslant \theta) \geqslant \tfrac{1}{2} \qquad (1.1.1)$$

In general, there may be many such values of θ that qualify as the median (Exercise 1.8.1). When this is the case, the ambiguity should be removed by specifying a rule for the selection of a specific median.

In this book, we assume that the median is unique and that F is absolutely continuous with a density function $f(x) = F'(x)$. We denote this class of distributions by

$$\Omega_0 = \left\{ F : F \text{ absolutely continuous and } F(0) = \tfrac{1}{2}, \text{ uniquely} \right\} \quad (1.1.2)$$

The sampling model consists of a random sample $X_1, \ldots, X_n$ of independent, identically distributed (i.i.d.) random variables, each distributed as $F(x - \theta)$, $F \in \Omega_0$. The first statistical inference problem that we consider is a test of:

$$H_0 : \theta = 0 \quad \text{versus}$$
$$\qquad\qquad\qquad\qquad\qquad\qquad\qquad\qquad (1.1.3)$$
$$H_A : \theta > 0.$$

This is the most general one-sided hypothesis that we need to consider. The test $H_0 : \theta = \theta_0$, θ_0 specified against $H_A : \theta > \theta_0$, can be reduced to (1.1.3) by noting that $Y_1 = X_1 - \theta_0, \ldots, Y_n = X_n - \theta_0$ is a sample from $F \in \Omega_0$ under H_0. Furthermore, we will usually discuss one-sided hypothesis testing because it is then generally clear how to develop the corresponding two-sided procedures. Finally note that both the null and alternative hypotheses are composite; H_0 only specifies that the sample comes from some arbitrary, absolutely continuous distribution with unique median 0, and H_A specifies an arbitrary distribution with median greater than 0.

Example 1.1.1. In the early 1950s G. V. T. Matthews and others carried out interesting experiments on bird navigation. See Matthews (1952, 1974) for an account of this research. In some of the early work with homing pigeons, the birds were trained to "home" from sizeable distances out along a specific training line. To determine whether the birds could find their way home systematically from unfamiliar points, experiments were conducted in which the birds were taken out on lines at 90° and 180° from the training line and released. One measurement of interest was the angle between the bird's line of flight as he disappeared over the horizon and the homing line. These error angles were measured between 0° and 180° so lines above or below the homing line were not distinguished. Let θ denote the population median error angle. If the birds are not homing, $\theta = 90°$. One research hypothesis specified that the birds were using the sun to navigate, and in fact $\theta < 90°$. Hence, we want to construct a test of $H_0 : \theta = 90°$ versus $H_A : \theta < 90°$. If we consider only the data on birds released on sunny days we cannot conclude that the birds are using the sun to navigate if we reject $H_0 : \theta = 90°$. We may only conclude the birds are homing. The data is given in Example 1.5.1. In Chapter 3 we introduce two-sample tests which can be used to compare birds released on sunny and cloudy days. In that case we randomly assign birds to sunny or cloudy days and compare the results.

1.2. THE SIGN TEST AND ITS DISTRIBUTION

In this section we introduce the simplest test possible for (1.1.3): the sign test. This test is one of the oldest statistical procedures, dating back to Arbuthnott's research in 1710 on whether the proportion of male births exceeded $\frac{1}{2}$ in London. Although the sign test is very simple, several of the later rank tests which we will study can be put into a sign test or counting form. For this reason we develop the properties of the sign test in detail. This will make the analysis of some aspects of the later rank tests almost routine.

Let

$$S = \#(X_i > 0), \qquad i = 1, \ldots, n$$

$$= \sum_{i=1}^{n} s(X_i) \tag{1.2.1}$$

where $s(x) = 1$ if $x > 0$ and 0 otherwise. The rule is to reject $H_0 : \theta = 0$ in favor of $H_A : \theta > 0$ if $S \geqslant k$. The critical value k is determined so that $P_{H_0}(S \geqslant k) = \alpha$, the significance level of the test. Hence we must first determine the distribution of S under H_0.

Under $H_0 : \theta = 0$, $s(X_1), \ldots, s(X_n)$ are i.i.d., each binomial with parameters 1 and $p = P(X > 0) = 1 - F(0) = 1/2$, written $B(1, 1/2)$. Hence, S is the sum of n, i.i.d. $B(1, 1/2)$ random variables and has a $B(n, 1/2)$ distribution. The critical value k is found in a binomial table.

Note that the distribution of S does not depend on which $F \in \Omega_0$ we are sampling from and hence k, the critical value, can be found without knowing F. It is in this sense that we say S is distribution free or nonparametric under $H_0 : \theta = 0$. On the other hand, under $H_A : \theta = \theta' > 0$, S still has a $B(n, p)$ distribution but now

$$p = P(X > 0) = 1 - F(-\theta') \tag{1.2.2}$$

which depends on F. Hence S is not distribution free under H_A.

Since, under either hypothesis, $S = \sum s(X_i)$ is the sum of i.i.d. $B(1, p)$ random variables with $\mathrm{Var}[s(X_i)] = p(1 - p) < \infty$ we can apply the Central Limit Theorem (Theorem A8 in the Appendix) to assert that

$$\frac{S - ES}{\sqrt{\mathrm{Var}\, S}} \tag{1.2.3}$$

has an approximate standard normal distribution, that is, a normal distribution with mean 0 and variance 1, provided $0 < p < 1$. This convergence in distribution may also be denoted by

$$\frac{S - ES}{\sqrt{\mathrm{Var}\, S}} \xrightarrow{D} Z \sim n(0, 1).$$

From (1.2.2), and since S has a $B(n, p)$ distribution, it follows that

$$ES = n\left[1 - F(-\theta)\right]$$
$$\mathrm{Var}\, S = n\left[1 - F(-\theta)\right]F(-\theta) \tag{1.2.4}$$

with $ES = n/2$ and $\mathrm{Var}\, S = n/4$, under $H_0 : \theta = 0$. These results can then be used to approximate the critical value k when a binomial table is not available.

$$\alpha = P(S \geqslant k)$$

$$= P\left(\frac{S - ES}{\sqrt{\mathrm{Var}\, S}} \geqslant \frac{k - n/2}{\sqrt{n/4}}\right)$$

$$\doteq 1 - \Phi\left(\frac{k - n/2}{\sqrt{n/4}}\right) \tag{1.2.5}$$

where $\Phi(\cdot)$ denotes the standard normal distribution function. Let Z_α denote the upper α percentile of the standard normal distribution, that is,

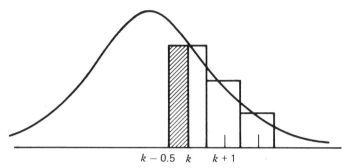

Figure 1.1.   Normal approximation to the binomial.

$\alpha = 1 - \Phi(Z_\alpha)$, then

$$\frac{k - n/2}{\sqrt{n/4}} \doteq Z_\alpha \qquad (1.2.6)$$

and $k \doteq n/2 + Z_\alpha\sqrt{n}\,/2$.

As can be seen from Fig. 1.1, the accuracy of the approximation can be improved by correcting for the discreteness of the binomial distribution. Hence (1.2.6) becomes, after subtracting 0.5 from k,

$$k \doteq n/2 + 0.5 + Z_\alpha\sqrt{n}\,/2. \qquad (1.2.7)$$

The corrected approximation to α in (1.2.5) is quite accurate for sample sizes as small as 3 or 4. This is because of the symmetry of the binomial distribution under $H_0 : \theta = 0$. Later, rank tests will be seen to have symmetric distributions under $H_0 : \theta = 0$ and again the corrected normal approximations for the significance level will be surprisingly accurate for small samples. Table 1.1 illustrates the approximation for the sign test. Sign test computations are illustrated in Example 1.5.1.

Table 1.1.   Normal Approximation to the Binomial, $p = 1/2$ and $n = 5$, with Continuity Correction

		k	
Function	0	1	3
$P(S \leqslant k)$	.03125	.1875	.5
$\Phi(\cdot)$	.0367	.1867	.5

1.3.   CONSISTENCY OF A STATISTICAL TEST

In this section we discuss the idea of consistency of a test. This corresponds to the idea of consistency or convergence in probability of an estimator and depends for its demonstration on Chebyshev's inequality (see Theorem A4 in the Appendix). Consistency of a test is an asymptotic property and thus reflects the behavior of the test for large sample sizes.

A test is consistent for some alternative when the power to detect that alternative tends to one as the sample size increases and the significance level is bounded away from zero. Any reasonable test will be consistent for some set of alternatives. We present a theorem that helps define that set.

In this book the null hypothesis is generally composite and we let Ω_{null} denote the set of distributions specified by the null hypothesis. Likewise we let Ω_{alt} denote the composite alternative. Then we are interested in testing

$$H_0 : G \in \Omega_{\text{null}} \quad \text{versus} \quad H_A : G \in \Omega_{\text{alt}} \qquad (1.3.1)$$

where G denotes the sampled distribution. We now suppose that the test statistic V_n, based on a sample size of n, satisfies

$$V_n \xrightarrow{P} \mu(G), \qquad \text{as} \quad n \to \infty, \qquad (1.3.2)$$

definition A1 in the Appendix, and that the function $\mu(\cdot)$ satisfies

$$\mu(G) = \mu_0 \qquad \forall G \in \Omega_{\text{null}}$$
$$> \mu_0 \qquad \forall G \in \Omega_c \subset \Omega_{\text{alt}}. \qquad (1.3.3)$$

Hence $\mu(G)$, the stochastic limit of the test statistic, separates the null hypothesis from a subclass of the alternative hypothesis. It is this ability of the test to distinguish the null from alternative hypotheses that determines its consistency. The class Ω_c defined by $\mu(G)$ in (1.3.3) is called the consistency class of V_n.

The nonparametric test statistics developed in this book are generally asymptotically normally distributed. Recall, for example, the sign test and (1.2.3). Hence is it possible to construct the critical value k_n in such a way that

$$\alpha_n = P_G(V_n \geq k_n) \to \alpha, \qquad \text{as} \quad n \to \infty, \quad \forall G \in \Omega_{\text{null}}. \qquad (1.3.4)$$

We will say a test that satisfies (1.3.4) is asymptotically size α.

The following theorem combines the stochastic convergence and separation in (1.3.2) and (1.3.3) with asymptotic normality to establish the asymptotic size and consistency of a test. The assumption of asymptotic normality is stronger than needed since consistency is basically a property of convergence in probability. The reader is referred to a paper by Lehmann (1951) which establishes consistency under weaker assumptions essentially requiring Chebyshev's inequality.

Theorem 1.3.1. Suppose V_n is a test statistic for (1.3.1) that rejects H_0 for large values and satisfies (1.3.2) and (1.3.3). Suppose further that there is a constant σ_0 such that

$$\sqrt{n}\,\frac{(V_n - \mu_0)}{\sigma_0} \xrightarrow{D} Z \sim n(0,1) \qquad \forall G \in \Omega_{\text{null}}. \tag{1.3.5}$$

Then there exists a sequence of critical values $\{k_n\}$ such that V_n is asymptotically size α and

$$P_G(V_n \geqslant k_n) \to 1 \qquad \forall G \in \Omega_c.$$

Proof. Let Z_α be the upper α percentile for the standard normal distribution and define

$$k_n = \mu_0 + Z_\alpha \frac{\sigma_0}{\sqrt{n}}. \tag{1.3.6}$$

Now

$$\alpha_n = P_G(V_n \geqslant k_n)$$

$$= P_G\left[\sqrt{n}\,\frac{(V_n - \mu_0)}{\sigma_0} \geqslant Z_\alpha\right],$$

$\alpha_n \to \alpha$ by (1.3.5) and V_n is asymptotically size α. Next fix $G^* \in \Omega_c$ and define

$$\epsilon = \frac{\mu(G^*) - \mu_0}{2}. \tag{1.3.7}$$

Now, by (1.3.3), $\epsilon > 0$ and for sufficiently large n, $k_n < \mu_0 + \epsilon$ because $k_n \to \mu_0$ from (1.3.6). Further, from (1.3.7), $\mu_0 = \mu(G^*) - 2\epsilon$; hence

$$k_n < \mu(G^*) - \epsilon. \tag{1.3.8}$$

Now $|V_n - \mu(G^*)| < \epsilon$ implies $V_n - \mu(G^*) > -\epsilon$, which implies $V_n > \mu(G^*) - \epsilon$, which implies $V_n \geqslant k_n$, the last implication following from

(1.3.8). Hence we have

$$P_{G^*}(|V_n - \mu(G^*)| < \epsilon) \leqslant P_{G^*}(V_n \geqslant k_n) \leqslant 1.$$

By (1.3.2), the left side tends to 1, hence $P_{G^*}(V_n \geqslant k_n) \to 1$. Since G^* was arbitrarily fixed in Ω_c the theorem is proved.

Example 1.3.1.   The consistency of the sign test.
   Let $\bar{S} = S/n$, where S is given in (1.2.1). From (1.2.4) and Chebyshev's inequality (Theorem A4),

$$\bar{S} \xrightarrow{P} \mu(F, \theta) = 1 - F(-\theta)$$

where $F \in \Omega_0$. Hence we have

$$\mu(F, \theta) = \tfrac{1}{2} \qquad \forall F \in \Omega_0, \quad \theta = 0$$

$$> \tfrac{1}{2} \qquad \forall F \in \Omega_0, \quad \theta > 0$$

and the sign test separates the null hypothesis $F \in \Omega_0$, $\theta = 0$, from the whole alternative $F \in \Omega_0$, $\theta > 0$. The consistency set for the sign test is the class of absolutely continuous distributions with unique positive median. The required asymptotic normality follows from (1.2.3) with $\mu_0 = \tfrac{1}{2}$ and $\sigma_0 = \tfrac{1}{2}$.

Any reasonable test should be consistent, so consistency does not provide a criteria for distinguishing among tests. If a test is not consistent for a reasonable set of alternatives, it should be rejected as defective; the next example illustrates a defective test.

Example 1.3.2.   Let $X_1, \ldots, X_n$ be a random sample from a Cauchy distribution with density

$$f(x - \theta) = \frac{1}{\pi\left[1 + (x - \theta)^2\right]}, \qquad -\infty < x < \infty.$$

The characteristic function is given by $\phi(t) = \exp\{-|t| + i\theta t\}$.
   Suppose, for testing $H_0 : \theta = 0$ versus $H_A : \theta > 0$, we reject H_0 if $\bar{X} \geqslant c$. But the characteristic function of $\bar{X}$ is

$$\phi_{\bar{x}}(t) = \left[\phi\left(\frac{t}{n}\right)\right]^n = \phi(t).$$

Hence $\bar{X}$ has exactly the same distribution as X_i, independent of n. This means the power function of $\bar{X}$ does not depend on n, the power cannot tend to 1 for any $\theta > 0$, and $\bar{X}$ does not provide a consistent test for any reasonable set of alternatives.

1.4. A MOST POWERFUL TEST

We next apply the Neyman–Pearson approach to construct a uniformly most powerful test for the one-sided alternative. At first glance it might appear surprising that an optimal test can be found in such a general setting. The fact is that there are very few size α tests for the composite null under consideration, so if one can be found, it may turn out to be optimal.

The null hypothesis simply states that we are sampling from $F \in \Omega_0$, and the alternative hypothesis states that we are sampling from $G(x) = K(x - \theta)$ where $K \in \Omega_0$ and $\theta > 0$. We use different notation for the null and alternative (F and K) to emphasize their composite nature. We can state the hypotheses in slightly different form:

$$H_0 : F(0) = \tfrac{1}{2}$$

$$H_A : 1 - G(0) > \tfrac{1}{2}.$$

Any distribution function G can be decomposed in the following way:

$$G(x) = P(X \leqslant x)$$

$$= P(X \leqslant x \mid X \leqslant 0)P(X \leqslant 0) + P(X \leqslant x \mid X > 0)P(X > 0).$$

Recall from (1.2.2) that $p = 1 - K(-\theta) = 1 - G(0)$, and define

$$g_-(x) = \frac{d}{dx} P(X \leqslant x \mid X \leqslant 0) \qquad \text{if} \quad x \leqslant 0$$
$$0 \qquad\qquad\qquad\qquad\qquad\qquad \text{otherwise}$$

and similarly for $g_+(x)$. Now the density of G can be written

$$g(x) = (1 - p)g_-(x) + pg_+(x). \tag{1.4.1}$$

This allows us to isolate the median, the parameter to be tested, from the shape of the distribution and to formulate the hypotheses in terms of p.

Example 1.4.1.   Let $g(x) = 1/3$ for $-1 < x < 2$ and 0 otherwise, then

$$G(x) = 0 \qquad\qquad x \leqslant -1$$
$$(x+1)/3 \qquad -1 < x < 2$$
$$1 \qquad\qquad 2 \leqslant x.$$

On $-1 < x < 0$, $P(X \leqslant x \,|\, X \leqslant 0) = x + 1$, and $g_-(x) = 1$ on $-1 < x < 0$ and 0 otherwise. Likewise, on $0 < x < 2$, $P(X \leqslant x \,|\, X > 0) = x/2$ and $g_+(x) = 1/2$ on $0 < x < 2$ and 0 otherwise. Now $p = 1 - G(0) = 2/3$. Hence $g(x)$ can be thought of as two uniform densities pieced together and weighted by $p = 2/3$.

In general we can specify the distribution being sampled by the triple (p, g_+, g_-); in the example we have $(2/3, g_+, g_-)$. The hypotheses become

$$H_0 : (\tfrac{1}{2}, g_+, g_-)$$

$$H_A : (p, h_+, h_-), \qquad p > \tfrac{1}{2}$$

where G and H are arbitrary distribution functions possessing densities.

To apply the Neyman–Pearson lemma and construct the uniformly most powerful test, we resort to the method of least favorable distributions. The strategy is outlined in the following four steps:

1. Fix an alternative distribution and try to choose that distribution in the composite null hypothesis that is hardest to distinguish from the fixed alternative. If the distribution is chosen correctly, it will be the least favorable distribution.
2. Construct the Neyman–Pearson best size α test for this simple versus simple testing problem.
3. Show that the test remains size α on the composite null hypothesis. If this is not possible, then the least favorable distribution was probably not chosen properly.
4. The final step consists in showing that the test does not depend on the fixed alternative; hence the test is uniformly most powerful. We now carry out this strategy for the one-sample location problem.

1. Fix $p = p' > 1/2$ and fix h_+, h_- so that the alternative is (p, h_+, h_-). The natural guess for a least favorable distribution is $(1/2, h_+, h_-)$. By using h_+ and h_- while setting $p = 1/2$, we hope to make it most

difficult to separate the null from the alternative hypothesis. Hence we will try

$$H_0 : (\tfrac{1}{2}, h_+, h_-) \quad \text{versus}$$

$$H_A : (p', h_+, h_-)$$

with $p' > 1/2$, h_+, h_- all specified.

Example 1.4.2. From Example 1.4.1, the anticipated least favorable distribution is given by

$$\begin{aligned}
\tfrac{1}{2} g_-(x) + \tfrac{1}{2} g_+(x) = \tfrac{1}{2} \qquad & -1 < x < 0 \\
\tfrac{1}{4} \qquad & 0 < x < 2 \\
0 \qquad & \text{otherwise}
\end{aligned}$$

2. The Neyman–Pearson lemma states that H_0 should be rejected when

$$\frac{\displaystyle\prod_{i=1}^{n} \left[\tfrac{1}{2} h_-(x_i) + \tfrac{1}{2} h_+(x_i) \right]}{\displaystyle\prod_{i=1}^{n} \left[(1 - p')h_-(x_i) + p'h_+(x_i) \right]} \leqslant k.$$

Suppose $x_{(i)}$ is the ith order statistic and that for the sample,

$$x_{(1)} < \cdots < x_{(t)} < 0 < x_{(t+1)} < \cdots < x_{(n)}.$$

We can then rewrite (1.4.1) as

$$\frac{(1/2)^n \displaystyle\prod_{1}^{t} h_-(x_{(i)}) \prod_{t+1}^{n} h_+(x_{(i)})}{(1 - p')^n \displaystyle\prod_{1}^{t} h_-(x_{(i)}) \prod_{t+1}^{n} \left(\frac{p'}{1 - p'} \right) h_+(x_{(i)})} \leqslant k,$$

and hence as

$$\frac{(1/2)^n}{(1 - p')^n} \left(\frac{1 - p'}{p'} \right)^{\#(X_i > 0)} \leqslant k.$$

Since $p' > 1/2$, after taking logarithms and letting k denote a generic constant, we see the test is equivalent to rejecting H_0 when $S \geqslant k$ where S is the sign test, (1.2.1).

3. The sign test is exact size α on the composite null hypothesis because of the distribution-free property.

4. The sign test is uniformly most powerful because the critical region remains the same for all $p' > 1/2$ and any other (h_-, h_+).

Hence we conclude that if we can make no shape assumption about the underlying population distribution then the sign test is uniformly most powerful. This means that for any fixed distribution with positive median, there is no other size α test with greater power. On the other hand, there are not very many size α tests for the composite null hypothesis considered here.

Definition 1.4.1. A size α test of $H_0 : G \in \Omega_{\text{null}}$ versus $H_A : G \in \Omega_{\text{alt}}$ with critical region $V \geqslant k$ is said to be unbiased if

$$P_G(V \geqslant k) \leqslant \alpha \qquad \text{for all} \quad G \in \Omega_{\text{null}}$$

$$P_G(V \geqslant k) \geqslant \alpha \qquad \text{for all} \quad G \in \Omega_{\text{alt}}$$

The sign test for testing $H_0 : \theta = 0$ versus $H_A : \theta > 0$ with $F \in \Omega_0$ is unbiased; see Exercise 1.8.3. In fact, the sign test is unbiased for testing the two-sided hypothesis $H_0 : \theta = 0$ versus $H_A : \theta \neq 0$ and is the uniformly most powerful unbiased test. For a complete discussion of uniformly most powerful unbiased tests, see Lehmann (1959, p. 147).

1.5. ESTIMATION

In this section we describe how to derive point and interval estimates of θ from hypothesis tests. We then apply the method to the sign test. Since several of the nonparametric tests can be put into a sign test or counting form, the results based on the sign test carry over immediately to these other procedures. This allows us to derive the point and interval estimates based on the Wilcoxon signed rank test and the Mann–Whitney–Wilcoxon rank sum test with virtually no effort.

The ideas are more easily introduced in terms of the one-sample t test.

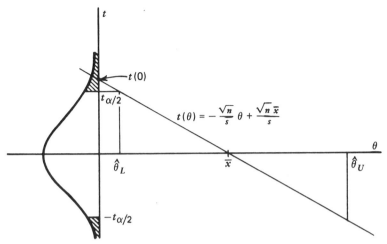

Figure 1.2. Graph of $t(\theta)$.

Let $x_1, \ldots, x_n$ be the observed sample values and define

$$t(\theta) = \sqrt{n} \left(\frac{\bar{x} - \theta}{s} \right)$$

$$= -\frac{\sqrt{n}}{s} \theta + \frac{\sqrt{n}}{s} \bar{x}$$

where s^2 is the usual unbiased estimate of σ^2. We have written $t(\theta)$ as a linear function of θ. In Fig. 1.2 the graph of $t(\theta)$ is drawn on a θ, t axis system. The horizontal axis then represents the parameter space and the vertical axis the sample space of the test statistic. We have also drawn the null distribution of $t(0)$, under the assumption of normality, on the vertical axis along with the critical points for the size α, two-sided test of $H_0 : \theta = 0$ versus $H_A : \theta \neq 0$.

The hypothesis test is carried out by checking to see where $t(\theta)$ crosses the vertical axis relative to the critical points; hence, check the value $t(0)$. The $(1 - \alpha)$ 100% confidence interval is found by inverting the acceptance region of the test. It is clearly marked by the interval on the horizontal axis from $\hat{\theta}_L$ to $\hat{\theta}_U$. Clearly $P_\theta(\hat{\theta}_L \leq \theta \leq \hat{\theta}_U) = P_\theta(|t(\theta)| \leq t_{\alpha/2}) = 1 - \alpha$. Note that $t(\hat{\theta}_L) = t_{\alpha/2}$ and $t(\hat{\theta}_U) = -t_{\alpha/2}$; so, for example, $\hat{\theta}_U = \bar{x} + t_{\alpha/2} s / n^{1/2}$.

The principle that we use to determine the point estimate corresponding to $t(\theta)$ is to select that value of θ which corresponds to the point of symmetry of the null distribution of $t(0)$. For reasonable tests this generally

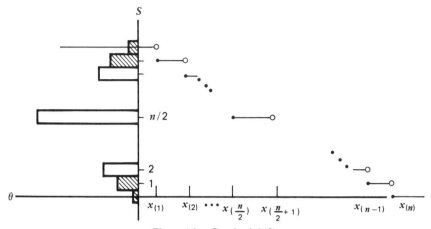

Figure 1.3. Graph of $S(\theta)$.

corresponds to the mode of the null distribution of the test statistic. In essence, we suggest estimating θ by that value which yields a test statistic value at the center of the null distribution, that is, a most likely value of the test statistic. For the case of $t(\theta)$ this occurs for $t(\theta) = 0$; hence $\hat{\theta} = \bar{x}$.

Hence all inference, both testing and estimation, can be linked together through this graphical representation. We present the estimate in a more formal fashion later; however, Fig. 1.2 carries the heuristic interpretation of these estimates.

For the sign test, define $S(\theta) = \#(X_i > \theta)$, $i = 1, \ldots, n$. This can also be written $S(\theta) = \#(X_{(i)} > \theta)$, $i = 1, \ldots, n$, where $X_{(i)}$ is the ith order statistic. In Fig. 1.3 we graph $S(\theta)$ versus θ for an even sample size n.

The null distribution of $S(0)$, the binomial, is constructed on the vertical axis, or sample space. Clearly, $S(\theta)$ is a nonincreasing step function of θ which steps down at each order statistic. Furthermore, the function $S(\theta)$ is continuous from the right. We thus have the following inequalities:

$$x_{(k+1)} \leqslant \theta \qquad \text{if and only if} \quad S(\theta) \leqslant n - k - 1$$
$$\theta < x_{(n-k)} \qquad \text{if and only if} \quad k + 1 \leqslant S(\theta).$$

Hence

$$X_{(k+1)} \leqslant \theta < X_{(n-k)} \qquad \text{if and only if} \quad k + 1 \leqslant S(\theta) \leqslant n - k - 1,$$

and it follows that

$$P_\theta\big(X_{(k+1)} \leqslant \theta < X_{(n-k)}\big) = P_\theta\big(k + 1 \leqslant S(\theta) \leqslant n - k - 1\big)$$
$$= 1 - P_\theta\big(S(\theta) \leqslant k\big) - P_\theta\big(S(\theta) \geqslant n - k\big).$$

Note that $P_\theta(S(\theta) \leqslant k) = P_0(S(0) \leqslant k)$. If we choose k such that $P(S \leqslant k)$ $= \alpha/2$ from the binomial table, then $[X_{(k+1)}, X_{(n-k)})$ is a $(1 - \alpha)$ 100% confidence interval for θ, independent of $F \in \Omega_0$. For simplicity, we will generally use the closed interval $[X_{(k+1)}, X_{(n-k)}]$. This interval still has $(1 - \alpha)$ 100% confidence by the continuity of the underlying distribution. Recall also that k can be approximated [see (1.2.6)] using the Central Limit Theorem.

Since the point of symmetry for the binomial null distribution is $n/2$ (when n is even) we seek a value $\hat{\theta}$ such that $S(\hat{\theta}) = n/2$. Any value $\hat{\theta}$ between $X_{(n/2)}$ and $X_{(n/2+1)}$ will work and by convention we take

$$\hat{\theta} = \frac{X_{(n/2)} + X_{(n/2+1)}}{2},$$

the median of the sample.

When the sample size is odd, say $n = 2r + 1$, the argument for the confidence interval remains the same. There are two middle bars in the binomial histogram and the estimate $\hat{\theta}$ is seen to be $X_{(r+1)}$, the unique sample median (Exercise 1.8.5).

Hence the sample median is the natural point estimate derived from the sign test, and a confidence interval constructed from the order statistics is the natural interval with its confidence coefficient given by the null distribution of the sign statistic. We now discuss a general formulation of the estimation problem due to Hodges and Lehmann (1963) and Lehmann (1963).

Definition 1.5.1.   Let $X_1, \ldots, X_n$ be a random sample from $F(x - \theta)$, $F \in \Omega_0$. Suppose V is a statistic for testing $H_0 : \theta = 0$ and define $V(\theta)$ by replacing X_i by $X_i - \theta$, $i = 1, \ldots, n$. Suppose that $V(\theta)$ is a nonincreasing function of θ and the null distribution of $V = V(0)$ is symmetric about μ_0, free of F. Define

$$\theta^* = \sup\{\theta : V(\theta) > \mu_0\}$$

$$\theta^{**} = \inf\{\theta : V(\theta) < \mu_0\} \tag{1.5.1}$$

$$\hat{\theta} = \frac{\theta^* + \theta^{**}}{2}.$$

The estimator $\hat{\theta}$ is called the Hodges–Lehmann estimator of θ.

If, in addition, we define

$$\hat{\theta}_L = \inf\{\theta : V(\theta) < C_1\}$$

$$\hat{\theta}_U = \sup\{\theta : V(\theta) > C_2\} \tag{1.5.2}$$

where $P(V \geqslant C_1) = P(V \leqslant C_2) = \alpha/2$. Then the interval $[\hat{\theta}_L, \hat{\theta}_U]$ is a $(1 - \alpha)$ 100% confidence interval based on V.

In the sign test example for n even, $\theta^* = X_{(n/2)}$ and $\theta^{**} = X_{(n/2+1)}$, whereas for $n = 2r + 1$ odd, $\theta^* = \theta^{**} = X_{(r+1)}$. The rank tests studied in this book are nonincreasing step functions of θ with graphs similar to Fig. 1.3. The null distributions of the rank tests are symmetric and nondecreasing (nonincreasing) on the left (right) of the point of symmetry. The point or points of maximum probability are called modal points. When n is even, $n/2$ is the unique modal point of S; when n is odd, $n = 2r + 1$, r and $r + 1$ are the two modal points. The Hodges–Lehmann estimate corresponds to these modal points and can be loosely thought of as a maximum probability estimate relative to the distribution of the test statistic.

When the critical values, C_1 and C_2, are integers, we can define $\hat{\theta}_L$ and $\hat{\theta}_U$ as the smallest and largest solutions such that $V(\hat{\theta}_L) = C_1 - 1$ and $V(\hat{\theta}_U) = C_2 + 1$.

We next present a theorem that shows the estimation procedures are translation statistics; so, if we add a constant to all the observations, we need only add that constant to the estimates. This translation property allows us to let $\theta = 0$ without loss of generality in the study of the distributional properties of the estimates.

Theorem 1.5.1. $\hat{\theta}$, $\hat{\theta}_L$, and $\hat{\theta}_U$ are translation statistics; that is, $\hat{\theta}(x_1 + a, \ldots, x_n + a) = \hat{\theta}(x_1, \ldots, x_n) + a$, and likewise for $\hat{\theta}_L, \hat{\theta}_U$.

Proof. Let V_a denote V computed on $x_1 + a, \ldots, x_n + a$ for some fixed a. Then $V_a(\theta)$ is computed on $x_1 - \theta + a, \ldots, x_n - \theta + a$ and hence $V_a(\theta) = V(\theta - a)$. Now,

$$\sup\{\theta : V_a(\theta) > \mu_0\} = \sup\{\theta : V(\theta - a) > \mu_0\} - a + a$$

$$= \sup\{\theta - a : V(\theta - a) > \mu_0\} + a$$

$$= \sup\{\delta : V(\delta) > \mu_0\} + a.$$

Hence $\theta^*(x_1 + a, \ldots, x_n + a) = \theta^*(x_1, \ldots, x_n) + a$. A similar argument works for θ^{**}, $\hat{\theta}_L$, and $\hat{\theta}_U$.

Example 1.5.1. Recall from Example 1.1.1 that experiments were conducted in which the birds were taken out on lines at 90° and 180° from the training line and released. One measurement of interest was the angle between the bird's line of flight, as it disappeared over the horizon, and the homing line. Table 1.2 gives these angles for 28 birds released on a sunny day. All angles are measured between 0° and 180° so lines above or below

Table 1.2. Angular Errors for Birds Released on a Sunny Day (Artificial Data)

6, 7, 9, 17, 18, 18, 22, 28, 32, 35, 36, 42, 42, 42, 48, 48,
51, 52, 53, 55, 56, 57, 58, 63, 72, 83, 91, 97

the homing line are not distinguished. Let θ denote the population median error angle. We make no shape assumption on the underlying population of error angles. If the birds are not homing then $\theta = 90°$; hence we will test $H_0 : \theta = 90°$ versus $H_A : \theta < 90°$. Using the sign test and corresponding estimates, we begin the investigation.

For testing $H_0 : \theta = 90°$ versus $H_A : \theta < 90°$ we find that $S = 2$ where $S = \#$ (observations $> 90°$). For $n = 28$ we reject H_0 at level $\alpha = .018$ when $S \leqslant 8$. Since $S = 2$ we easily reject H_0 at this level. The estimated error angle $\hat{\theta}$ is the median of the sample and $\hat{\theta} = 45°$. A 91% confidence interval for θ is given by $[X_{(10)}, X_{(19)}) = [35, 53)$ since $P(S \leqslant 9) = .045$. Hence the sign test and companion estimates provide strong evidence that on sunny days the birds are systematically tending toward the homing line. The major portion of the study involved two sample comparisons in which birds were released on both sunny and cloudy days in order to see if the birds used the sun to navigate. See Exercise 1.8.11.

If there are observations in the sample equal to the null hypothesized value they are set aside when calculating the sign test and the sample size is reduced accordingly. These values are not excluded for purposes of estimation. For a further discussion and other alternatives for handling this problem, see Lehmann (1975, p. 123).

1.6. STABILITY

Thus far we have discussed the statistical properties of the sign test. Robustness of statistical methods can be thought of as describing the stability of the procedure itself. In particular, we wish to avoid statistical procedures that can be unduly influenced by a small fraction of the data. The t test and its companion estimate $\overline{X}$ are examples of procedures that have high power and efficiency at a specific model (normal) but are also highly unstable. Only one observation suffices to alter the values of t and $\overline{X}$ by an arbitrarily large amount. The sign test and the sample median, on the other hand, are quite stable. It is possible to alter a few observations without changing either the test or the estimate. In the rest of this section we describe more formally some aspects of stability and apply these ideas to the statistical methods considered above.

Definition 1.6.1. Given an estimate $\hat{\theta}$ of θ suppose there exists an integer a such that

   (a) $X_{(a+1)} \leqslant \hat{\theta} \leqslant X_{(n-a)}$.
   (b) For any fixed values of $x_{(a+2)}, \ldots, x_{(n)}$, if $x_{(a+1)} \to -\infty$, then $\hat{\theta} \to -\infty$.
   (c) For any fixed values of $x_{(1)}, \ldots, x_{(n-a-1)}$, if $x_{(n-a)} \to +\infty$, then $\hat{\theta} \to +\infty$.

If a^* is the smallest such integer, then $\hat{\theta}$ can tolerate at most a^* bad observations. The tolerance is defined as

$$\tau_n = \frac{a^*}{n}.$$

The idea of tolerance was introduced by Hodges (1967) and is related to the idea of breakdown of an estimator. For a discussion of breakdown, see Hampel (1974) and Huber (1981). In many cases, $\lim \tau_n = \tau$, and the asymptotic tolerance is the breakdown value. All other things being equal, we would like to have an estimator with high tolerance.

Example 1.6.1. The sample mean $\overline{X}$ has 0 tolerance because $X_{(1)} \leqslant \overline{X} \leqslant X_{(n)}$, so that $a^* + 1 = 1$ and $a^* = 0$. If $n = 2r$, the tolerance of the median is $(r - 1)/2r$; and if $n = 2r + 1$, the tolerance is $r/(2r + 1)$, and in either case the asymptotic tolerance $\tau = .5$. The α-trimmed mean is the mean of the middle $n - 2[\alpha n]$ observations after $[\alpha n]$ observations have been trimmed from both ends of the ordered sample. Since $a^* = [\alpha n]$, the asymptotic tolerance of the trimmed mean is $\tau = \alpha$.

Definition 1.6.2. Suppose V is a test statistic such that we reject $H_0 : \theta = 0$ for $H_A : \theta > 0$ when $V \geqslant k$. Suppose there exists an integer a such that for given, fixed values of $x_{a+2}, \ldots, x_n$ values of $x_1, \ldots, x_{a+1}$ can be chosen to force $V < k$. Let a^* be the smallest such integer, then the tolerance to acceptance is defined to be

$$\tau_n(\text{accept}) = \frac{a^*}{n}.$$

Since V cannot be forced to accept with a^* or fewer observations, it can tolerate at most a^* bad observations. Furthermore, for a given sample, we can control V if we can alter $a^* + 1$ sample values.

Similarly, the tolerance to rejection is defined by the smallest integer b^* such that, for any fixed values of $x_{b^*+2}, \ldots, x_n$, we can choose values $x_1, \ldots, x_{b^*+1}$ to force $V \geqslant k$. Then $\tau_n(\text{reject}) = b^*/n$. A similar definition of testing tolerance is given by Ylvisacker (1977) and discussed further by Rieder (1982).

Example 1.6.2. The sign test rejects $H_0 : \theta = 0$ for $H_A : \theta > 0$ when $S \geqslant k$. If we fix $n - k + 1$ observations less than zero, then no matter what the values of the remaining observations, it follows that $S < k$. Hence $a^* = n - k$ and $\tau_n(\text{accept}) = (n - k)/n$. Substitute the approximate value of k from (1.2.7) to get

$$\tau_n(\text{accept}) \doteq \frac{1}{2} - Z_\alpha \frac{1}{2\sqrt{n}} - \frac{1}{2n}$$

$$\tau(\text{accept}) = \lim \tau_n(\text{accept}) = \frac{1}{2} .$$

Hence the tolerance to acceptance of the sign test depends on the significance level α and converges to the tolerance of the median from below. See Exercises 1.8.6 for $\tau_n(\text{reject})$.

In Exercise 1.8.7 it is pointed out that for the t statistic $\tau_n(\text{accept}) = 0$. Ylvisacker (1977) shows that $\tau_n(\text{reject}) = [k^2/(n + k^2)] - 1/n$, where k is the critical value of the t test.

Just as in the case with estimation, it is desirable to have tests with high tolerance to both acceptance and rejection. Then the test will not be controlled by a small fraction of the data. Table 1.3 lists some numerical values of the tolerances of the sign and t tests.

We next consider the effect on the estimate of tossing a single observation into the sample.

Definition 1.6.3. If the estimate $\hat{\theta}_n$ is based on a sample of n observations and an additional observation with value x is introduced, the differential

Table 1.3. Tolerance of Sign and t Tests.
$\alpha = .05$

n	$\tau_n(\text{accept})$		$\tau_n(\text{reject})$	
	t	S	t	S
10	0	.20	.15	.70
13	0	.23	.11	.69
18	0	.28	.08	.67
30	0	.33	.06	.63
100	0	.40	.02	.59
∞	0	.50	0	.50

effect on the estimate is measured by the sensitivity curve:

$$SC(x) = (n + 1)(\hat{\theta}_{n+1} - \hat{\theta}_n).$$

The Princeton robustness study (Andrews et al., 1972) included a stylized version of the sensitivity curve. Nineteen expected normal order statistics were taken as the base "sample" and the additional x was varied to determine the stylized sensitivity. See Section 5E of the Princeton study for the graphs of stylized sensitivity curves.

Hampel (1974) points out that when the sensitivity curve is properly normalized, in the limit it corresponds to the influence curve. The influence curve measures the influence on the population characteristic being estimated (e.g., population mean or median) of a contaminating point mass in the underlying distribution.

The definition of the influence curve for an estimator depends on being able to represent the estimator as a functional evaluated at the empirical cumulative distribution function (cdf). For example, the mean and median of a cdf, $H(x)$, can be written $A(H) = \int xh(x)\,dx = \int x\,dH(x)$ and $M(H) = H^{-1}(\frac{1}{2})$. See the Appendix for a discussion of the Stieltjes integral $\int g(x)\,dH(x)$. We suppose the mean $A(H)$ exists and when the median is not unique we will take $M(H) = \inf\{x : H(x) \geqslant 1/2\}$. The natural estimates of $A(H)$ and $M(H)$ are $A(H_n) = \int x\,dH_n(x) = n^{-1}\sum x_i$, the sample mean, and $M(H_n) = H_n^{-1}(1/2)$, the sample median.

In order to measure the influence of a point mass at y, Hampel (1974) suggests evaluating the Gateaux derivative of the functional $T(H)$ in the direction of a cdf that assigns probability 1 to the value y. This derivative is just the ordinary derivative of $T(H_t)$ with respect to the real variable t evaluated at $t = 0$, where $H_t(x) = (1 - t)H(x) + t\delta_y(x)$, $\delta_y(x) = 0$ if $x < y$, and 1 if $x \geqslant y$. See Huber (1981) for more discussion. It measures how quickly the functional changes in the direction of contamination.

Definition 1.6.4. The influence curve for the functional $T(\cdot)$ is

$$\Omega(y) = \frac{d}{dt}T\big[(1 - t)H + t\delta_y\big]\bigg|_{t=0}.$$

From the point of view of robustness it is desirable to have estimates with bounded sensitivity and influence curves. This means that a single observation cannot have an arbitrarily large effect on the estimate. Another desirable property, perhaps better reflected in the influence curve, is continuity. Jumps or discontinuities in the sensitivity or influence curves indicate local instability at the jump point. For example, an estimate with such a sensitivity curve may be adversely effected by round-off error at these points.

Example 1.6.3. Without loss of generality, suppose we have a sample of n observations, and by good fortune the sample mean and median are both 0. It is easy to verify that the sample mean has $SC(x) = x$. Hence the sensitivity is linear and unbounded. This is just another way of saying that a single observation can change the sample mean by an arbitrarily large amount.

Now suppose $n = 2r$, even, and $x_{(1)} < \cdots < x_{(r)} < 0 < x_{(r+1)} < \cdots < x_{(n)}$. The sensitivity curve for the median is given by

$$SC(x) = (n + 1)x_{(r)}, \qquad x \leqslant x_{(r)}$$
$$(n + 1)x, \qquad x_{(r)} \leqslant x \leqslant x_{(r+1)}$$
$$(n + 1)x_{(r+1)}, \qquad x_{(r+1)} \leqslant x,$$

and the graph is given in Fig. 1.4.

Example 1.6.4. Let $H_t(x) = (1 - t)H(x) + t\delta_y(x)$, and then $A(H_t) = \int x \, dH_t(x) = \int x \, d[(1 - t)H(x) + t\delta_y(x)]$. Hence

$$\frac{d}{dt} A(H_t) = \int x \, d\delta_y(x) - \int x \, dH(x)$$

$$= y - A(H).$$

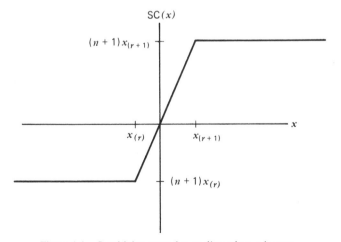

Figure 1.4. Sensitivity curve for median when n is even.

Hence the mean functional $A(H)$ has an unbounded, linear influence curve, similar to the sensitivity curve. In fact, if we take the mean of $H(\cdot)$ to be 0, then $\Omega(y) = y$.

Now, note that $1/2 = H_t(H_t^{-1}(\frac{1}{2})) = H(H_t^{-1}(\frac{1}{2})) + t[\delta_y(H_t^{-1}(\frac{1}{2})) - H(H_t^{-1}(\frac{1}{2}))]$. Hence, if $h(\cdot)$ is the probability density function (pdf) of $H(\cdot)$ we have,

$$
\begin{aligned}
0 &= \frac{d}{dt} H_t\left(H_t^{-1}\left(\frac{1}{2}\right)\right)\Bigg|_{t=0} \\
&= h\left(H^{-1}\left(\frac{1}{2}\right)\right) \frac{d}{dt} H_t^{-1}\left(\frac{1}{2}\right)\Bigg|_{t=0} + \delta_y\left(H^{-1}\left(\frac{1}{2}\right)\right) - H\left(H^{-1}\left(\frac{1}{2}\right)\right)
\end{aligned}
$$

and

$$
\frac{d}{dt} H_t^{-1}\left(\frac{1}{2}\right)\Bigg|_{t=0} = \frac{\frac{1}{2} - \delta_y\left(H^{-1}\left(\frac{1}{2}\right)\right)}{h\left(H^{-1}\left(\frac{1}{2}\right)\right)}.
$$

Using the definition of $\delta_y(\cdot)$ and supposing $H(\cdot)$ has median 0 without loss of generality, we have

$$
\Omega(y) = \begin{cases} -\dfrac{1}{2h(0)} & y \leqslant 0 \\[2ex] \dfrac{1}{2h(0)} & y > 0. \end{cases}
$$

This can be compared to the sensitivity curve in Fig. 1.4. The influence curve $\Omega(y)$ is the limiting case of the sensitivity curve and has a jump discontinuity at the population median. Both sensitivity and influence curves are bounded, showing the extreme insensitivity of the median to outliers.

In cases where the estimator $\hat{\theta}$ is a functional $T(\cdot)$ evaluated at the empirical cdf, we have $\hat{\theta} = T(F_n)$ and $\theta = T(F)$. The influence function $\Omega(y)$ often provides a representation suggesting the asymptotic distribution of the estimator. Huber (1981, Section 2.5) points out that, under regularity conditions,

$$
\sqrt{n}\left(T(F_n) - T(F)\right) = \frac{1}{\sqrt{n}} \sum_{i=1}^{n} \Omega(X_i) + o_p(1) \tag{1.6.1}
$$

where $o_p(1)$ tends to 0 in probability. The Central Limit Theorem applies to the first term on the right side; and, since $n^{1/2}(\hat{\theta} - \theta) = n^{1/2}(T(F_n) - T(F))$, we have

$$\sqrt{n}\,(\hat{\theta} - \theta) \xrightarrow{D} Z \sim n\left(0, \int_{-\infty}^{\infty} \Omega^2(x)\,dF(x)\right), \qquad (1.6.2)$$

when the influence function is centered so that $\int_{-\infty}^{\infty} \Omega(x)\,dF(x) = 0$.

A rigorous development requires a careful analysis of the remainder term $o_p(1)$ for each estimator. However, (1.6.2) is an excellent heuristic guide which generally anticipates the correct asymptotic distribution. From Example 1.6.4, if the sample comes from $H(x - \theta)$, the influence curve for the median yields $\int \Omega^2(x)\,dH(x) = 1/[4h^2(0)]$. Hence we would at least need the density existing and positive at 0. Then (1.6.2) suggests that if $\hat{\theta}$ = med X_i then

$$\sqrt{n}\,(\hat{\theta} - \theta) \xrightarrow{D} Z \sim n\left(0, \frac{1}{4h^2(0)}\right). \qquad (1.6.3)$$

For an extensive discussion of this approach see Serfling (1980, Chapter 6).

1.7. SUMMARY AND EFFECT OF DEPENDENCE IN DATA

If we suppose that our samples come from an absolutely continuous distribution with a unique median θ, (1.1.2), then the simple sign test, (1.2.1), is the uniformly most powerful size α test of $H_0 : \theta = 0$ versus $H_A : \theta > 0$. The distribution theory under both null and alternative hypotheses is given by the binomial distribution. Under $H_0 : \theta = 0$, the distribution of the sign test does not depend on the distribution sampled, that is, it is distribution free. The sign statistic is expressible as the sum of i.i.d. random variables under both hypotheses, and the Central Limit Theorem can be applied to approximate needed probabilities and critical values. In addition to being optimal, the sign test is consistent and unbiased for any alternative distribution with nonzero median. It also has positive tolerance to both acceptance and rejection and is thus not unduly effected by a small portion of the data.

Natural point and interval estimates can be easily derived from the sign test. The point estimate is the sample median and the interval estimate is defined by appropriate order statistics. The confidence coefficient is determined by the null binomial distribution of the sign test. The median is a robust estimate; it has positive tolerance and bounded sensitivity and influence curves. In Section 6.2 the sign test is extended to the one-sample multivariate location model.

Throughout this chapter we have assumed that the data consists of independent observations from a location model. Most of the properties of the sign statistic depend quite heavily on the independence assumption. To illustrate what happens when independence fails, we consider a simple serial correlation model: $X_1, X_2, \ldots$ are normally distributed with mean θ, variance 1, and correlation $\rho_{ij} = \rho$ if $j = i + 1$ and 0 otherwise. The correlation ρ is restricted to $-.5 \leqslant \rho \leqslant .5$; see Scheffé (1959, Section 10.2). Hence only immediate neighbors are correlated. We further suppose that, for $i = 1, 2, \ldots,$ (X_i, X_{i+1}) has a bivariate normal distribution with means θ, variances 1, and correlation ρ. If we did not suspect serial correlation in the data, for large n, a nominal 5% test based on S would reject $H_0: \theta = 0$ in favor of $H_A: \theta > 0$ if $S \geqslant n/2 + 1.645(n)^{1/2}/2$. Here we have used (1.2.6) and ignored the continuity correction for large n.

Under the null hypothesis $H_0: \theta = 0$ and assuming nonzero serial correlation we have

$$ES = E \sum s(X_i)$$

$$= nP(X > 0) = \frac{n}{2}$$

$$\text{Var } S = E\left\{ \sum \left(s(X_i) - \tfrac{1}{2}\right) \right\}^2 \tag{1.7.1}$$

$$= n \operatorname{Var}(s(X_1)) + 2(n-1)\operatorname{Cov}(s(X_1), s(X_2))$$

$$= \frac{n}{4} + 2(n-1)\left\{ P(X_1 > 0, X_2 > 0) - \frac{1}{4} \right\},$$

since $\operatorname{Var} s(X_1) = E(s(X_1))^2 - \frac{1}{4} = \frac{1}{2} - \frac{1}{4} = \frac{1}{4}$. Hence the $\operatorname{Var} S$ is altered and includes $P(X_1 > 0, X_2 > 0)$ which shows that S is no longer distribution free under H_0.

In Exercise 1.8.14 you are asked to show that $P(X_1 > 0, X_2 > 0) = 1/4 + (1/2\pi)\sin^{-1}\rho$. Hence we have

$$\text{Var } S = \frac{n}{4} + \frac{n-1}{\pi} \sin^{-1}\rho. \tag{1.7.2}$$

In Exercise 1.8.15 you are asked to verify the conditions of Theorem A16 and hence show that $(S - ES)/(\operatorname{Var} S)^{1/2}$ is limiting $n(0, 1)$.

We can now approximate the true level of the test and compare it to the nominal 5% value which we assumed to be true. Let α_T denote the true signficance level, then for large n,

$$\alpha_T = P\left(S \geqslant \frac{n}{2} + 1.645\frac{\sqrt{n}}{2}\right)$$

$$\doteq 1 - \Phi\left(\frac{1.645\sqrt{n}/2}{\sqrt{\mathrm{Var}\,S}}\right)$$

$$\doteq 1 - \Phi\left(\frac{1.645}{\sqrt{1 + (4/\pi)\sin^{-1}\rho}}\right). \tag{1.7.3}$$

Again, ignoring the serial correlation, the nominal 5% test based on the sample mean rejects $H_0: \theta = 0$ when $n^{1/2}\overline{X} \geqslant 1.645$. Under $H_0: \theta = 0$ and assuming the serial correlation model, $E(n^{1/2}\overline{X}) = 0$, $\mathrm{Var}(n^{1/2}\overline{X}) = 1 + 2(n-1)\rho/n$ and $n^{1/2}\overline{X}$ is normally distributed since it is a linear combination of normally distributed random variables. The true significance level is α_T,

$$\alpha_T = P(\sqrt{n}\,\overline{X} \geqslant 1.645)$$

$$= P\left[\frac{\sqrt{n}\,\overline{X}}{\sqrt{1 + 2(n-1)\rho/n}} \geqslant \frac{1.645}{\sqrt{1 + 2(n-1)\rho/n}}\right]$$

$$= 1 - \Phi\left(\frac{1.645}{\sqrt{1 + 2\rho}}\right). \tag{1.7.4}$$

In Table 1.4 we record the true significance level, for various values of ρ, for nominal 5% sign and $\overline{X}$ tests.

Table 1.4. True Significance Level for 5% S and $\overline{X}$ Tests

Test	$-.49$	$-.4$	$-.3$	$-.2$	$-.1$	0	.1	.2	.3	.4	.49
$\overline{X}$	.000	.000	.005	.017	.033	.05	.067	.082	.097	.109	.121
S	.003	.009	.018	.028	.039	.05	.061	.071	.081	.092	.100
T^a	.000	.000	.006	.018	.033	.05	.067	.081	.095	.107	.119

[a] T is the Wilcoxon signed rank test; see (2.7.12).

Hence both tests have true levels that can be far from the nominal level. The sign test not only is no longer distribution free under H_0 but is effected almost but not quite as much as the $\overline{X}$ test. When the correlation is positive, we tend to have "too many" rejections since large observations pull others up with them. Hence, when $\rho = .4$, the tests reject H_0 about 10% of the time rather than the expected 5%. The moral is clear: even the most robust procedures may be highly unreliable under the most simple models of dependence in the data. Gastwirth and Rubin (1971) reach the same conclusion on more general autoregressive models. The same authors, in 1975, studied the behavior of robust estimators for models with dependent data.

1.8. EXERCISES

1.8.1.   Construct a distribution that has many medians.

1.8.2.   For $n = 10$, $\alpha = .0547$

      a.   Construct the sign test for $H_0 : \theta = 0$ versus $H_A : \theta > 0$.

      b.   Find the exact power for $n(1,1)$ and double exponential with $f(x) = 2^{-1}\exp\{-|x - 1|\}$.

      c.   Find the approximate power at $n(1,1)$ using the normal approximation with continuity correction.

      d.   Suppose F is $n(1,1)$ and find the power of the test based on $\overline{X}$.

1.8.3.   Prove that the sign test is strictly unbiased for $H_0 : \theta = 0$ versus $H_A : \theta > 0$, $F \in \Omega_0$; that is, show that for admissible values of α, $P_{H_0}(S \geqslant k) = \alpha$ and $P_G(S \geqslant k) > \alpha$ for any G in Ω_{alt}. Hint: Use the beta integral representation of the binomial distribution.

1.8.4.   Suppose V rejects $H_0 : \theta = 0$ versus $H_A : \theta > 0$ when $V \geqslant k$. Let $V(\theta)$ be V computed on $X_1 - \theta, \ldots, X_n - \theta$. If $P_{H_0}(V \geqslant k) = \alpha$ and $V(\theta)$ is nonincreasing in θ, then prove that $P_G(V \geqslant k) \geqslant \alpha$ for all G in Ω_{alt}. Hence V provides an unbiased test.

1.8.5.   Construct the graph of $S(\theta)$ for n odd and indicate the point and interval estimates.

1.8.6.   Find the tolerance to rejection for the sign test and show how it converges, as n increases, to the tolerance of the median.

1.8.7.   Show that the t test has 0 tolerance to acceptance.

1.8.8.   Construct the sensitivity curve for the median when n is odd.

1.8.9. Construct the sensitivity curve for the trimmed mean defined in Example 1.6.1.

1.8.10. Suppose F is symmetric about 0 and define

$$T(F) = (1 - 2\alpha)^{-1} \int_{\alpha}^{1-\alpha} F^{-1}(t)\, dt,$$

then $T(F) = 0$, the point of symmetry. The α trimmed mean $\overline{X}_\alpha = T(F_n)$. Show the influence curve is given by

$$\Omega(y) = \begin{cases} (1 - 2\alpha)^{-1} F^{-1}(\alpha), & y < F^{-1}(\alpha) \\ (1 - 2\alpha)^{-1} y, & F^{-1}(\alpha) \leqslant y \leqslant F^{-1}(1 - \alpha) \\ (1 - 2\alpha)^{-1} F^{-1}(1 - \alpha), & y > F^{-1}(1 - \alpha). \end{cases}$$

Hence when sampling from $F(x - \theta)$, $F \in \Omega_s$, by (1.6.2) we have $n^{1/2}(\overline{X}_\alpha - \theta)$ is asymptotically normally distributed with mean 0 and variance:

$$\sigma^2 = \frac{1}{(1 - 2\alpha)^2} \left\{ \int_a^b t^2 f(t)\, dt + 2\alpha \left[F^{-1}(a) \right]^2 \right\}$$

where $b = -a = F^{-1}(\alpha)$, and Ω_s defined in 2.1.1.

1.8.11. In Example 1.5.1, error angle measurements were given for 28 birds released on sunny days. The following data was taken on 13 birds released on cloudy days (error angle in degrees): 8, 10, 38, 43, 45, 57, 73, 76, 83, 105, 112, 126, 141. Construct an approximate 5% sign test for $H_0 : \theta = 90°$ versus $H_A : \theta < 90°$ and carry out the test on the data. Also construct the point estimate of θ and an approximate 90% confidence interval for θ. As in Example 1.5.1, θ represents the population median error angle.

1.8.12. Define $S^* = \#(X_i > 0) - \#(X_i < 0)$. Find the mean, variance, and distribution of S^*. Let $\text{sgn}(x) = 1$, 0, or -1 as $x > 0$, $= 0$, or < 0, respectively, then $S^* = \sum_1^n \text{sgn}(X_i)$. Find the relationship between S and S^*. Find the limiting distribution of S^* and describe how to construct a confidence interval for θ based on S^*.

1.8.13. Suppose $X_1, X_2, \ldots, X_{2n}$ are independent observations such that X_i has cdf $F(x - \theta_i)$ where F is symmetric about 0. For testing $H_0 : \theta_1 = \cdots = \theta_{2n}$ versus $H_A : \theta_1 \leqslant \cdots \leqslant \theta_{2n}$ with at least one strict inequality, we consider the Cox–Stuart (1955) sign test for trend:

$$S = \sum_{i=1}^{n} s(X_{n+i} - X_i), \qquad s(x) = 1 \text{ if } x > 0 \text{ and } 0 \text{ otherwise.}$$

a. Define $p_i = P(X_{n+i} > X_i)$, $i = 1, 2, \dots$ and $q_i = 1 - p_i$. Let $Z_i = s(X_{n+i} - X_i) - p_i$ and use Theorem A6 to prove that

$$\frac{S - \sum_1^n p_i}{\sqrt{\sum_1^n p_i q_i}} \xrightarrow{D} Z \sim n(0,1)$$

provided $\sum_1^\infty p_i q_i$ is a divergent series.

b. Discuss the small sample and asymptotic distribution of S under H_0.

c. Show that $\bar{S} = S/n$ provides a consistent test provided $[1/(n)^{1/2}]\sum_1^n (p_i - 1/2) \to +\infty$. In particular, show the test is consistent if $\theta_{n+i} - \theta_i = \Delta > 0$ for all $i = 1, 2, \dots, 2n$ and $n = 1, 2, \dots$. Hint: Use the asymptotic normality under alternatives given in part (a).

1.8.14. Suppose (X, Y) has a bivariate normal distribution with means 0, variances 1, and correlation ρ. Show that $P(X \leq 0, Y \leq 0) = P(X \geq 0, Y \geq 0) = \frac{1}{4} + (1/2\pi)\sin^{-1}\rho$. Hint: Transform to polar coordinates.

1.8.15. Suppose $X_1, X_2, \dots$ are normally distributed with mean 0, variance 1, and correlation $\text{corr}(X_i, X_j) = \rho$ if $j = i + 1$ and 0 otherwise. Show that Theorem A16 applies and that $n^{1/2}(\bar{S} - 1/2)$ is asymptotically normally distributed with mean 0 and variance $\sigma^2 = 1/4 + 2\{P(X_1 > 0, X_2 > 0) - 1/4\}$, where $\bar{S} = n^{-1}\sum s(X_i)$.

The One-Sample Location Model with a Symmetric, Continuous Distribution

2.1. INTRODUCTION

In the first chapter we found that if there is no shape assumption made on the underlying population, then the sign test is uniformly most powerful size α for a one-sided hypothesis about the median. Hence, without further assumptions, there is no reason to introduce additional statistical tests. If we make more restrictive assumptions, such as normality of the underlying population, then we can again seek optimal procedures. For the location problem, these normal model procedures are based on the mean. In this chapter, we stop short of normality and introduce the assumption of symmetry on the underlying population, then consider how this information can be used to develop additional statistical procedures, such as rank tests.

Typically, the distribution theory for a rank test is more complicated than that for the sign test. Under the null hypothesis, the rank test statistic can be represented as a sum of independent but not identically distributed random variables. The independence is lost when we consider the alternative hypothesis. This necessitates the presentation of theorems beyond the Central Limit Theorem to handle the asymptotics. We also introduce asymptotic efficiency as a means of comparing statistical procedures. Efficiency is a local measure and is strictly valid only in a neighborhood of the null hypothesis. Although not as informative as a power curve, efficiency is much easier to work with. Empirical studies are quoted to show that in many practical situations the efficiency results are relevant for small sample sizes.

A powerful motivation for the symmetry assumption is provided by the paired data design presented in the following example:

Example 2.1.1.   Suppose the pair of random variables (T, C), representing a treatment and control response, respectively, have a joint distribution function $F(t, c)$. Assume that the treatment and control are assigned independently and at random to the subjects of the experiment. Then the null hypothesis of no difference between treatment and control implies that $F(t, c) = F(c, t)$.

Next, we introduce $X = T - C$, the usual form for the data analysis. This reduces the problem to a one-sample location model in which we formulate hypotheses concerning $G(x) = P(X \leqslant x)$.

Under the null hypothesis that $F(t, c) = F(c, t)$ it follows that $P(T - C \leqslant x) = P(C - T \leqslant x) = P(-(T - C) \leqslant x)$ and hence that X and $-X$ have the same distribution. This means that under the null hypothesis, the difference $X = T - C$ is symmetrically distributed about 0.

If the alternative hypothesis specifies that the treatment effect adds a constant to the control then we will test $H_0 : \theta = 0$ versus $H_A : \theta > 0$ where θ is the center of the sampled symmetric population of differences.

There may of course be perfectly good reasons for assuming symmetry of the underlying population distribution in the one-sample location model. The paired data design just provides one source of such examples. Another approach, advocated by many data analysts, is to transform the data. However, the transformation is often selected on the basis of the data, and this makes it difficult to interpret the significance levels and confidence coefficients.

We now introduce the subclass of Ω_s of symmetric distributions, centered at 0. Let

$$\Omega_s = \{ F : F \in \Omega_0 \text{ and } F(x) = 1 - F(-x) \}. \qquad (2.1.1)$$

Hence X (or F) is said to be symmetric about 0 and 0 is the unique median and mean (when it exists). The sampling model is then given by $X_1, \ldots, X_n$, a random sample from $F(x - \theta)$, $F \in \Omega_s$. Hence θ is the unique median and mean (when it exists) at the center of the distribution.

In this sampling model, we assume that the experimental effect is experienced solely in a change of location. This may not always be the case, and we will see later that if the effect is to introduce asymmetry, the rank tests may still tend to reject the null hypothesis (Example 2.5.1). However, unless further qualified, the hypotheses under test are:

$$H_0 : \theta = 0$$

$$H_A : \theta > 0$$

with $F \in \Omega_s$.

Example 2.1.2. Rosenzweig et al. (1972) describe experiments, carried out in the 1960s, to determine the effects of environment on brain anatomy. The hypothesis of such an effect can be traced back to an Italian anatomist, Gaetano Malacarne, working in the 1780s. In the more recent experiments three male rats from each of 12 litters are randomly assigned to a standard laboratory cage, an enriched cage containing a variety of toys, and an inpoverished environment in which the rats lived in isolation. Various measures of brain weight and enzymatic activity were taken. For this example, the weight gain of the cortex over a specific period of time was considered. If we compare rats in an impoverished environment to rats in an enriched environment, we have a paired-data experiment. The pairs are naturally formed by litter mates with the same genetic makeup. Let $X(Y)$ denote the impoverished (enriched) measurement, then the basic random variable of interest is $D = Y - X$. Under the null hypothesis of no difference in the effects due to the two environments, D has a distribution symmetric about 0. If we let θ denote the center of the distribution of D, then the experiment yields 12 observations $D_1, \ldots, D_{12}$ to test $H_0 : \theta = 0$ versus $H_A : \theta > 0$. Data for this example is presented in Example 2.3.1.

2.2. THE WILCOXON SIGNED RANK TEST

The sign test $S = \sum s(X_j)$ uses only information in the sign of the observation; no metric information on how far the observation is from zero is incorporated into the test. However, for a distribution that is symmetric about 0, the vector of absolute values of the observations is a sufficient statistic (see Lehmann, 1959, p. 56). Absolute value is just the distance from 0; so, for symmetric distributions, it would seem reasonable to try to use this information.

In general, the rank of a quantity Z_i among $Z_1, \ldots, Z_n$ is the number of items $Z_k \leqslant Z_i$, $k = 1, \ldots, n$. Hence the rank of Z_i is its position in the ordered set $Z_{(1)} < \cdots < Z_{(n)}$.

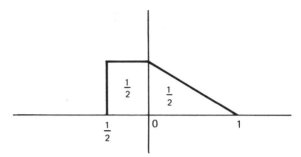

Figure 2.1. A right-skewed distribution with median 0.

Returning to the sampling model $X_1, \ldots, X_n$ i.i.d. $F(x - \theta)$, $F \in \Omega_s$, we rank $|X_1|, \ldots, |X_n|$ and define a statistic T as the sum of ranks of the positive sample items among the absolute values. We will reject $H_0 : \theta = 0$ in favor of $H_A : \theta > 0$ if $T \geqslant k$, where k is determined by the null distribution of T.

The statistic T was proposed by Wilcoxon (1945) and is referred to as the Wilcoxon signed rank statistic. When $\theta > 0$ and the symmetric distribution is shifted to the right, positive observations tend to be farther from 0 than the negative observations. Hence T tends to be large and rejects $H_0 : \theta = 0$. In contrast to the sign statistic, T takes into account, through the ranks, the relative distances from 0.

As mentioned previously, it may happen that the median is 0 but the distribution is skewed to the right. An example is given in Fig. 2.1. From the figure it is easy to see that T will tend to be large even though the median is 0. Hence the symmetry assumption is necessary for an unambiguous interpretation of large values of T. If the population median is known, then T provides a test of symmetry.

Define

$$W_j = 1 \qquad \text{if } |X|_{(j)} \text{ corresponds to a positive observation}$$

$$\qquad 0 \qquad \text{otherwise,}$$

(2.2.1)

where $|X|_{(1)} < \cdots < |X|_{(n)}$ are the ordered absolute values. Then

$$T = \sum_{j=1}^{n} jW_j = \sum_{j=1}^{n} R_j s(X_j) \qquad (2.2.2)$$

is the Wilcoxon signed rank statistic where R_j is the rank of $|X_j|$. In order to study the distribution theory involving T we need further notation.

Definition 2.2.1. If R_j is the rank of $|X_j|$ then $|X_j| = |X|_{(R_j)}$. The antirank is then defined by D_j such that $|X_{D_j}| = |X|_{(j)}$. Hence D_j labels the X which corresponds to the jth ordered absolute value.

From the definition of D_j and from (2.2.1) it follows that

$$W_j = s(X_{D_j}) \tag{2.2.3}$$

where $s(x) = 1$ if $x > 0$ and 0 otherwise.

Theorem 2.2.1. Under the null hypothesis $H_0 : \theta = 0$, $F \in \Omega_s$, $s(X_1)$, . . . , $s(X_n)$ and the vector $(R_1, \ldots, R_n)$ are mutually independent.

Proof. Since $(s(X_i), |X_i|)$, $i = 1, \ldots, n$ are independent pairs we first need to show that $s(X_i)$ and $|X_i|$ are independent. Consider

$$P(s(X_i) = 1, |X_i| \leqslant x) = P(0 \leqslant X_i \leqslant x)$$

$$= F(x) - F(0)$$

$$= F(x) - \tfrac{1}{2}$$

$$= (1/2)(2F(x) - 1)$$

$$= P(s(X_i) = 1) \cdot P(|X_i| \leqslant x).$$

Similarly for $P(s(X_i) = 0, |X_i| \leqslant x)$. Hence $s(X_i)$ and $|X_i|$ are independent. Since $(R_1, \ldots, R_n)$ is a function of $|X_1|, \ldots, |X_n|$, the theorem follows.

Just as the signs are independent of the ranks, they are also independent of the antiranks, and we have $s(X_1), \ldots, s(X_n)$ and $(D_1, \ldots, D_n)$ are mutually independent.

Theorem 2.2.2. Under the null hypothesis $H_0 : \theta = 0$, $F \in \Omega_s$, W_1, . . . , W_n are independent, identically distributed with

$$P(W_i = 0) = P(W_i = 1) = 1/2$$

Proof.   Let $D = (D_1, \ldots, D_n)$ and $d = (d_1, \ldots, d_n)$, then using (2.2.3) and Theorem 2.2.1 we have:

$$P(W_1 = w_1, \ldots, W_n = w_n)$$

$$= \sum_d P\big(s(X_{D_1}) = w_1, \ldots, s(X_{D_n}) = w_n \,|\, D = d\big) P(D = d)$$

$$= \sum_d P\big(s(X_{d_1}) = w_1, \ldots, s(X_{d_n}) = w_n\big) P(D = d)$$

$$= \left(\tfrac{1}{2}\right)^n \sum_d P(D = d)$$

$$= (1/2)^n.$$

Hence $P(W_1 = w_1, \ldots, W_n = w_n) = \prod_1^n P(W_i = w_i)$, $P(W_i = w_i) = 1/2$, and the theorem follows.

Hence $T = \sum j W_j$ is a linear combination of i.i.d. $B(1, 1/2)$ random variables under $H_0 : \theta = 0$, $F \in \Omega_s$. This results in fairly straight forward distribution theory under H_0 and shows that T is distribution free under H_0. Although a closed formula for the distribution of T is not available, the following example illustrates the simplicity of the distribution.

Example 2.2.1.   For sample size 4, the possible ranks of the absolute values are 1, 2, 3, 4 with the associated signs attached independently with probability $1/2$. We list a few of the $2^4 = 16$ configurations.

	Ranks				
	1	2	3	4	Value of T
Signs	+	+	+	+	10
	−	+	+	+	9
	+	−	+	+	8
	+	+	−	+	7
	+	+	+	−	6
	−	−	+	+	7
	−	+	−	+	6
	−	+	+	−	5

Since each configuration of signs has probability $1/2^4 = 1/16$, we can enumerate the probabilities for T. For example, $P(T = 10) = 1/16$, $P(T = 6) = 2/16$, and so on. If we agree to reject $H_0: \theta = 0$ if $T \geqslant 9$, then the size of the test is $\alpha = P(T \geqslant 9) = 1/8$.

The mean and variance of T, under H_0, are easily derived from Theorem 2.2.2 as (see Exercise 2.10.1):

$$E_{H_0} T = n(n + 1)/4$$

$$\text{Var}_{H_0} T = n(n + 1)(2n + 1)/24.$$

(2.2.4)

In fact, the moment-generating function for T is easy to derive from Theorem 2.2.2.

Example 2.2.2. Under $H_0: \theta = 0$, $F \in \Omega_s$, the moment-generating function of T is given by:

$$M(t) = \frac{1}{2^n} \prod_{j=1}^{n} (1 + e^{tj}).$$

(2.2.5)

Proof. From the definition of $M(t)$ and Theorem 2.2.2 we have

$$M(t) = E(e^{tT})$$

$$= E e^{t \Sigma j W_j}$$

$$= \prod E e^{tj W_j}.$$

Now $E e^{tj W_j} = e^0/2 + e^{tj}/2 = (1 + e^{tj})/2$, and the result follows.

The moments, given by (2.2.4), could be derived from (2.2.5). Since the probability function for T cannot be given in closed form, the moment-generating function provides an alternative method for constructing the probabilities of T. Hence if $M(t) = a_0 e^{0t} + a_1 e^t + a_2 e^{2t} + \cdots$, then $P(T = j) = a_j$. In the following example we show how these probabilities can be developed in a systematic way.

Example 2.2.3. We begin with $n = 2$, and thus

$$M(t) = (1/2^2)(1 + e^t)(1 + e^{2t}) = (1/2^2)(1 + e^t + e^{2t} + e^{3t}).$$

We now list the powers and the coefficients.

$$
\begin{array}{cccccl}
 & 0 & 1 & 2 & 3 & \text{Powers} \\
1/4 \quad x & 1 & 1 & 1 & 1 & \text{Coefficients.}
\end{array}
\tag{2.2.6}
$$

Thus, $P(T = 0) = P(T = 1) = \cdots = P(T = 3) = 1/4$. For $n = 3$, we have

$$
M(t) = (1/2^3)(1 + e^t)(1 + e^{2t})(1 + e^{3t}),
$$

and the powers and coefficients are developed from (2.2.6) as

$$
\begin{array}{cccccccccl}
 & 0 & 1 & 2 & 3 & | & 4 & 5 & 6 & & \text{Powers} \\
 & 1 & 1 & 1 & 1 & & & & & \\
 & & & & 1 & & 1 & 1 & 1 & \\
1/8 \quad x & 1 & 1 & 1 & 2 & & 1 & 1 & 1 & & \text{Coefficients.}
\end{array}
\tag{2.2.7}
$$

The vertical dividing line in the first line of (2.2.7) indicates where additional powers for the next sample size are added. For $n = 4$, we have

$$
M(t) = (1/2^4)(1 + e^t) \ldots (1 + e^{4t}).
$$

Hence, from (2.2.7), we have

$$
\begin{array}{ccccccccccccl}
 & 0 & 1 & 2 & 3 & 4 & 5 & 6 & | & 7 & 8 & 9 & 10 & \text{Powers} \\
 & 1 & 1 & 1 & 2 & 1 & 1 & 1 & & & & & & \\
 & & & & 1 & 1 & 1 & & 2 & 1 & 1 & 1 & & \\
1/16 \quad x & 1 & 1 & 1 & 2 & 2 & 2 & 2 & & 2 & 1 & 1 & 1 & \text{Coefficients}
\end{array}
\tag{2.2.8}
$$

and, for example, $P(T = 7) = 2/16 = 1/8$.

The pattern of this example should now be obvious. A computer can easily be programmed to develop tables of the distribution of T under H_0. Further, the example provides evidence for the symmetry of the distribution of T (see Exercise 2.10.2). A recurrence formula for the distribution of T is given in Exercise 3.7.3.

As in the case of the sign test, when the sample size is large or when a table of the distribution is not available, it is important to have a normal approximation for the distribution of T under H_0. From the independence of $W_1, \ldots, W_n$, Theorem 2.2.2, we can still rely on central limit theory. However, T is a linear function of $W_1, \ldots, W_n$ and the usual Central Limit Theorem must be extended. The extension that we use is proved in Theorem A9 of the Appendix. In Exercise 2.10.4 you are asked to apply

Theorem A9 to T and show that

$$\frac{T - n(n+1)/4}{\sqrt{\dfrac{n(n+1)(2n+1)}{24}}}$$

has a standard normal limiting distribution. Hence if we reject $H_0 : \theta = 0$ for $H_A : \theta > 0$ when $T \geqslant c$, then the size α critical value can be approximated (with continuity correction) by

$$c \doteq \frac{n(n+1)}{4} + .5 + Z_\alpha \sqrt{\frac{n(n+1)(2n+1)}{24}} \qquad (2.2.9)$$

where Z_α is the upper α percentile of the standard normal distribution.

The accuracy of the approximation is usually increased by using an Edgeworth approximation; see Cramér (1946, Sections 12.6 and 17.7). If V has a symmetric distribution, the approximation is given by

$$P(V \leqslant v) \doteq \Phi(t) - \tfrac{1}{24}\lambda_4(t^3 - 3t)\psi(t) \qquad (2.2.10)$$

where $\lambda_4 = \{ E(V - EV)^4 / [E(V - EV)^2]^2 \} - 3$, called the excess, is a function of the kurtosis, $\psi(\cdot)$ is the standard normal pdf, and $t = (v - EV)/(\operatorname{Var} V)^{1/2}$. (If V is discrete, t usually contains a correction for continuity.) For the Wilcoxon signed rank statistic, Fellingham and Stoker (1964) show

$$P(T \leqslant k) \doteq \Phi(t) + \left[\frac{(3n^2 + 3n - 1)}{10n(n+1)(2n+1)} \right](t^3 - 3t)\psi(t) \qquad (2.2.11)$$

where $t = (k + .5 - ET)/(\operatorname{Var} T)^{1/2}$. Example 2.2.2 can be used to find the necessary moments. Bickel (1974) has shown this is a rigorous asymptotic expansion taken to terms of order n^{-1}. See Table 2.1.

Table 2.1. Normal and Edgeworth Approximations to the Null Distribution of T, with Continuity Correction and $n = 5$

	\multicolumn{8}{c}{k}								
	0	1	2	3	4	5	6	7
$P(T \leqslant k)$	.031	.062	.094	.156	.219	.312	.406	.500
Normal	.029	.053	.089	.140	.212	.295	.394	.500
Edgeworth	.027	.055	.096	.152	.227	.309	.402	.500

There are various other approaches for establishing the limiting normality of T. Several are based on the moments or moment-generating function. See the papers by Haigh (1971) and Noether (1970).

2.3. POINT AND INTERVAL ESTIMATES BASED ON THE WILCOXON SIGNED RANK STATISTIC

Recall Section 1.5 in which we defined, for the sign test, $S(\theta) = \# X_i > \theta$, $i = 1, \ldots, n$. The Hodges–Lehmann estimate and the interval estimate of θ based on S then follow easily from the graph of $S(\theta)$, which is a nonincreasing step function with steps at the ordered sample values. In this section, we derive a counting form for the Wilcoxon signed rank statistic which enables us to easily construct the corresponding point and interval estimates.

Definition 2.3.1. Given a random sample $X_1, \ldots, X_n$ the $n(n+1)/2$ Walsh averages are defined by $(X_i + X_j)/2$, $i \leqslant j$. This name was given to the pairwise averages by John Tukey in reference to work of John Walsh. Tukey (1949) then went on to develop the following representation.

Theorem 2.3.1. The Wilcoxon signed rank statistic T defined in (2.2.2) can be written as

$$T = \#\left(\frac{X_i + X_j}{2} > 0\right), \qquad i \leqslant j.$$

Hence T is the number of positive Walsh averages.

Proof. Let $X_{i_1}, \ldots, X_{i_p}$ denote the p positive sample items; then T is the sum of ranks of these items among the absolute values.

Draw a circle with center at the origin and radius X_{i_1}, as in Fig. 2.2. Then the rank of X_{i_1} is equal to the number of sample points in the circle

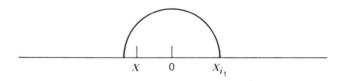

Figure 2.2. Counting positive Walsh averages.

including X_{i_1} since we are ranking the distance to 0. Every average formed by a sample point in the circle and X_{i_1} is positive. Hence the rank of X_{i_1} is just the number of positive Walsh averages formed by X_{i_1} and sample points less than or equal to X_{i_1}. If this procedure is repeated systematically for $i_2, \ldots, i_p$ then the sum of ranks is seen to be identical to the number of positive Walsh averages.

We now see that

$$T(\theta) = \#\left(\frac{X_i + X_j}{2} > \theta \right), \qquad i \leqslant j, \tag{2.3.1}$$

is a nonincreasing step function with steps at the Walsh averages. It follows at once from Definition 1.4.1 and Exercise 1.8.4 that T provides an unbiased test of $H_0 : \theta = 0$ versus $H_A : \theta > 0$, $F \in \Omega_s$.

From the signlike or counting structure of T, and the symmetry of its null distribution (Exercise 2.10.2) we can at once write down the Hodges–Lehmann estimate of θ (see Section 1.5) as

$$\hat{\theta} = \operatorname*{med}_{i \leqslant j}\left(\frac{X_i + X_j}{2} \right), \tag{2.3.2}$$

the median of the Walsh averages. Further, if $W_{(1)} \leqslant \cdots \leqslant W_{(N)}$, $N = n(n+1)/2$, are the ordered Walsh averages and $P(T \leqslant a) = \alpha/2 = P(T \geqslant N - a)$, then

$$[W_{(a+1)}, W_{(N-a)}) \tag{2.3.3}$$

is the $(1 - \alpha)$ 100% confidence interval for θ based on T. From (2.2.9), a can be approximated (with continuity correction) by

$$a \doteq \frac{n(n+1)}{4} - .5 - Z_{\alpha/2}\sqrt{\frac{n(n+1)(2n+1)}{24}} \ .$$

If n is of moderate size, the number of Walsh averages will be quite large. It is then not practical to compute the estimate or confidence interval by hand. The Minitab statistical computing system contains the commands WTEST and WINT which provide the Wilcoxon signed rank test, point estimate, and confidence interval.

The next theorem shows that under most sampling situations $\hat{\theta}$ is an unbiased estimate of θ. Later in Theorem 2.6.5, we will show that $\hat{\theta}$ is approximately normally distributed.

Theorem 2.3.2. If $F \in \Omega_s$, then the distribution of $\hat{\theta}$ is symmetric about θ.

Proof. We first observe that $P_\theta(\hat{\theta} - \theta < x) = P_0(\hat{\theta} < x)$ by Theorem 1.5.1. Hence we take $\theta = 0$ without loss of generality. Now $X = (X_1, \ldots, X_n)$ and $-X = (-X_1, \ldots, -X_n)$ have the same joint distribution since $F \in \Omega_s$. If we write $\hat{\theta}(X)$ for med$(X_i + X_j)/2$, $i \leqslant j$, then $\hat{\theta}(X)$ and $\hat{\theta}(-X)$ have the same distribution, and $\hat{\theta}(-X) = -\hat{\theta}(X)$ implies that $\hat{\theta}$ and $-\hat{\theta}$ have the same distribution. Hence the theorem follows.

One drawback of the median of the Walsh averages is the large number of Walsh averages that must be computed and ordered. For example, if $n = 50$, then there are $n(n + 1)/2 = 1275$ Walsh averages. Estimates, such as the sample median and Galton's estimate (see Exercise 2.10.7), require less computation and have positive tolerance. If they have good statistical efficiency properties they could provide an attractive alternative to the median of the entire set of Walsh averages. We discuss the statistical properties in later sections of the chapter.

In summary, when sampling from a symmetric distribution with point of symmetry θ, T, the sum of ranks of the positive items among the absolute values, provides a natural test of $H_0 : \theta = 0$ versus $H_A : \theta > 0$. This test utilizes information in the symmetry by ranking the distances of the observations from the origin. The test is distribution free under H_0 and the underlying hope is that it provides more power for detecting symmetric shift alternatives than the sign test. This problem of power and efficiency is taken up in Sections 2.5 and 2.6. We have further seen that, just as in the sign test case, natural point and interval estimates based on T are available. Under the model $F \in \Omega_s$, $\hat{\theta}$ is an unbiased estimate of θ, and T provides an unbiased test of $H_0 : \theta = 0$ versus $H_A : \theta > 0$. In the next section we consider the stability properties of the Wilcoxon procedures.

Example 2.3.1. In Example 2.1.2 we considered littermate rats randomly assigned to an enriched or impoverished environment. The measurement taken was the weight in milligrams of the cortex after a fixed period of time. There were 12 pairs and the data is given in Table 2.2. We suppose that the distribution of differences is symmetrically distributed about θ, and

Table 2.2. Cortex Weight (mg) (Artificial Data)

	\multicolumn{12}{c	}{Pair}
Environment	1	2	3	4	5	6	7	8	9	10	11	12
Enriched	689	663	653	740	699	690	685	718	742	651	687	679
Impoverished	657	646	642	650	698	621	647	689	652	661	612	678
Difference	32	17	11	90	1	69	38	29	90	-10	75	1

we wish to test $H_0: \theta = 0$ versus $H_A: \theta > 0$ based on the 12 observed differences. If the significance level is taken to be $\alpha = .01$, then we reject $H_0: \theta = 0$ if $T \geqslant 68$ [the normal approximation, (2.2.9), yields $T \geqslant 70$]. The observed value of T is 75; hence we reject $H_0: \theta = 0$ and claim, at $\alpha = .01$, that the cortex weight of rats raised in an enriched environment is significantly greater than that of rats raised in an impoverished environment. The point estimate of θ is $\hat{\theta} = 36.5$, the median of the 78 Walsh averages. A 94.5% confidence interval for θ is determined by the 15th and the 64th ordered Walsh averages: $[11.0, 59.5)$.

In practice it is often necessary to deal with zeros (observations that are equal to the null hypothesized value) and with ties among the absolute values. As in the case of the sign test, we set aside the zeros and reduce the sample size accordingly before computing the Wilcoxon signed rank test. Zeros are not excluded for the purposes of estimation. Observations that have the same absolute value are assigned the average rank for that set of observations. These midranks are then used in the calculation of T. This corresponds to counting a zero Walsh average as $1/2$. There are other methods for handling ties and none is superior to the others in all situations. The midrank method is the one most commonly used in practice; Conover (1973) discusses and compares several of the methods. See also Lehmann (1975) and Pratt (1959). Noether (1967) shows that, when the underlying population is discrete, the closed confidence interval always has confidence coefficient bounded below by the stated confidence coefficient.

2.4. STABILITY PROPERTIES OF RANK TESTS AND ESTIMATES

In Exercise 2.10.5 you are asked to compute the testing tolerance of T to both acceptance and rejection and to show that the asymptotic tolerance in both cases is .29. Hence T provides a test with positive tolerance, between the highly tolerant sign test and the intolerant t test.

We now turn to a general study of tolerance of estimators which are constructed from the Walsh averages in a special way (see Hodges, 1967). Note that the sample median is the median of the set of Walsh averages restricted by $i = j$ (see Definition 2.3.1).

Let B be a subset of indices (i, j) contained in the set $\{(i, j): i \leqslant j\}$ and define the estimate $\hat{\theta}$ by

$$\hat{\theta} = \underset{(i,j) \in B}{\text{med}} \ \frac{X_{(i)} + X_{(j)}}{2} \tag{2.4.1}$$

where $X_{(1)} < \cdots < X_{(n)}$ are the ordered sample values. For example,

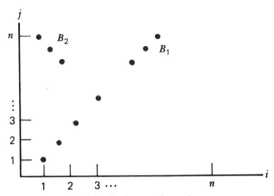

Figure 2.3.   Index sets for estimates.

$B = \{(i, j) : i = j\}$ defines the sample median, and $B = \{(i, j) : i \leqslant j\}$ defines the median of the Walsh averages. Another estimate, attributed to Galton by Hodges (1967), is defined by $B = \{(i, j) : i + j = n + 1\}$. A convenient way to visualize the index set B is given in Fig. 2.3. The set $\{(i, j) : i \leqslant j\}$ is represented by the lattice points in the upper triangle. The sets B_1 and B_2, defining the sample median and Galton's estimate, respectively, are shown. Other estimates are illustrated later.

If the pair (i, j) in B implies that $(n - j + 1, n - i + 1)$ is also in B, the set B is said to be symmetric. The following theorem shows that a symmetric index set produces unbiased estimates of θ when sampling from a symmetric population.

Theorem 2.4.1.   If the sampling model is given by $F(x - \theta)$ with $F \in \Omega_s$ and if B is symmetric, then the estimate $\hat{\theta}$, (2.4.1), is symmetrically distributed about θ.

Proof.   Without loss of generality take $\theta = 0$. From the form of the joint density of pairs of order statistics, it is easy to see that $(X_{(i)}, X_{(j)})$ and $(-X_{(n-j+1)}, -X_{(n-i+1)})$ have the same distribution. Hence $\hat{\theta}(X)$ based on $X = (X_1, \ldots, X_n)$ has the same distribution as $\hat{\theta}(-X)$. But $\hat{\theta}(-X) = -\hat{\theta}(X)$; hence $\hat{\theta}$ and $-\hat{\theta}$ have the same distribution, and $\hat{\theta}$ is symmetrically distributed about 0.

Theorem 2.4.2.   Suppose B is symmetric. Then the tolerance of $\hat{\theta}$, (2.4.1), is a/n, where a is the largest integer such that

$$\tfrac{1}{2}\{ \# B + 1\} \leqslant \# K_{a+1}, \tag{2.4.2}$$

where $K_t = \{(i, j) : t \leqslant i \leqslant j\}$ and $\#$ denotes the number of elements in the set.

Proof. We will prove the result for $\# B = 2q$, even. The odd case is left as Exercise 2.10.6.

Now a is the largest integer such that

$$(\tfrac{1}{2})[2q + 1] = q + 1/2 \leqslant \# K_{a+1}.$$

Hence $q + 1 \leqslant \# K_{a+1}$. This shows that $\# K_{a+1}$ exceeds half the $\# B$. Further $\# K_{a+2} \leqslant q$ because of the maximal property of a.

If $(i, j) \in K_{a+1}$, then $a + 1 \leqslant i \leqslant j$ and

$$X_{(a+1)} \leqslant \frac{X_{(i)} + X_{(j)}}{2}.$$

Hence more than half of the Walsh averages defined by B are bounded below by $X_{(a+1)}$, and it follows that

$$X_{(a+1)} \leqslant \underset{B}{\text{med}} \, \frac{X_{(i)} + X_{(j)}}{2} = \hat{\theta},$$

as required in Definition 1.6.1.

Now fix $X_{(a+2)}, \ldots, X_{(n)}$ and let $X_{(a+1)} \to -\infty$. If $i \leqslant a + 1$, then as $X_{(a+1)} \to \infty$,

$$\frac{X_{(i)} + X_{(j)}}{2} \leqslant \frac{X_{(a+1)} + X_{(j)}}{2} \to -\infty.$$

Since $K_{a+2} = \{(i, j) : a + 2 \leqslant i \leqslant j\}$, $i \leqslant a + 1$ implies that (i, j) is in the complement of K_{a+2}. Hence all Walsh averages defined by the complement of K_{a+2} tend to $-\infty$. We have seen that $\# K_{a+2} \leqslant q$; so the number of elements in the complement of K_{a+2} is at least $q + 1$, more than half the $\# B$. This means that $\hat{\theta}$ is contained in the complement of K_{a+2}, and hence $\hat{\theta} \to -\infty$ as $X_{(a+1)} \to -\infty$, as required by Definition 1.6.1.

The rest of the conditions in Definition 1.6.1 follow from the symmetry of B.

Example 2.4.1. Let $B = \{(i, j) : i \leqslant j\}$, so that $\hat{\theta}$ is the median of the Walsh averages, (2.3.2). In this case $\# B = n(n + 1)/2$. The set $K_{a+1} = \{(i, j) : a + 1 \leqslant i \leqslant j\}$, and hence $\# K_{a+1} = (n - a) + (n - a - 1) + \cdots + 1 = (n - a)(n - a + 1)/2$. The theorem shows that we must find the largest a such that

$$\frac{1}{2}\left\{\frac{n(n+1)}{2}+1\right\} \leqslant \frac{(n-a)(n-a+1)}{2}. \tag{2.4.3}$$

This reduces to the quadratic inequality

$$a^2 - (2n+1)a + (n^2 + n - 2)/2 \geqslant 0,$$

and a is the greatest integer in

$$\frac{2n + 1 - \sqrt{(2n^2 + 2n + 5)}}{2}. \tag{2.4.4}$$

The other solution of the quadratic yields a value of a outside the range 1 to n. If we disregard the noninteger character of (2.4.4), the tolerance (Definition 1.6.1) can be written as

$$\tau_n \doteq 1 + \frac{1}{2n} - \frac{1}{2n}\sqrt{2n^2 + 2n + 5},$$

and

$$\tau = \lim \tau_n = 1 - \frac{1}{\sqrt{2}} \doteq .29.$$

We see, as with the sign test and the sample median, that the testing tolerances of the Wilcoxon signed rank test (Exercise 2.10.5) converge to the asymptotic estimation tolerance of the Hodges–Lehmann estimate, the median of the Walsh averages. The convergence of τ_n to τ is quite rapid; for example, some values of (n, τ_n) are $(4, .286)$, $(6, .297)$, $(10, .300)$, $(20, .298)$.

We complete this section with a discussion of the sensitivity and influence curves of the median of the Walsh averages. The results of the following example can be compared to Example 1.6.4.

Example 2.4.2. We consider $\hat{\theta} = \text{med}(X_i + X_j)/2$, $i \leqslant j$, (2.3.2), the Hodges and Lehmann estimate of θ, derived from the Wilcoxon signed rank statistic. We suppose that $\hat{\theta}$ is based on a sample of size n from $H(x) = F(x - \theta)$, $F \in \Omega_s$. To construct the influence curve we need to represent $\hat{\theta}$ as a functional evaluated at the empirical cdf, $H_n(x)$. This functional is implicitly defined by setting the cdf of $(X_i + X_j)/2$ equal to

$1/2$. To simplify the discussion, we consider $\hat{\theta}^* = \text{med}(X_i + X_j)/2$, $i, j = 1$, $2, \ldots, n$.

The cdf of $(X_i + X_j)/2$ is given by

$$P\left(\frac{X_i + X_j}{2} \leqslant z\right) = \int_{-\infty}^{\infty}\int_{-\infty}^{2z-u} h(x)h(u)\,dx\,du$$

$$= \int_{-\infty}^{\infty} H(2z - u)h(u)\,du.$$

Next define the functional $T(H)$, the median of the distribution of $(X_i + X_j)/2$, implicitly by

$$\int_{-\infty}^{\infty} H(2T(H) - u)h(u)\,du = 1/2. \tag{2.4.5}$$

First we will show $\hat{\theta}^* = T(H_n)$ and then we will derive the influence curve. Replace H by H_n in (2.4.5) to get

$$\int_{-\infty}^{\infty} H_n(2T(H_n) - u)\,dH_n(u) = 1/2$$

and

$$\frac{1}{n}\sum_{i=1}^{n} H_n(2T(H_n) - x_i) = 1/2.$$

Now $H_n(2T(H_n) - x_i) = \sum_{j=1}^{n} I(x_j \leqslant 2T(H_n) - x_i)/n$, where $I(\cdot)$ is the indicator function. Hence we have

$$\frac{1}{n^2}\sum_{i=1}^{n}\sum_{j=1}^{n} I\left(\frac{x_i + x_j}{2} \leqslant T(H_n)\right) = \frac{1}{2},$$

and this defines $T(H_n)$ as the median of the n^2 averages $(x_i + x_j)/2$, $i, j = 1, 2, \ldots, n$.

We now apply Definition 1.6.4 to find the influence curve. With $H_t = (1 - t)H + t\delta_y = H + t(\delta_y - H)$, we have, from (2.4.5),

$$\int_{-\infty}^{\infty}\{H(2T(H_t) - u) + t[\delta_y(2T(H_t) - u) - H(2T(H_t) - u)]\}$$

$$d[H(u) + t(\delta_y(u) - H(u))] = \frac{1}{2}.$$

Expand this expression by multiplying out the integrand and integrator and then differentiate with respect to t. Differentiation and integration can be interchanged by applying Theorem A18 since the integrand is bounded and differentiable. The result of evaluating at $t = 0$ is

$$\frac{d}{dt} T(H_t)\bigg|_{t=0} 2 \int_{-\infty}^{\infty} h(2T(H) - u) \, dH(u) + \int_{-\infty}^{\infty} H(2T(H) - u) \, d\delta_y(u)$$

$$- \int_{-\infty}^{\infty} H(2T(H) - u) \, dH(u) + \int_{-\infty}^{\infty} \delta_y(2T(H) - u) \, dH(u)$$

$$- \int_{-\infty}^{\infty} H(2T(H) - u) \, dH(u) = 0.$$

Since $H(x) = F(x - \theta)$, $F \in \Omega_s$, without loss of generality we let $\theta = 0$. This means that in the foregoing equation we can take $T(H) = 0$, just as we let the mean and median be 0 in Example 1.6.5.

Now $F(-u) = 1 - F(u)$, $f(-u) = f(u)$ and so $\int F(-u) \, dF(u) = 1/2$, $\int \delta_y(-u) \, dF(u) = F(-u) = 1 - F(u)$, and $\int F(-u) \, d\delta_y(u) = F(-y) = 1 - F(y)$. The equation foregoing now becomes

$$\frac{d}{dt} T(F_t)\bigg|_{t=0} 2 \int f^2(u) \, du + (1 - F(y)) - 1/2 + (1 - F(y)) - 1/2 = 0$$

and

$$\Omega(y) = \frac{d}{dt} T(F_t)\bigg|_{t=0}$$

$$= \frac{2F(y) - 1}{2 \int F^2(u) \, du}$$

$$= \frac{F(y) - 1/2}{\int f^2(u) \, du}. \tag{2.4.6}$$

Hence the influence curve is the cdf, centered and scaled. This means that $\hat{\theta}*$ (and $\hat{\theta}$) has a bounded, continuous influence curve and is robust. Now from (2.4.6), since $\int [F(y) - 1/2]^2 \, dF(y) = 1/12$, we have $\sigma^2 = \int \Omega^2(y) \, dF(y) = 1/12\{\int f^2(y) \, dy\}^2$. Hence (1.6.2) suggests $n^{1/2}(\hat{\theta} - \theta)$ is asymptotically $n(0, \sigma^2)$.

The stylized sensitivity curve for $\hat{\theta}$ is a bounded, nondecreasing step function, much like an empirical cdf that has been centered and scaled. For large n the sensitivity curve looks like (2.4.6); see the Princeton robustness study (Andrews et al., 1972, Section 5E).

2.5.   GENERAL ASYMPTOTIC THEORY FOR THE WILCOXON SIGNED RANK STATISTIC

We now consider a random sample $X_1, \ldots, X_n$ from an arbitrary, continuous distribution $H(x)$. Later we will want to consider the location model $H(x) = F(x - \theta)$. First, we develop the mean and variance of T and establish the consistency of the test based on T. Next, we point out that under alternative hypotheses, T can no longer be represented as the sum of independent random variables. This means that the Central Limit Theorem cannot be applied directly. In order to find the limiting distribution, it is necessary to project T onto the class of sums of independent random variables. The strategy is then to apply the Central Limit Theorem to the sum and show that the difference between T and its projection tends to zero in probability. After establishing the asymptotic distribution of T in general, we will consider the asymptotic power. Power considerations motivate the development of asymptotic efficiency which will then be developed in the next section.

The mean and variance of T, in general, are determined by the following parameters which depend on the underlying distribution:

$$p_1 = P(X_1 > 0)$$

$$p_2 = P(X_1 + X_2 > 0)$$

$$p_3 = P(X_1 + X_2 > 0, X_1 > 0) \tag{2.5.1}$$

$$p_4 = P(X_1 + X_2 > 0, X_1 + X_3 > 0).$$

In Exercise 2.10.8 it is pointed out that $p_3 = (p_2 + p_1^2)/2$; hence only p_1, p_2, and p_4 are needed. The development is more natural, however, when we include p_3.

Theorem 2.5.1.

$$ET = np_1 + \frac{n(n-1)}{2} p_2$$

$$\tag{2.5.2}$$

$$\operatorname{Var} T = np_1(1 - p_1) + \frac{n(n-1)}{2} p_2(1 - p_2)$$

$$+ 2n(n-1)(p_3 - p_1 p_2) + n(n-1)(n-2)(p_4 - p_2^2).$$

Proof. Define, in accordance with Theorem 2.3.1,

$$
T_{ij} = \begin{cases} 1 & \text{if } \dfrac{(X_i + X_j)}{2} > 0 \\ 0 & \text{otherwise,} \end{cases}
\tag{2.5.3}
$$

and hence $T = \sum\sum T_{ij}$, where the double summation is over subscripts $i \leqslant j$. We then have

$$
ET = \sum_{i \leqslant j}\sum ET_{ij}
$$

$$
= nET_{11} + \frac{n(n-1)}{2} ET_{12}.
$$

Now, $ET_{11} = P(X_1 > 0) = p_1$ and $ET_{12} = P(X_1 + X_2 > 0) = p_2$.
Next, we have

$$
\text{Var } T = \text{Var}\left(\sum_{i<j}\sum T_{ij} + \sum_k T_{kk} \right).
$$

Consider these sums written out:

$$
T_{12} + T_{13} + \cdots + T_{1n} + T_{23} + \cdots + T_{2n} + \cdots + T_{jj+1} + \cdots
$$

$$
+ T_{jn} + \cdots + T_{n-1n} + T_{11} + \cdots + T_{nn}.
\tag{2.5.4}
$$

The variance of a sum is the sum of the variances plus twice the sum of the covariances. There are three types of covariances we need to consider: $\text{Cov}(T_{11}, T_{12})$, $\text{Cov}(T_{12}, T_{13})$, $\text{Cov}(T_{12}, T_{34})$. From the independence of $X_1, \ldots, X_n$, we first note that

$$
\text{Cov}(T_{12}, T_{34}) = 0.
$$

The first type, $\text{Cov}(T_{11}, T_{12})$, results from three matching subscripts. There are n ways to choose T_{jj} (represented by T_{11}). Then T_{jk} or T_{kj} appears $n - j$ times in the jth block of (2.5.4) and $j - 1$ times, once in each block before the jth, for a total of $n - j + j - 1 = n - 1$. Hence there are $n(n - 1)$ covariances of the type $\text{Cov}(T_{11}, T_{12})$.

The second type, $\text{Cov}(T_{12}, T_{13})$, results from two matching subscripts. There are $n(n - 1)/2$ ways to choose T_{jk}. In the jth block of (2.5.4) there are $n - j - 1$ T_{jk}'s and there are one each of T_{ij} in the preceding $j - 1$ blocks for a total of $(n - j - 1) + (j - 1) = n - 2$. Hence, there are $n(n - 1)(n - 2)/2$ covariances of the type $\text{Cov}(T_{12}, T_{13})$.

We can thus write the variance of T as

$$\text{Var } T = n \text{ Var } T_{11} + \frac{n(n-1)}{2} \text{Var } T_{12}$$

$$+ 2\left[n(n-1)\text{Cov}(T_{11}, T_{12}) + \frac{n(n-1)(n-2)}{2} \text{Cov}(T_{12}, T_{13}) \right].$$

To complete the argument, note that $\text{Var } T_{12} = ET_{12}^2 - (ET_{12})^2 = p_2 - p_2^2$, and $\text{Cov}(T_{11}, T_{12}) = ET_{11}T_{12} - ET_{11}ET_{12} = p_3 - p_1 p_2$, and similarly $\text{Var } T_{11} = p_1 - p_1^2$ and $\text{Cov}(T_{12}, T_{13}) = p_4 - p_2^2$.

Example 2.5.1. Define the class of absolutely continuous distribution functions Ω_p by $H \in \Omega_p$ if $H(x) + H(-x) \leq 1$ for all x with strict inequality for some interval of x values. We refer to Ω_p as the class of stochastically positive distributions since, for $x > 0$, $P(X \geq x) = 1 - H(x) \geq H(-x) = P(X \leq -x)$, and consider the consistency of the Wilcoxon signed rank test on Ω_p. Let $X_1, \ldots, X_n$ be a random sample from a distribution, $G(x)$, and let $\bar{T} = T/n(n+1)$, then from (2.5.2):

$$E_G \bar{T} = \frac{1}{n+1} p_1 + \frac{1}{2}\left(\frac{n-1}{n+1} \right)p_2 \to \frac{1}{2} p_2$$

$$(2.5.5)$$

$$\text{Var}_G \bar{T} \to 0.$$

Hence $\bar{T}$ satisfies (1.3.2) with $\mu(G) = p_2/2$. Now,

$$2\mu(G) = p_2 = P(X_1 + X_2 > 0) = \int_{-\infty}^{\infty} \int_{-x_2}^{\infty} g(x_1)g(x_2)\, dx_1\, dx_2$$

$$= \int_{-\infty}^{\infty} \left[1 - G(-x_2)\right] g(x_2)\, dx_2.$$

If $G \in \Omega_s$ then $G(x) = 1 - G(-x)$ and $p_2 = \int_{-\infty}^{\infty} G(x)g(x)\, dx = 1/2$. If $G \in \Omega_p$ then $G(x) < 1 - G(-x)$ for some interval and $p_2 > 1/2$. From Section 2.2., we have the required asymptotic normality, and hence we can apply Theorem 1.3.1 to assert that $\bar{T}$ is consistent for stochastically positive alternatives.

Note the location model, given by $G(x) = F(x - \theta)$, $F \in \Omega_s$, is an example of a stochastically positive distribution when $\theta > 0$. Hence the Wilcoxon signed rank test is consistent for symmetric shift alternatives.

If the treatment acts to alter the symmetry of the control population and leaves the median at 0, a stochastically positive distribution may result. Then T will still tend to fall in the critical region and reject $H_0: \theta = 0$.

Hence it is important to specify carefully the hypotheses under test. The Wilcoxon signed rank test could just as well be considered a test for symmetry if the location is fixed. See Fig. 2.1 for an example.

We now turn to the asymptotic distribution of T. Under $H_0 : \theta = 0$, $F \in \Omega_s$, we saw in Theorem 2.2.2 that T is a linear combination of independent random variables, and a form of the Central Limit Theorem could be applied to establish the limiting normality. In the following simple example, we show that this independence breaks down under the alternative hypothesis.

Example 2.5.2. Suppose we have a sample X_1, X_2 from a uniform distribution on $(-1, 2)$. Then the sample comes from $F(x - \theta)$, $F \in \Omega_s$, $\theta > 0$.

From the definition of W_1 and W_2, (2.2.1), we have $W_1 = 0$ and $W_2 = 1$ if and only if $-1 < Y_1 < 0 < -Y_1 < Y_2 < 2$ where Y_1 and Y_2 denote the order statistics. Hence

$$P(W_1 = 0, W_2 = 1) = \int_{-1}^{0} \int_{-y_1}^{2} 2! \frac{1}{9} \, dy_2 \, dy_1 = \frac{3}{9} .$$

Likewise $P(W_1 = 0, W_2 = 0) = 1/9$, $P(W_1 = 1, W_2 = 0) = 1/9$ and $P(W_1 = 1, W_2 = 1) = 4/9$. The marginal probabilities are $P(W_1 = 0) = 4/9$, $P(W_1 = 1) = 5/9$, $P(W_2 = 0) = 2/9$, $P(W_2 = 1) = 7/9$, and so, for example,

$$P(W_1 = 0)P(W_2 = 1) \neq P(W_1 = 0, W_2 = 1).$$

In order to deal with T in general, we resort to the method of projection. Hajek (1968) has an excellent discussion of the technique. Essentially, the statistic T is projected, via conditional expectations given the X_i's, onto the class of sums of independent random variables. The asymptotic normality of the projection follows from the Central Limit Theorem, and the asymptotic normality of T follows when the difference between the projection and T is small in probability. Hence we find the Central Limit Theorem remains the main tool for determining the asymptotic normality. We proceed by introducing, more formally, the idea of a projection and then the idea of smallness in probability.

Theorem 2.5.2. (Projection) Suppose $X_1, \ldots, X_n$ is a random sample from an arbitrary distribution, $H(x)$. Let $V = V(X_1, \ldots, X_n)$ be a random variable such that $EV = 0$. If $W = \sum_{i=1}^{n} p_i(X_i)$, then $E(V - W)^2$ is minimized by choosing the function $p_i(x)$ as:

$$p_i^*(x) = E(V \mid X_i = x). \tag{2.5.6}$$

The random variable $V_p = \sum_{i=1}^{n} p_i^*(X_i)$ is called the projection of V and

$$E(V - V_p)^2 = \operatorname{Var} V - \operatorname{Var} V_p. \tag{2.5.7}$$

Proof. By adding and subtracting V_p we have

$$E(V - W)^2 = E(V - V_p)^2 + E(V_p - W)^2 + 2E(V - V_p)(V_p - W).$$

The cross product term is

$$E(V - V_p)(V_p - W)$$

$$= E \sum_{i=1}^{n} \left[p_i^*(X_i) - p_i(X_i) \right] \left[V - V_p \right]$$

$$= \sum EE \left\{ \left[p_i^*(X_i) - p_i(X_i) \right] \left[V - V_p \right] \mid X_i \right\}$$

$$= \sum E \left\{ \left[p_i^*(X_i) - p_i(X_i) \right] E \left[V - V_p \mid X_i \right] \right\}, \tag{2.5.8}$$

and

$$E \left[V - V_p \mid X_i \right] = E \left[V - p_i^*(X_i) - \sum_{j \neq i} p_j^*(X_j) \mid X_i \right].$$

But from the definition, (2.5.6),

$$E \left[V - p_i^*(X_i) \mid X_i \right] = 0.$$

Further,

$$E \left[p_j^*(X_j) \mid X_i \right] = E p_j^*(X_j)$$

$$= EE(V \mid X_j)$$

$$= EV = 0.$$

Hence (2.5.8) is 0, and

$$E(V - W)^2 = E(V - V_p)^2 + E(V_p - W)^2$$

which is minimized by choosing $W = V_p$.
If we choose $W \equiv 0$, then (2.5.7) also follows and the proof is complete.

It should be noted that the same proof works if $X_1, \ldots, X_n$ are independent but not identically distributed. See Hajek and Sidak (1967, p. 59) for an example in which the random variables are dependent. In general the $p_i^*(\cdot)$ will be different for different i. In most of the examples in this book, however, they are identical because the X_i's are identically distributed.

Example 2.5.3. The projection method is illustrated by considering the sample variance for a sample of size n from a distribution with mean μ and variance σ^2. The major application is given in Example 2.5.5.

$$S^2 = \frac{1}{n-1} \sum (X_i - \bar{X})^2$$

$$= \frac{1}{n(n-1)} \left[(n-1) \sum X_i^2 - 2 \sum\sum_{i<j} X_i X_j \right]$$

$$= \frac{1}{\binom{n}{2}} \sum\sum_{i<j} \frac{(X_i - X_j)^2}{2} . \tag{2.5.9}$$

Since $ES^2 = \sigma^2$, define $V = S^2 - \sigma^2$. Then

$$E\left[\frac{(X_j - X_k)^2}{2} \mid X_i = x \right] = \begin{cases} E\dfrac{(x - X_k)^2}{2} & \text{if } i = j \text{ or } k \\ E\dfrac{(X_j - X_k)^2}{2} & \text{otherwise} \end{cases}$$

$$= \begin{cases} \dfrac{x^2 - 2\mu x + \mu^2 + \sigma^2}{2} & \text{if } i = j \text{ or } k \\ \sigma^2 & \text{otherwise} \end{cases} \tag{2.5.10}$$

$$p_i^*(x) = E(V \mid X_i = x) = E\left\{ \frac{1}{\binom{n}{2}} \sum_{j=1}^{n-1} \sum_{k=j+1}^{n} \left[\frac{(X_j - X_k)^2}{2} - \sigma^2 \right] \mid X_i = x \right\}$$

$$= \frac{(n-1)}{\binom{n}{2}} \frac{(x^2 - 2\mu x + \mu^2 - \sigma^2)}{2} . \tag{2.5.11}$$

The $n-1$ factor occurs because if $j = i$, then $i = j < k$ generates $n - i$ terms, and if $k = i$, then $j < k = i$ generates $i - 1$ terms for a total of $n - i + i - 1 = n - 1$.

Hence the projection V_p of the centered sample variance $S^2 - \sigma^2$ is

$$V_p = \frac{n-1}{\binom{n}{2}} \sum_{i=1}^{n} \left[\frac{X_i^2 - 2\mu X_i + \mu^2 - \sigma^2}{2} \right]$$

$$= \frac{1}{n} \sum_{i=1}^{n} \left[(X_i - \mu)^2 - \sigma^2 \right] \qquad (2.5.12)$$

which is an average of independent, identically distributed random variables. Provided $EX^4 < \infty$, the Central Limit Theorem implies that $n^{1/2}V_p$ is asymptotically normally distributed.

The strength of the projection method lies in showing next that $n^{1/2}(S^2 - \sigma^2)$ and $n^{1/2}V_p$ have the same limiting distribution. We will need the following result.

Theorem 2.5.3. Suppose W_n has a limiting $n(0, \sigma^2)$ distribution and $E(U_n - W_n)^2 \to 0$. Then U_n has the same $n(0, \sigma^2)$ limiting distribution as W_n.

Proof. Define $R_n = U_n - W_n$, then in accordance with Theorem A3(a), we can write $U_n = W_n + R_n$. Note that by Chebyshev's inequality, Theorem A4,

$$P(|R_n| \geq \epsilon) \leq \frac{ER_n^2}{\epsilon^2} = \frac{E(U_n - W_n)^2}{\epsilon^2} \to 0.$$

Hence R_n converges to 0 in probability, and the theorem follows from Theorem A3(a) with $c = 0$.

In applications, W_n is the projection of U_n so that by Theorem 2.5.2, $E(U_n - W_n)^2 = \text{Var}\, U_n - \text{Var}\, W_n$, and it is necessary then to show this difference tends to zero.

Example 2.5.4. We complete the previous example now by showing that $n^{1/2}(S^2 - \sigma^2)$ is approximately normally distributed.

We define $\mu_k = E(X - \mu)^k$ and note from (2.5.12) that

$$\text{Var}\sqrt{n}\, V_p = \mu_4 - \sigma^4.$$

From Cramér (1946, p. 348)

$$\text{Var}\sqrt{n}\, S^2 = \mu_4 - \left[\frac{n-3}{n-1} \right] \sigma^4. \qquad (2.5.13)$$

Because, from (2.5.7), $E(n^{1/2}S^2 - n^{1/2}V_p)^2 = \text{Var}\, n^{1/2}S^2 - \text{Var}\, n^{1/2}V_p$, which tends to zero, it follows from Theorem 2.5.3 that $n^{1/2}(S^2 - \sigma^2)$ has the same limiting distribution as $n^{1/2}V_p$, namely, $n(0, \mu_4 - \sigma^4)$.

In the next example we apply these theorems to determine the asymptotic distribution of T. The main result is stated at the end of the example in Theorem 2.5.5. The strategy involves four steps: find the projectic of V, say V_p; argue the asymptotic normality of V_p; find the Var V and \ V_p; and show Var V − Var $V_p \to 0$.

Example 2.5.5. Let $X_1, \ldots, X_n$ be a random sample from an arbitrary distribution with absolutely continuous cdf $H(x)$. We continue to exploit the counting form for T, defined in (2.5.3), and write:

$$T = \sum_{i<j}\sum T_{ij} + \sum_k T_{kk}. \tag{2.5.14}$$

From (2.5.2) define $V = T - ET$, hence

$$V = T - \frac{n(n-1)}{2}\, p_2 - np_1$$

$$= \sum_{i<j}\sum (T_{ij} - p_2) + \sum_k (T_{kk} - p_1). \tag{2.5.15}$$

1. We first show that the projection V_p is given by

$$V_p = (n-1)\sum \left[1 - H(-X_i) - p_2\right] + \sum (T_{kk} - p_1). \tag{2.5.16}$$

The second term on the right side of (2.5.15) is just the centered sign statistic. Since it is already a sum of independent, identically distributed random variables, it is its own projection. We consider then the first term on the right side of (2.5.15).

From the definition, (2.5.3),

$$E(T_{ij} - p_2 \mid X_k = x) = \begin{cases} P(X > -x) - p_2 & \text{if } k = i \text{ or } j \\ 0 & \text{otherwise} \end{cases}$$

$$= \begin{cases} 1 - H(-x) - p_2 & \text{if } k = i \text{ or } j \\ 0 & \text{otherwise.} \end{cases} \tag{2.5.17}$$

Hence

$$E\left[\sum_{i<j}\sum (T_{ij} - p_2 \mid X_k = x)\right] = (n-1)\left[1 - H(-x) - p_2\right],$$

since there are only $n - 1$ nonzero terms in the sum. When $i = k$ then j can assume the $n - k$ values greater than $i = k$ and when $j = k$ then i can assume $k - 1$ values less than k for a total of $n - k + k - 1 = n - 1$ values. Now (2.5.16) follows from Theorem 2.5.2.

We next show that $n^{-3/2}V_p$ is asymptotically distributed as $n(0, p_4 - p_2^2)$.

2.   From (2.5.16) we are led to consider

$$\frac{1}{n^{3/2}} V_p = \frac{n-1}{n^{3/2}} \sum \left[1 - H(-X_i) - p_2 \right] + \frac{1}{n^{3/2}} \sum (T_{kk} - p_1)$$

$$= \frac{n-1}{n^{3/2}} V_p' + \frac{1}{n^{3/2}} \sum (T_{kk} - p_1). \qquad (2.5.18)$$

We first consider the term $n^{-1/2}V_p' = n^{-1/2}\sum[1 - H(-X_i) - p_2]$. Let $Y_i = 1 - H(-X_i)$, then

$$EY_i = \int_{-\infty}^{\infty} (1 - H(-x))h(x)\,dx$$

$$= \int_{-\infty}^{\infty} \int_{-x}^{\infty} h(y)h(x)\,dy\,dx$$

$$= P(X_1 > -X_2) = p_2$$

given in (2.5.1), where X_1, X_2 are i.i.d. $H(x)$. Furthermore

$$EY_i^2 = \int (1 - H(-x))^2 h(x)\,dx \leqslant 1 < \infty.$$

Hence the Central Limit Theorem implies that $n^{-1/2}V_p'$ is asymptotically normally distributed with mean 0 and variance given by

$$\mathrm{Var}(1 - H(-X)) = E(1 - H(-X))^2 - p_2^2$$

$$= \int_{-\infty}^{\infty} (1 - H(-X))^2 h(x)\,dx - p_2^2$$

$$= \int_{-\infty}^{\infty} \int_{-x}^{\infty} \int_{-x}^{\infty} h(y)h(z)h(x)\,dy\,dz\,dx - p_2^2$$

$$= P(X_1 + X_2 > 0, X_1 + X_3 > 0) - p_2^2$$

$$= p_4 - p_2^2.$$

Now write (2.5.18) as

$$\frac{1}{n^{3/2}} V_p = \frac{1}{n^{1/2}} V_p' - \frac{1}{n^{3/2}} V_p' + \frac{1}{n^{3/2}} \sum (T_{kk} - p_1).$$

The second two terms are $n^{-1/2}$ times the averages of i.i.d. random variables. Since $Y_i - p_2$ and $T_{kk} - p_1$ both have means 0 and variances bounded above by 1, Chebyshev's inequality implies that they tend to zero in probability. Hence Theorem 2.5.3 implies that $n^{-3/2}V_p$ is asymptotically normally distributed with mean 0 and variance $p_4 - p_2^2$. In fact, $\mathrm{Var}(n^{-3/2}V_p) \to p_4 - p_2^2$.

   3.   In order to show $n^{-3/2}V = n^{-3/2}(T - ET)$ has the same asymptotic distribution as $n^{-3/2}V_p$, we consider (by Theorem 2.5.2)

$$E\left(n^{-3/2}V - n^{-3/2}V_p\right)^2 = \frac{1}{n^3}\,\mathrm{Var}\, V - \frac{1}{n^3}\,\mathrm{Var}\, V_p.$$

Now, by (2.5.2),

$$\frac{1}{n^3}\,\mathrm{Var}\, V \to p_4 - p_2^2,$$

and since $\mathrm{Var}\, n^{-3/2}V_p$ has already been shown in part 2 to converge to $p_4 - p_2^2$, Theorem 2.5.3 implies that $n^{-3/2}(T - ET)$ is asymptotically normally distributed with mean 0 and variance $p_4 - p_2^2$.

We summarize the result in the following theorem.

Theorem 2.5.4. If $H(x)$ is an arbitrary, continuous distribution function such that $0 < H(0) < 1$, then $(T - ET)/(\mathrm{Var}\, T)^{1/2}$ is asymptotically $n(0, 1)$ with $\mathrm{Var}\, T$ given by (2.5.2).

Proof.   Note that $p_4 - p_2^2 = \mathrm{Var}(1 - H(-X)) \geqslant 0$. This variance is 0 only if the distribution of $-X$ is constant on the support of X. As long as $0 < H(0) < 1$ this cannot happen; hence $p_4 - p_2^2 > 0$.
   Hence we have

$$\frac{T - ET}{\sqrt{n^3(p_4 - p_2^2)}}$$

is asymptotically $n(0, 1)$ and the theorem follows since

$$\frac{\mathrm{Var}\, T}{n^3(p_4 - p_2^2)} \to 1.$$

In the location model, $H(x) = F(x - \theta)$, $F \in \Omega_s$. Hence, for $\theta > 0$, $H(0) = F(-\theta)$, and if $0 < F(\theta) < 1$, Theorem 2.5.4 implies that $(T - ET)$ $/(\text{Var } T)^{1/2}$ has a standard normal limiting distribution with ET and Var T given by (2.5.2).

That we need 0 in the support of X is made intuitively reasonable by thinking of a distribution $K(x)$ such that the support has been shifted to the positive axis. In this case $T = n(n + 1)/2$ with probability 1, and this degenerate random variable has variance 0. Thus, the condition is needed to insure that T has a nondegenerate distribution.

Exercise 2.10.35 describes another application of the Projection Theorem. In general, the Projection Theorem provides the basis for the asymptotic distribution theory of a large class of statistics called U statistics. These statistics can be expressed as symmetric functions of the observations, and many estimates and test statistics can be written as U statistics. For further reading see Randles and Wolfe (1979, Chapter 3), Puri and Sen (1971, Chapter 3) or Lehmann (1975, Section 5 of the Appendix).

We are now in a position to make a tentative comparison of the sign test and the Wilcoxon signed rank test. The fact that T is not distribution free under alternative hypotheses (Example 2.5.2) makes it extremely difficult to compute exact probabilities of T in general. We discuss these probabilities in more detail in the next chapter. In the meantime, Theorem 2.5.4 provides the basis for approximate power calculations. The following paragraph will show that even the asymptotic approach leads to computational difficulties in the location model. These considerations motivate giving up power as a means of comparing tests in favor of asymptotic efficiency.

Recall that we let $X_1, \ldots, X_n$ be a random sample from $F(x - \theta)$, $F \in \Omega_s$. Since this is to be an asymptotic comparison of S and T, we put T into a form that eliminates the higher-order terms from its mean and variance.

As in Example 2.5.1, let

$$\overline{T} = \frac{1}{n(n + 1)} T.$$

Then, from (2.5.2), we have

$$E\overline{T} \to \frac{p_2}{2}$$

$$\text{Var}\sqrt{n}\ \overline{T} \to (p_4 - p_2^2),$$

and, from Theorem 2.5.4,

$$\frac{\sqrt{n}\left(\overline{T}-p_2/2\right)}{\sqrt{\left(p_4-p_2^2\right)}} \tag{2.5.19}$$

is approximately $n(0,1)$. Hence, for large n, we need to compute p_2 and p_4 in order to approximate the power.

For S, the sign statistic, from (1.2.3) and defining $\overline{S}=S/n$, we have

$$\frac{\sqrt{n}\left(\overline{S}-p_1\right)}{\sqrt{p_1(1-p_1)}} \tag{2.5.20}$$

is approximately $n(0,1)$. So, in addition to p_2 and p_4, we need to compute p_1.

In Exercise 2.10.9 you are asked to compute p_1, p_2, p_4 for the uniform distribution on $(-1,2)$, and hence find the approximate power for the two tests. If this is done for various sample sizes, we see two things emerge. On the one hand $\overline{T}$ appears to be more powerful than $\overline{S}$, and on the other, both have power approaching 1 as the sample sizes become large. This last point reflects the consistency of the tests.

This sort of asymptotic power comparison is not very satisfying. We need to assume large sample sizes in order to neglect the higher-order terms in the means and variances and have an accurate normal approximation; but then the power is generally close to one for these sample sizes. The parameter p_4 is especially troublesome to compute and may be impossible for many choices of F. The following example illustrates one more case where it can, with some effort, be calculated.

Example 2.5.6. Suppose $X_1, \ldots, X_n$ is a random sample from $n(\theta, \sigma^2)$. Then

$$p_1 = P\left(\frac{X-\theta}{\sigma} > \frac{-\theta}{\sigma}\right) = 1 - \Phi\left(\frac{-\theta}{\sigma}\right) = \Phi\left(\frac{\theta}{\sigma}\right)$$

$$p_2 = P\left(\frac{X_1+X_2-2\theta}{\sqrt{2}\,\sigma} > \frac{-2\theta}{\sqrt{2}\,\sigma}\right) = \Phi\left(\frac{\sqrt{2}\,\theta}{\sigma}\right)$$

$$p_4 = P\left(\frac{X_1+X_2-2\theta}{\sqrt{2}\,\sigma} > \frac{-\sqrt{2}\,\theta}{\sigma}, \frac{X_1+X_3-2\theta}{\sqrt{2}\,\sigma} > \frac{-\sqrt{2}\,\theta}{\sigma}\right).$$

Now a straightforward calculation shows that $E(X_1 + X_2)(X_1 + X_3)$ $= \sigma^2 + 4\theta^2$ and $\rho = 1/2$. Hence p_4 can be found in tables of the bivariate normal distribution with means 0, variances 1, and correlation $1/2$.

As an example we take $n = 10$, $\alpha = .0527$ so the critical region is determined by $T \geqslant 44$. In this case the power at $\theta = \sigma$ is approximately $\Phi(1.475) = .93$ or $\Phi(1.586) = .94$, corrected for continuity. The exact value, calculated by Klotz (1963), is .89. Hence the power is overestimated when we neglect the higher-order terms. For these calculations $p_1 = .8413$, $p_2 = .9213$, and $p_4 = .8657$, and when the full expressions for the mean and variance of T are used with the normal approximation, it is seen that the power is approximated by .90.

The approach of the power to 1 can be partly compensated for by choosing the alternative close to the null hypothesis. However, for any fixed alternative, this will ultimately fail for a consistent test. In the next calculation, we see that, for alternatives suitably close to the null hypothesis, it will no longer be necessary to compute the troublesome p_4. This sort of approximation will then motivate the introduction of asymptotic efficiency which is a local (i.e., close to the null hypothesis) measure of power.

Example 2.5.7. Let $X_1, \ldots, X_n$ be a sample from $F(x - \theta)$, $F \in \Omega_s$. For testing $H_0 : \theta = 0$ versus $H_A : \theta > 0$ with T, the approximate critical value c is given by (2.2.9). Hence, for a fixed alternative, the power is approximated by

$$P(T \geqslant c) \doteq 1 - \Phi\left(\frac{c - ET}{\sqrt{\mathrm{Var}\, T}} \right), \qquad (2.5.21)$$

with ET and $\mathrm{Var}\, T$ given by (2.5.2). Now, we write

$$\frac{n(n+1)}{4} - np_1 - \frac{n(n-1)}{2} p_2 = \frac{n(n-1)}{2}\left(\frac{1}{2} - p_2 \right) + n\left(\frac{1}{2} - p_1 \right),$$

and hence

$$c - ET = Z_\alpha \sqrt{\frac{n(n+1)(2n+1)}{24}} + \frac{n(n-1)}{2}\left(\frac{1}{2} - p_2 \right) + n\left(\frac{1}{2} - p_1 \right).$$

$$(2.5.22)$$

We now make a further approximation of p_1 and p_2. We expand these functions of θ about $\theta = 0$, retain the linear terms, and neglect the higher-order terms. This approximation is valid for small values of θ. Recall

$$p_1 = P(X > 0)$$

$$= 1 - F(-\theta)$$

$$\doteq 1 - F(0) + \theta f(0)$$

$$= \tfrac{1}{2} + \theta f(0)$$

and $\tfrac{1}{2} - p_1 \doteq -\theta f(0)$. Likewise,

$$p_2 = P(X_1 + X_2 > 0)$$

$$= 1 - F^*(-2\theta)$$

$$\doteq \tfrac{1}{2} + 2\theta f^*(0),$$

where F^* is the convolution distribution, the cdf of $X_1 + X_2$, where X_1, X_2 are i.i.d. $F(x)$, and $1/2 - p_2 \doteq -2\theta f^*(0)$. Furthermore, by continuity if θ is close to 0,

$$\operatorname{Var} T \doteq \frac{n(n+1)(2n+1)}{24},$$

its value under $H_0: \theta = 0$. Substitution in (2.5.22) and (2.5.21) yields

$$P(T \geqslant c) \doteq 1 - \Phi\left\{ Z_\alpha - \frac{n(n-1)\theta f^*(0) + n\theta f(0)}{\sqrt{\dfrac{n(n+1)(2n+1)}{24}}} \right\}. \quad (2.5.23)$$

Note that p_4 is not necessary for the calculation of (2.5.23). We next consider in more detail the convolution density f^*. If we suppose sufficient regularity to pass derivatives through the integrals, then we have

$$F^*(t) = \int_{-\infty}^{\infty} \int_{-\infty}^{t-x_2} f(x_1) f(x_2) \, dx_1 \, dx_2$$

$$= \int_{-\infty}^{\infty} F(t - x_2) f(x_2) \, dx_2$$

and

$$f^*(t) = \int_{-\infty}^{\infty} f(t - x_2) f(x_2) \, dx_2.$$

Hence, since $F \in \Omega_s$ and $f(-x) = f(x)$,

$$f^*(0) = \int_{-\infty}^{\infty} f^2(x)\, dx. \qquad (2.5.24)$$

If F is a $n(0, \sigma^2)$ distribution, then it is easy to see that

$$f(0) = \frac{1}{\sqrt{2\pi}\,\sigma}$$

$$f^*(0) = \frac{1}{2\sqrt{\pi}\,\sigma}$$

and

$$P(T \geqslant c) \doteq 1 - \Phi\left[Z_\alpha - \frac{\left(\dfrac{n(n-1)}{2} + \dfrac{n}{\sqrt{2}}\right)}{\sqrt{\dfrac{n(n+1)(2n+1)}{24}}} \, \frac{\theta}{\sigma\sqrt{\pi}} \right]. \qquad (2.5.25)$$

From Example 2.5.6, $n = 10$, $\alpha = .0527$, $Z_\alpha = 1.645$, and $\theta = \sigma$ yield a value of .91 for (2.5.25).

The approximation (2.5.23) is valid for values of θ close to 0. Note that if $\theta > 0$ is fixed as n tends to infinity,

$$P(T \geqslant c) \to 1 - \Phi(-\infty) = 1.$$

Hence if we wish to stabilize the power away from 1, we need to let θ approach 0 as n increases. For large n, using (2.5.24), (2.5.23) can be written

$$P(T \geqslant c) \doteq 1 - \Phi\left(Z_\alpha - \sqrt{12}\left(\int f^2(x)\, dx \right)\theta\sqrt{n} \right) \qquad (2.5.26)$$

and hence, if we take the sequence $\theta_n = a / n^{1/2}$,

$$1 - \Phi\left(Z_\alpha - \sqrt{12}\left(\int f^2(x)\, dx \right)a \right) \qquad (2.5.27)$$

can be considered the asymptotic local power of the Wilcoxon signed rank test.

In Exercise 2.10.10, you are asked to find a similar expression for the sign test. Then it is possible to make a local, asymptotic power comparison

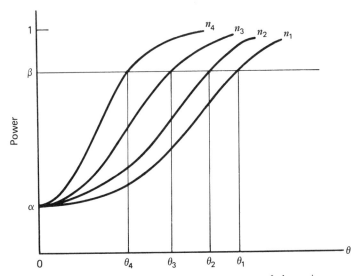

Figure 2.4. Stabilizing the power along a sequence of alternatives.

of T and S for various underlying distributions. In that exercise, you will see that, for an underlying normal population, T is more powerful than S; but the reverse is true for an underlying double exponential (Laplace) distribution. We can thus conclude that even though the sign test appears to utilize very little information in the sample, it can be quite powerful even for symmetric populations.

The preceding approximations provide a heuristic analysis in terms of local power. It is clear that to stabilize the power away from one, it is necessary to consider the power along a sequence of alternatives $\{\theta_n\}$. The sequence suggested by the examples is $\theta_n = a/n^{1/2}$ which approaches zero at the rate of $1/n^{1/2}$. The next section makes these ideas more formal and replaces the idea of local, asymptotic power by asymptotic relative efficiency. See Fig. 2.4.

2.6. ASYMPTOTIC RELATIVE EFFICIENCY

The following development of asymptotic efficiency is due to Pitman and was presented in a summer course that he taught at Columbia in 1949. Noether (1955) gives the first published version. We begin with a definition of efficiency in the case of finite sample sizes.

Definition 2.6.1. Let $V_n^{(i)}$, $i = 1, 2$, be size α tests of $H_0 : \theta = 0$ versus

$H_A : \theta > 0$ based on $X_1, \ldots, X_n$, a random sample from $F(x - \theta)$, F fixed in Ω_0.

For given θ and β, $\alpha < \beta < 1$, let $n^{(i)}$, $i = 1, 2$, be the sample size required for

$$P_{\theta, F}\left(V_n^{(i)} \geqslant k^{(i)} \right) = \beta.$$

Then the efficiency of $V_n^{(1)}$ relative to $V_n^{(2)}$ is $n^{(2)} / n^{(1)}$.

Hence, if $n^{(2)} / n^{(1)} = .5$, then $V_n^{(1)}$ requires about twice as many observations as $V_n^{(2)}$ to obtain the same power. Note that we would need the distribution of $V_n^{(i)}$, $i = 1, 2$, under the alternative hypothesis in order to compute the efficiency. If we only consider large sample sizes to avoid this problem, then we again have the problem of comparing two consistent tests, both with approximate power equal to one. Recall the discussion of Example 2.5.7 with Fig. 2.4. We have the following definition of asymptotic relative efficiency.

Definition 2.6.2. Let $V_n^{(i)}$, $i = 1, 2$, provide two tests for $H_0 : \theta = 0$ versus $H_A : \theta > 0$ based on $X_1, \ldots, X_n$ a random sample from $F(x - \theta)$, F fixed in Ω_0. Suppose the tests are asymptotically size α, that is,

$$P_{H_0}\left(V_n^{(i)} \geqslant k_n^{(i)} \right) \to \alpha, \qquad i = 1, 2.$$

For fixed β, $\alpha < \beta < 1$, suppose $\{\theta_j\}$ is a sequence of alternatives such that $\theta_j \to 0$ with the corresponding sequences $\{n_j^{(i)}\}$, $i = 1, 2$, of sample sizes such that

$$P_{\theta_j}\left(V^{(i)} \geqslant k^{(i)} \right) \to \beta, \qquad i = 1, 2.$$

Here we have suppressed the subscript $n_j^{(i)}$ on $V^{(i)}$. Now at θ_j, $n_j^{(2)} / n_j^{(1)}$ is the efficiency of $V^{(1)}$ relative to $V^{(2)}$, and when the limit

$$e_{12} = \lim_{j \to \infty} \frac{n_j^{(2)}}{n_j^{(1)}} \tag{2.6.1}$$

exists and is independent of $\{\theta_j\}$, α, and β, it is called the asymptotic relative efficiency of $V^{(1)}$ relative to $V^{(2)}$.

As we shall see, this asymptotic efficiency is easy to calculate and the limit is often approached quite rapidly. This means that e_{12} gives a good indication of the ratio of sample sizes necessary to attain the same level and power for alternatives close to the null hypothesis. Note also this efficiency, as expected, does depend on the F fixed in Ω_0.

Example 2.6.1. In this example we give still another heuristic comparison of T and S. Under the same assumptions and notation of Example 2.5.7, we see from (2.5.26) that the Wilcoxon signed rank test has approximate power given by

$$1 - \Phi\left(Z_\alpha - \sqrt{12} \int f^2(x)\,dx\,\theta\sqrt{n}\right).$$

If β, $\alpha < \beta < 1$, is fixed, then the requirement that the power be approximately $1 - \Phi(Z_\beta) = \beta$ means

$$Z_\beta \doteq Z_\alpha - \sqrt{12} \int f^2(x)\,dx\,\theta\sqrt{n}.$$

Solving for n, we have

$$n \doteq \frac{(Z_\alpha - Z_\beta)^2}{\theta^2 12\left(\int f^2(x)\,dx\right)^2}. \qquad (2.6.2)$$

If the sequence of alternatives is $\{\theta_j\}$, then n_j can be defined by (2.6.2). Likewise, from Exercise 2.10.11, for the sign test, we have

$$n \doteq \frac{(Z_\alpha - Z_\beta)^2}{\theta^2 4f^2(0)}. \qquad (2.6.3)$$

Hence, for the same α, β, and $\{\theta_j\}$, the ratio of sample sizes

$$\frac{n_T}{n_S} \doteq \frac{1}{3}\left(\frac{f(0)}{\int f^2(x)\,dx}\right)^2 \qquad (2.6.4)$$

yields the asymptotic efficiency. If f is the standard normal density, we have $e_{12} = 2/3$. Hence, when sampling from a normal distribution with alternatives close to the null hypothesis, the Wilcoxon signed rank test requires about $2/3$ as many observations as the sign test to attain the same level and power.

We now state a set of regularity conditions, due to Pitman, that make the calculation of efficiency quite easy in many cases. We suppose that V_n is a statistic for testing $H_0: \theta = 0$ versus $H_A: \theta > 0$ with critical region $V_n \geqslant k_n$.

1. V_n provides a consistent test.
2. There exist sequences $\{\mu_n(\theta)\}$ and $\{\sigma_n(\theta)\}$ such that $(V_n - \mu_n(\theta))/\sigma_n(\theta)$ is asymptotically $n(0,1)$, uniformly in a neighborhood of $\theta = 0$.
3.

$$\frac{d}{d\theta}\mu_n(\theta)\bigg|_{\theta=0} = \frac{d}{d\theta}\mu_n(0) = \mu'_n(0) \quad \text{exists.}$$

4. For a sequence $\{\theta_n\}$ such that $\theta_n \to 0$ as $n \to \infty$

$$\frac{\sigma_n(\theta_n)}{\sigma_n(0)} \to 1 \quad \text{and} \quad \frac{\mu'_n(\theta_n)}{\mu'_n(0)} \to 1, \quad \text{as} \quad n \to \infty.$$

5.

$$\frac{\mu'_n(0)}{\sqrt{n}\,\sigma_n(0)} \to c > 0.$$

Definition 2.6.3. The quantity c is called the efficacy of the test based on V_n. For large n, it measures the rate of change in standard units of the "asymptotic" mean of V_n at the null hypothesis. A test with relatively large efficacy c is responding rapidly to alternatives and might be expected to have good local power properties.

The conditions 3 and 4 are smoothness conditions on the parameters of V_n as functions of θ. Often, the statistic is constructed so that $V_n \overset{P}{\to} \mu(\theta)$ or $E_\theta V_n \to \mu(\theta)$ and $n \operatorname{Var} V_n \to \sigma^2(\theta)$ so that we can take $\mu_n(\theta) = \mu(\theta)$ for all n and $\sigma_n(\theta) = \sigma(\theta)/n^{1/2}$. Then condition 2 becomes: $n^{1/2}(V_n - \mu(\theta))/\sigma(\theta)$ is asymptotically $n(0,1)$, uniformly in θ near 0; and the efficacy is easily computed from the asymptotic parameters $\mu(\theta)$ and $\sigma(\theta)$ as $c = \mu'(0)/\sigma(0)$.

The conditions 1 and 2 are asymptotic distribution conditions. The uniform convergence to normality, condition 2, in a neighborhood of the null hypothesis requires slightly stronger central limit theorems. An important tool for establishing the uniform convergence is the Berry–Esseen Theorem A14 and is illustrated on the sign test.

Example 2.6.2. Let $X_1, \ldots, X_n$ be a random sample from $F(x - \theta)$, $F \in \Omega_0$, with $f(0) < \infty$. For testing $H_0 : \theta = 0$ versus $H_A : \theta > 0$, we use the sign test defined in (1.2.1) as $S = \sum s(X_i)$ with $p(\theta) = P(X > 0) = 1 - F(-\theta)$ defined in (1.2.2).

In the notation of the Pitman conditions, we let $\mu_n(\theta) = np(\theta)$, $\sigma_n^2(\theta) = np(\theta)(1 - p(\theta))$. The smoothness conditions are then just smoothness conditions on the underlying distribution function. We apply the Berry–Esseen Theorem A14 to $s(X_1), \ldots, s(X_n)$, then

$$\sigma^2(\theta) = p(\theta)(1 - p(\theta)) = p(\theta)q(\theta)$$

$$\rho^3(\theta) = E|s(X_i) - p(\theta)|^3$$

$$= p(\theta)q(\theta)\left[p^2(\theta) + q^2(\theta)\right].$$

Hence we see, for a constant K,

$$\frac{K\rho^3}{\sqrt{n}\,\sigma^3} = \frac{K}{\sqrt{n}}\,\frac{p(\theta)q(\theta)\left[p^2(\theta) + q^2(\theta)\right]}{(p(\theta)q(\theta))^{3/2}}$$

$$\leqslant \frac{K}{\sqrt{n}}\,\frac{1}{\sqrt{p(\theta)q(\theta)}}. \qquad (2.6.5)$$

Now suppose, in the neighborhood of $\theta = 0$, that the variance $\sigma^2(\theta) = p(\theta)q(\theta) \geqslant M > 0$; that is, it is bounded away from 0. Then

$$\frac{K\rho^3}{\sqrt{n}\,\sigma^3} \leqslant \frac{K}{\sqrt{n}} \cdot \frac{1}{\sqrt{M}} \to 0,$$

uniformly in θ in the neighborhood. Hence the sign test converges uniformly to normality.

The condition that $p(\theta)q(\theta) \geqslant M > 0$ in a neighborhood is not unreasonable since at $\theta = 0$, $p(0)q(0) = 1/2$, and we have a continuous distribution function F. In typical applications we will need an upper bound on the third absolute moment and a positive lower bound on the second moment. See Theorem A15 in the Appendix. When this is the case, we will be able to show the projection (Theorem 2.5.2) converges to normality uniformly, and a uniform version of Slutsky's Theorem A3 will complete the argument. Bounding the moments from above is generally not a difficulty for rank tests since their moments are in terms of probabilities, which are already bounded above by 1. (See Theorem 2.5.1 for an example.) Hence essentially the only new point to be considered is bounding the variance away from 0, which occurs when the asymptotic variance is continuous in θ near 0 and positive at $\theta = 0$. We now proceed with the development of asymptotic relative efficiency.

Theorem 2.6.1. Suppose V_n satisfies the regularity conditions 1 through 5 and provides an asymptotically size α test of $H_0 : \theta = 0$ versus $H_A : \theta > 0$. Let $\theta_n = \theta/n^{1/2}$ for $\theta > 0$, fixed. Then the asymptotic power is given by

$$\lim_{n \to \infty} P_{\theta_n}(V_n \geq k_n) = 1 - \Phi(Z_\alpha - \theta c) \qquad (2.6.6)$$

where

$$\lim_{n \to \infty} P_0(V_n \geq k_n) = 1 - \Phi(Z_\alpha) = \alpha, \qquad (2.6.7)$$

and c is the efficacy defined in condition 5.

Proof. Using the regularity conditions and Theorem A1, we have, for $\epsilon > 0$ and for n sufficiently large,

$$\left| P_\theta(V_n \geq k_n) - 1 + \Phi\left(\frac{k_n - \mu_n(\theta)}{\sigma_n(\theta)} \right) \right| < \epsilon, \qquad (2.6.8)$$

for all θ in a neighborhood of 0. Next we expand $\mu_n(\theta)$ about 0 to get

$$\mu_n(\theta) = \mu_n(0) + \theta\mu_n'(\theta^*),$$

where $0 < \theta^* < \theta$. Hence the argument of Φ becomes

$$\frac{k_n - \mu_n(\theta)}{\sigma_n(\theta)} = \frac{k_n - \mu_n(0)}{\sigma_n(\theta)} - \frac{\theta\mu_n'(\theta^*)}{\sigma_n(\theta)}. \qquad (2.6.9)$$

By the uniform convergence condition 2, for sufficiently large n, $\theta_n = \theta/n^{1/2}$ will be in the neighborhood of 0 and can be inserted into (2.6.9). Multiplying and dividing the right side of (2.6.9) by $\sigma_n(0)$ and replacing θ by θ_n yields

$$\frac{k_n - \mu_n(0)}{\sigma_n(0)} \left(\frac{\sigma_n(0)}{\sigma_n(\theta_n)} \right) - \frac{\theta}{\sqrt{n}} \frac{\mu_n'(\theta_n^*)}{\sigma_n(0)} \cdot \left(\frac{\sigma_n(0)}{\sigma_n(\theta_n)} \right),$$

where $0 < \theta_n^* < \theta_n$. We now have, from (2.6.7), that

$$\frac{k_n - \mu_n(0)}{\sigma_n(0)} \left(\frac{\sigma_n(0)}{\sigma_n(\theta_n)} \right) \to Z_\alpha$$

and

$$\frac{\theta}{\sqrt{n}} \frac{\mu_n'(\theta_n^*)}{\sigma_n(0)} \left(\frac{\sigma_n(0)}{\sigma_n(\theta_n)} \right) \to \theta c.$$

Hence (2.6.6) follows upon replacing θ by θ_n in (2.6.8).

Thus we can formally approximate the power of a consistent test near the null hypothesis. This was the point of the discussion of Example 2.5.7. The power approximation in (2.6.6) requires the asymptotic variance under H_0 and the asymptotic mean under H_A to compute the efficacy c. The results of Theorem 2.6.1 can be expressed in a different form. The limit, (2.6.7), can be rewritten, with $k_n = \mu_n(0) + t\sigma_n(0)$, as

$$P_0\left(\frac{V_n - \mu_n(0)}{\sigma_n(0)} \leqslant t \right) \to \Phi(t)$$

or $[V_n - \mu_n(0)]/\sigma_n(0) \overset{D_0}{\to} Z \sim n(0, 1)$, where D_0 denotes convergence in distribution when $\theta = 0$. Likewise (2.6.6) becomes, with $\theta_n = \theta/n^{1/2}$,

$$P_{\theta_n}\left(\frac{V_n - \mu_n(0)}{\sigma_n(0)} \leqslant t \right) \to \Phi(t - \theta c)$$

or $[V_n - \mu_n(0)]/\sigma_n(0) \overset{D_{\theta_n}}{\to} Z \sim n(\theta c, 1)$. Hence the asymptotic distribution of $[V_n - \mu_n(0)]/\sigma_n(0)$ changes only in the mean, from 0 to θc, when the alternative is allowed to converge to 0. We now develop the asymptotic efficiency of two tests.

Theorem 2.6.2. Let $V_n^{(i)}$, $i = 1, 2$, provide two tests of $H_0: \theta = 0$ versus $H_A: \theta > 0$. Suppose they both satisfy the regularity conditions 1–5 and further satisfy the conditions of Definition 2.6.2. Then the asymptotic relative efficiency of $V_n^{(1)}$ relative to $V_n^{(2)}$ is

$$e_{12} = \lim_{j \to \infty} \frac{n_j^{(2)}}{n_j^{(1)}} = \frac{c_1^2}{c_2^2}, \tag{2.6.10}$$

where c_i is the efficacy of $V_n^{(i)}$, $i = 1, 2$.

Proof. For the moment we will suppress the superscript and consider V_n. The power converges according to Definition 2.6.2, and hence the standardized critical value

$$\frac{k_n - \mu_n(\theta_n)}{\sigma_n(\theta_n)} = \frac{k_n - \mu_n(0) - \theta_n \mu_n'(\theta_n^*)}{\sigma_n(\theta_n)} \to \text{constant}.$$

Multiply and divide by $n^{1/2}$ to get

$$\frac{k_n - \mu_n(0)}{\sigma_n(\theta_n)} - \frac{\sqrt{n}\,\theta_n \mu_n'(\theta_n^*)}{\sqrt{n}\,\sigma_n(\theta_n)} \to \text{constant}$$

and

$$\sqrt{n}\,\theta_n = \text{constant} + o(1). \qquad (2.6.11)$$

If we let θ denote this constant, then we see that the convergence of the power implies the form $\theta_n = \theta/n^{1/2} + o(1)/n^{1/2}$.

Now from Theorem 2.6.1, we have, for $i = 1, 2$,

$$P_{\theta_{n_j}}\left(V_n^{(i)} \geqslant k_n^{(i)}\right) \to 1 - \Phi\left(Z_\alpha - \theta^{(i)} c_i\right).$$

Since these powers must be the same,

$$\theta^{(1)} c_1 = \theta^{(2)} c_2.$$

Furthermore, from (2.6.11), since the sequence must be the same (see Definition 2.6.2),

$$\frac{\theta^{(1)}}{\sqrt{n_j^{(1)}}} + \frac{o(1)}{\sqrt{n_j^{(1)}}} = \frac{\theta^{(2)}}{\sqrt{n_j^{(2)}}} + \frac{o(1)}{\sqrt{n_j^{(2)}}}$$

which can be rewritten as

$$\sqrt{\frac{n_j^{(2)}}{n_j^{(1)}}}\,(1 + o(1)) = \frac{\theta^{(2)}}{\theta^{(1)}} + o(1)$$

$$= \frac{c_1}{c_2} + o(1).$$

Now $n_j^{(2)}/n_j^{(1)} \to c_1^2/c_2^2$, which completes the proof.

Example 2.6.3. Let $X_1, \ldots, X_n$ be a random sample from $F(x - \theta)$, $F \in \Omega_s$. The sign statistic S, (1.2.1), has efficacy

$$c_s = 2f(0). \tag{2.6.12}$$

This follows immediately from Definition 2.6.3 with $\mu_n(\theta) = ES = n(1 - F(-\theta))$, $\mu_n'(0) = nf(0)$ and $\sigma_n^2(0) = n/4$.

The Wilcoxon signed rank statistic T, (2.2.2), has efficacy

$$c_T = \sqrt{12} \int f^2(x) \, dx. \tag{2.6.13}$$

To see this, first note from (2.5.2), expressing p_1 and p_2 in terms of F, we have

$$\mu_n(\theta) = n(1 - F(-\theta)) + \frac{n(n-1)}{2} \int (1 - F(-x - \theta)) f(x - \theta) \, dx$$

$$= n(1 - F(-\theta)) + \frac{n(n-1)}{2} \int (1 - F(-y - 2\theta)) f(y) \, dy. \tag{2.6.14}$$

Hence

$$\mu_n'(0) = nf(0) + n(n-1) \int f^2(x) \, dx.$$

Now using $\sigma_n^2(0)$ given in (2.2.4) and Definition 2.6.3, the equation for c_T follows. The differentiation under the integral in (2.6.14) is justified by Theorem A17 in the Appendix. We only need to suppose that F is absolutely continuous and $\int f^2(x) \, dx < \infty$. The uniform convergence to normality, condition 2, is discussed in Exercise 2.10.12.

The t statistic, $t = n^{1/2} \overline{X} / S$ where $\overline{X}$, S are the sample mean and standard deviation, respectively. The efficacy is

$$c_t = \frac{1}{\sigma_f} \tag{2.6.15}$$

where $\sigma_f^2 = \int x^2 f(x) \, dx$ is the variance of F. This follows from $\mu_n(\theta) = n^{1/2} \theta / \sigma_f$ and $\sigma_n(0) = 1$ and applying the definition of efficacy.

We are now ready to list the Pitman efficiency equations for the various tests. Recall the efficiency of $V^{(1)}$ relative to $V^{(2)}$ is $e(T^{(1)}, T^{(2)}) = c_1^2 / c_2^2$

by (2.6.10). Hence we have

$$e(S, T) = \frac{f^2(0)}{3\left(\int f^2(x)\,dx\right)^2} \tag{2.6.16}$$

$$e(S, t) = 4\sigma_f^2 f^2(0) \tag{2.6.17}$$

$$e(T, t) = 12\sigma_f^2 \left(\int f^2(x)\,dx\right)^2. \tag{2.6.18}$$

It is pointed out in Exercise 2.10.13 that these efficiencies are scale invariant. Hence the value of the efficiency is independent of the value of σ_f^2, the variance of the underlying distribution.

Example 2.6.4. Let $X_1, \ldots, X_n$ be a random sample from $F_\epsilon(x) = (1 - \epsilon)\Phi(x) + \epsilon\Phi(x/3)$. This corresponds to sampling from a distribution that is a mixture of $n(0, 1)$ and $n(0, 9)$. For small values of ϵ, most of the observations come from $n(0, 1)$ with occasional large observations from $n(0, 9)$. This model is called the contaminated normal model. It was first treated in detail by Tukey (1960) and provides a model that is very difficult to distinguish from the normal model for small to moderate values of ϵ. Furthermore, contamination may have its greatest impact for small values of ϵ. For example, Tukey points out that the two distributions contribute equal amounts to the variance of $f_\epsilon(x)$ when $\epsilon = .10$.

We will compute the efficiency of T relative to t for this model. We have

$$f_\epsilon(x) = (1 - \epsilon)\phi(x) + \epsilon \frac{1}{3}\phi\left(\frac{x}{3}\right)$$

where $\phi(x)$ is the $n(0, 1)$ density function. Hence,

$$\int f_\epsilon^2(x)\,dx = \frac{(1 - \epsilon)^2}{2\sqrt{\pi}} + \frac{\epsilon^2}{6\sqrt{\pi}} + \frac{\epsilon(1 - \epsilon)}{\sqrt{5}\sqrt{\pi}}$$

$$\sigma_{f_\epsilon}^2 = 1 + 8\epsilon$$

$$e(T, t) = \frac{3(1 + 8\epsilon)}{\pi}\left[(1 - \epsilon)^2 + \frac{\epsilon^2}{3} + \frac{2\epsilon(1 - \epsilon)}{\sqrt{5}}\right]^2 \tag{2.6.19}$$

Table 2.3. Efficiency of T Relative to t

ϵ	0	.01	.03	.05	.08	.10	.15
$e(T,t)$	0.955	1.009	1.108	1.196	1.301	1.373	1.497

Table 2.3 illustrates (2.6.19). Note that when $\epsilon = 0$, we are simply sampling from a normal distribution and the efficiency of T relative to t is 0.955. This means that the loss of efficiency due to using the rank test T rather than the optimal t test is negligible. Furthermore, for mild contamination, such as 5%, the rank test T can be substantially more efficient (20%) than the t test. In fact, this example shows that the t test loses its optimality quite rapidly as we move from the normal model into a neighborhood of the normal model.

In Exercise 2.10.14, you are asked to compute the efficiencies $e(T,t)$ for various models. The implications of these results are that T is highly efficient for a broad selection of models. When the tails of the underlying distribution are sufficiently heavy, say close to a double exponential or Laplace distribution, S is more efficient than either t or T. In the examples considered, in which the underlying distribution has tails heavier than a normal distribution, t is never very efficient.

The examples and exercises compare the rank test T to the t test for several specific underlying distributions and two fairly broad families of models. The question still remains whether there are underlying distributions for which the t test is highly superior to T. The answer is in the negative as the next theorem due to Hodges and Lehmann (1956) shows.

Theorem 2.6.3. Let $X_1, \ldots, X_n$ be a random sample from $F(x - \theta)$, $F \in \Omega_s$. Then

$$\inf_{\Omega_s} e(T,t) = 0.864 \qquad (2.6.20)$$

This theorem shows that no matter what the underlying distribution, the efficiency of T relative to t is never less than 0.864. We have already seen in Exercise 2.10.14 that $e(T,t)$ can be made arbitrarily large.

Proof. From (2.6.18), $e(T,t) = 12\sigma_f^2(\int f^2(x)\,dx)^2$. If $\sigma_f^2 = \infty$ then clearly $e(T,t) > 0.864$, hence we can restrict attention to $F \in \Omega_s$ such that $\sigma_f^2 < \infty$. Furthermore, by Exercise 2.10.13, $e(T,t)$ is scale invariant; so without loss of generality we will only consider $F \in \Omega_s$ such that $\sigma_f^2 = 1$. In the following argument we suppress the dummy variable of integration.

The problem, then, is to minimize $\int f^2$ subject to $\int f = \int x^2 f = 1$ and $\int x f = 0$. This is equivalent to minimizing

$$\int f^2 + 2b \int x^2 f - 2ba^2 \int f \qquad (2.6.21)$$

where a, b are positive constants to be determined later. We now write (2.6.21) as

$$\int \left[f^2 + 2b(x^2 - a^2)f \right] = \int_{|x| \leqslant a} \left[f^2 + 2b(x^2 - a^2)f \right]$$

$$+ \int_{|x| > a} \left[f^2 + 2b(x^2 - a^2)f \right]. \quad (2.6.22)$$

First complete the square on the first term on the right side of (2.6.22) to get

$$\int_{|x| \leqslant a} \left[f + b(x^2 - a^2) \right]^2 - \int_{|x| \leqslant a} b^2(x^2 - a^2)^2. \qquad (2.6.23)$$

Now (2.6.22) is equal to the two terms of (2.6.23) plus the second term on the right side of (2.6.22). We can now write down the density f that minimizes (2.6.21).

If $|x| > a$ take $f(x) = 0$, since $x^2 > a^2$, and if $|x| \leqslant a$ take $f(x) = b(a^2 - x^2)$, since the integral in the first term of (2.6.23) is nonnegative.

We now determine the values of a and b from the side conditions. From $\int f = 1$ we have

$$\int_{-a}^{a} b(a^2 - x^2) \, dx = 1$$

which implies that $a^3 b = \frac{3}{4}$. Further, from $\int x^2 f = 1$ we have

$$\int_{-a}^{a} x^2 b(a^2 - x^2) \, dx = 1$$

from which $a^5 b = \frac{15}{4}$. Hence solving for a and b yields $a = 5^{1/2}$ and $b = [3(5)^{1/2}]/100$. Now,

$$\int f^2 = \int_{-\sqrt{5}}^{\sqrt{5}} \left[\frac{3\sqrt{5}}{100} (5 - x^2) \right]^2 dx = \frac{3\sqrt{5}}{25} ,$$

and to complete the proof of (2.6.20), note that

$$\inf_{\Omega_s} e(T, t) = 12\left(\frac{3\sqrt{5}}{25}\right)^2 = \frac{108}{125} = 0.864.$$

In summary, we see from the examples, exercises, and the last theorem that

$$e(T, t) = 0.955 \quad \text{for underlying normal model}$$
$$= 1.50 \quad \text{for underlying Laplace model}$$
$$\geqslant 0.864 \quad \text{for all underlying symmetric models.}$$

This would seem to suggest the price of an optimal test at the normal model is too high to pay. We would be better off using the suboptimal rank test which is more stable over the whole class of symmetric models.

The two major criticisms of this type of efficiency are that it provides an asymptotic comparison, valid only for very large samples, and it is a local comparison, valid only in a neighborhood of the null hypothesis. The latter simply points to the fact that a single number such as efficiency is not sufficient to describe the relative behavior of tests over the entire alternative. Bahadur (1967) efficiency does provide an efficiency curve similar to a power curve. For many cases of interest, as the alternative approaches the null hypothesis, the Bahadur efficiency approaches the Pitman efficiency. See Klotz (1965) for further reading on the Bahadur efficiency and a comparison of S, T, and t. Klotz shows that the Bahadur efficiency of T relative to t rises from 0.955 to around 0.98 at a location of $0.75\sigma_f$ and then decreases down to around 0.60 for very large location values.

The former criticism is considered by Klotz (1963) in which he studies the efficiency of T relative to t for finite n. For T we fix a sample size n, significance level α, and underlying distribution F, and a location θ. In the case of an underlying normal distribution it is possible to find the exact power of T. (The general problem of computing finite sample power of rank tests is discussed in some detail in Section 3.3 of the next chapter.) Next calculate the power of t for the given α and θ to obtain sample sizes n' and $n' + 1$ for which the power of t brackets the power of T. We then interpolate linearly to define $n^* = \delta n' + (1 - \delta)(n' + 1)$ the sample size for t.

The finite efficiency of T relative to t is taken to be n^*/n. We illustrate Klotz's results in Table 2.4. Klotz has other tables similar to this one. The results all indicate that the asymptotic efficiency is reflected quite accurately in finite samples from a normal distribution with common signifi-

Table 2.4. Finite Efficiency of T Relative to t for $n(0.25, 1)$ Distribution

n	α	Efficiency
5	.0625	0.986
8	.05469	0.980
10	.05273	0.968

cance levels and moderate sizes of θ. Arnold (1965) obtained similar results for nonnormal shift alternatives.

Another check on the asymptotic results is provided by Monte Carlo simulations of the power of the t, sign, and Wilcoxon signed rank tests. In Randles and Hogg (1973) and Randles and Wolfe (1979, p. 116) the power is simulated for the uniform, normal, logistic, double exponential, and Cauchy distributions. In most cases the empirical power for samples of size 10, 15, and 20 is consistent with the results predicted by the Pitman efficiency. For example, in the cases of the double exponential, the sign test has the best Pitman efficacy, and hence should have the highest power for alternatives near the null hypothesis. This is indeed the case for $\theta = 0.2\sigma$; however, for nonlocal alternatives the Wilcoxon signed rank test is better. Hence there is additional information to be gained from an empirical power study.

Using Theorem 2.6.1, we next show that the Hodges–Lehmann estimator has an asymptotic normal distribution. This is illustrated on the Hodges–Lehmann estimator derived from the Wilcoxon signed rank test in Example 2.6.5 and confirms the asymptotic distribution suggested in Example 2.4.2. It is interesting to note that it is the asymptotic distribution of the test under a sequence of alternatives that determines the asymptotic distribution of the estimator. We begin with a preliminary result.

Theorem 2.6.4. Suppose V is a statistic satisfying the conditions of Definition 1.5.1 and that $\hat{\theta}$ is the corresponding estimator. Then

$$P(V(a) < \mu_0) \leqslant P(\hat{\theta} < a) \leqslant P(V(a) \leqslant \mu_0).$$

Proof. From Definition 1.5.1 we have $\theta^{**} < a$ implies that $V(a) < \mu_0$ and hence $\theta^{**} \leqslant a$. These inequalities imply

$$P(\theta^{**} < a) \leqslant P(V(a) < \mu_0) \leqslant P(\theta^{**} \leqslant a).$$

Since θ^{**} is a continuous random variable (see Hodges and Lehmann,

1963), we have $P(\theta^{**} < a) = P(V(a) < \mu_0)$. Likewise, $P(\theta^* < a) = P(V(a) \leqslant \mu_0)$.

Since $\theta^* \leqslant \hat{\theta} \leqslant \theta^{**}$, if $\theta^{**} < a$ then $\hat{\theta} < a$ and $\theta^* < a$. Hence $P(\theta^{**} < a) \leqslant P(\hat{\theta} < a) \leqslant P(\theta^* < a)$ and finally $P(V(a) < \mu_0) \leqslant P(\hat{\theta} < a) \leqslant P(V(a) \leqslant \mu_0)$.

Theorem 2.6.5. Let $\hat{\theta}$ be the Hodges–Lehmann estimator corresponding to a statistic V which satisfies the Pitman conditions 1–5 with efficacy c. Then

$$\lim_{n\to\infty} P\left[\sqrt{n}\,(\hat{\theta} - \theta) < a\right] = \Phi(ac),$$

that is, $n^{1/2}(\hat{\theta} - \theta)$ is asymptotically $n(0, c^{-2})$.

Proof. By Theorem 1.5.1 we have

$$P_\theta\left[\sqrt{n}\,(\hat{\theta} - \theta) < a\right] = P_0(\sqrt{n}\,\hat{\theta} < a)$$

$$= P_0\left(\hat{\theta} < \frac{a}{\sqrt{n}}\right).$$

From the previous theorem we can write,

$$P_0\left(V\left(\frac{a}{\sqrt{n}}\right) < \mu_0\right) \leqslant P_0\left(\hat{\theta} < \frac{a}{\sqrt{n}}\right) \leqslant P_0\left(V\left(\frac{a}{\sqrt{n}}\right) \leqslant \mu_0\right). \quad (2.6.24)$$

Now μ_0 is the point of symmetry of the null distribution of V, hence $P_0(V(0) < \mu_0)$ converges to 0.5 which corresponds, in Theorem 2.6.1, to $Z_{.5} = 0$. Further,

$$P_0\left(V\left(\frac{a}{\sqrt{n}}\right) < \mu_0\right) = P_{-a/\sqrt{n}}(V(0) < \mu_0)$$

which converges to $\Phi(ac) = \Phi(a/(c^{-2})^{1/2})$ by Theorem 2.6.1 The same limit exists for the right side of (2.6.24), and hence $P_\theta(n^{1/2}(\hat{\theta} - \theta) < a)$ converges to $\Phi(ac)$.

The asymptotic efficiency of two asymptotically normal estimators is generally defined to be the reciprocal ratio of their asymptotic variances. Hence, if $\hat{\theta}_i$, $i = 1, 2$, are asymptotically $n(0, \sigma_i^2)$, $i = 1, 2$, then the efficiency of $\hat{\theta}_1$ relative to $\hat{\theta}_2$ is

$$e(\hat{\theta}_1, \hat{\theta}_2) = \frac{\sigma_2^2}{\sigma_1^2}. \quad (2.6.25)$$

It follows at once from Theorem 2.6.5 that the efficiency of two Hodges–Lehmann estimators derived from Pitman regular tests is identical to the Pitman efficiency of the respective tests. Hence efficiency properties of tests are inherited by the estimators.

Example 2.6.5. We now list the asymptotic distributions for $\text{med}\,X_i$, $\text{med}_{i<j}(X_i + X_j)/2$, and $\bar{X}$ and note their relative efficiencies. These are the Hodges–Lehmann estimators corresponding to S, T, and t, respectively. Hence, from Example 2.6.3, they are asymptotically normally distributed with asymptotic variances $1/4f^2(0)$, $1/12(\int f^2(x)\,dx)^2$, and σ_f^2, respectively. For example, from Theorem 2.6.5 it follows that

$$e\left(\operatorname*{med}_{i\leqslant j}\frac{(X_i + X_j)}{2},\bar{X}\right) = 12\sigma_f^2\left(\int f^2(x)\,dx\right)^2$$

$$= e(T,t)$$

$$\geqslant 0.864.$$

2.7. ASYMPTOTIC LINEARITY OF THE WILCOXON SIGNED RANK STATISTIC, SUMMARY AND EFFECTS OF DEPENDENCE

In this section we discuss the approximate linearity of $T(\theta)$, (2.3.1), as a function of θ. This idea will be made precise in Theorem 2.7.1. The approximate linearity of $T(\theta)$ enables us to study the asymptotic length of the confidence interval derived from T. Further, we can offer a heuristic development of the asymptotic distribution of $\hat{\theta}$ and the asymptotic local power of T. Hence the approximate linearity of $T(\theta)$ ties the point and interval estimates and test together nicely. Asymptotic linearity is crucial to the development of the distribution theory of rank tests and estimates in the linear model. At the end of this section we summarize the properties of the Wilcoxon procedures and briefly discuss the impact of lack of independence in the data.
We will work with

$$\bar{T}(\theta) = \frac{1}{n(n+1)}\,T(\theta)$$

$$= \frac{1}{n(n+1)}\sum\sum_{i\leqslant j} T_{ij}(\theta) \qquad (2.7.1)$$

where $T_{ij}(\theta) = 1$ if $(X_i + X_j)/2 > \theta$ and 0 otherwise. To simplify the discussion we will suppose, without loss of generality, that the true value of θ is 0. Later we will state the results for a true value θ_0. Note that, from Example 2.5.1, since $F \in \Omega_s$,

$$E_0 \overline{T}(\theta) = E_{-\theta} \overline{T}(0) \to \frac{1}{2} p_2(-\theta) \qquad (2.7.2)$$

where

$$\frac{1}{2} p_2(-\theta) = \frac{1}{2} \int_{-\infty}^{\infty} [1 - F(-x + \theta)] f(x + \theta) dx$$

$$= \frac{1}{2} \int_{-\infty}^{\infty} [1 - F(-x + 2\theta)] f(x) dx$$

$$= \frac{1}{2} \int_{-\infty}^{\infty} F(x - 2\theta) f(x) dx.$$

Further, the $\mathrm{Var}_0 \overline{T}(\theta) = \mathrm{Var}_{-\theta} \overline{T}(0) \to 0$ as $n \to \infty$; hence $\overline{T}(\theta)$ converges in probability (when the true value of θ is 0) to $p_2(-\theta)/2$. This can be written $\overline{T}(\theta) = p_2(-\theta)/2 + o_p(1)$ where $o_p(1)$ are the small order terms that tend to 0 in probability (under $\theta = 0$) when n increases. In general we will use the following notation: $o_p(\delta)$ means that $o_p(\delta)/\delta$ converges to 0 in probability as δ tends to 0.

If we can differentiate (2.7.2) with respect to θ under the integral, then

$$p_2(-\theta) = p_2(0) + \theta p_2'(0) + o(\theta)$$

where $o(\theta)/\theta \to 0$ as $\theta \to 0$, and $p_2'(0) = -2 \int f^2(x) dx$. Hence, for small θ and large n, with high probability

$$\overline{T}(\theta) \doteq \frac{1}{4} - \theta \int f^2(x) dx,$$

and this suggests that $\overline{T}(\theta)$ is "approximately" linear in θ with slope $-\int f^2(x) dx$.

Before we provide a rigorous statement of this approximate linearity, we develop a technical lemma.

Theorem 2.7.1. Let $V_n(b) = U_n(b) + c_n b$ where $U_n(b)$ is monotone in b and $|c_n| \leqslant c < \infty$. Suppose that for each b, $V_n(b) \overset{P}{\to} 0$ as $n \to \infty$. Then for any $B > 0$ and any $\epsilon > 0$, as $n \to \infty$

$$P\left\{ \sup_{|b| \leqslant B} |V_n(b)| > \epsilon \right\} \to 0.$$

Proof.   Let $\epsilon > 0$, $\gamma > 0$ be given. Partition $[-B, B]$ into d intervals: $-B = b_0 < b_1 < \cdots < b_d = B$ such that $|b_i - b_{i-1}| \leqslant \epsilon/2c$ for $i = 1, \ldots, d$.

Since there are finitely many b_i, we can find an integer N such that $n \geqslant N$ implies

$$P\left\{ \max_i |V_n(b_i)| < \frac{\epsilon}{2} \right\} \geqslant 1 - \gamma.$$

Without loss of generality suppose $U_n(b)$ is nonincreasing. If $|b| \leqslant B$ then $b_{i-1} \leqslant b \leqslant b_i$ for some i. First suppose $V_n(b) \geqslant 0$. Then $|V_n(b)| = V_n(b)$ and we can write

$$|V_n(b)| = U_n(b) + c_n b$$

$$\leqslant U_n(b_{i-1}) + c_n b_{i-1} + c_n(b - b_{i-1})$$

$$\leqslant |V_n(b_{i-1})| + |c_n|(b - b_{i-1})$$

$$\leqslant \max_i |V_n(b_{i-1})| + \frac{\epsilon}{2}.$$

A similar argument applies when $V_n(b) < 0$, hence

$$\sup_{|b| \leqslant B} |V_n(b)| \leqslant \max_i |V_n(b_i)| + \frac{\epsilon}{2}.$$

Finally, for $n \geqslant N$,

$$P\left\{ \sup_{|b| \leqslant B} |V_n(b)| > \epsilon \right\} \leqslant P\left\{ \max_i |V_n(b_i)| + \frac{\epsilon}{2} > \epsilon \right\}$$

$$\leqslant P\left\{ \max_i |V_n(b_i)| > \frac{\epsilon}{2} \right\} < \gamma,$$

and this completes the proof.

We now formalize the ideas of approximate linearity for the Wilcoxon signed rank statistic.

Theorem 2.7.2. Suppose that $F \in \Omega_s$, and $\int_{-\infty}^{\infty} f^2(x)\,dx < \infty$. Suppose the true value $\theta = 0$. Then, denoting probabilities computed under $\theta = 0$ by $P_0(\cdot)$, for $\epsilon > 0$ and $B > 0$,

$$\lim_{n \to \infty} P_0 \left\{ \sup_{|b| \leqslant B} \left| \sqrt{n} \left[\overline{T}(b/\sqrt{n}) - \overline{T}(0) \right] + b \int_{-\infty}^{\infty} f^2(x)\,dx \right| > \epsilon \right\} = 0.$$

$$(2.7.3)$$

Proof.   In Theorem 2.7.1 let $U_n(b) = n^{1/2}[\overline{T}(b/n^{1/2}) - \overline{T}(0)]$ and $c_n \equiv \int f^2(x)\,dx$, then we must first show that $U_n(b) \overset{P}{\to} -b \int f^2(x)\,dx$. Now

$$U_n(b) = \frac{-\sqrt{n}}{n(n+1)} \# \frac{X_i + X_j}{2} \in (0, b/\sqrt{n}\,]$$

and

$$EU_n(b) = \frac{-\sqrt{n}}{n(n+1)} \left\{ \frac{n(n-1)}{2} \left[F^*(b/\sqrt{n}) - F^*(0) \right] \right.$$

$$\left. + n \left[F(b/\sqrt{n}) - F(0) \right] \right\},$$

where $F^*(t) = P((X_1 + X_2)/2 \leqslant t) = \int_{-\infty}^{\infty} F(2t - x)f(x)\,dx$. Hence

$$EU_n(b) \sim -\sqrt{n} \left[F^*(b/\sqrt{n}) - F^*(0) \right]/2.$$

Provided the derivative exists, we have

$$EU_n(b) \sim \frac{-b}{2} \frac{\left[F^*(b/\sqrt{n}) - F^*(0) \right]}{b/\sqrt{n}} \to \frac{-b}{2} \frac{d}{dt} F^*(t) \Big|_{t=0}.$$

Since F^* is a convolution of two absolutely continuous cdfs Theorem A17 implies that it is also absolutely continuous with pdf at 0 given by:

$$\frac{d}{dt} F^*(t) \Big|_{t=0} = 2 \int f(2t - x)f(x)\,dx \Big|_{t=0} = 2 \int f^2(x)\,dx.$$

This shows that $EU_n(b) \to -b \int f^2(x)\,dx$.

Next we must consider the $\mathrm{Var}\, U_n(b)$. Let $I_{ij} = 1$ if $(X_i + X_j)/2 \in (0, b/n^{1/2}]$ and 0 otherwise, then

$$\mathrm{Var}\, U_n(b) \sim \frac{n}{n^2(n+1)^2} \mathrm{Var}\left(\sum\sum_{i<j} I_{ij}\right).$$

Now using the argument in Theorem 2.5.1,

$$\mathrm{Var}\left(\sum\sum_{i<j} I_{ij}\right) = \frac{n(n-1)}{2}\mathrm{Var}\, I_{12} + n(n-1)(n-2)\mathrm{Cov}(I_{12}, I_{13}),$$

hence $\mathrm{Var}\, U_n(b) \sim \mathrm{Cov}(I_{12}, I_{13})$ and

$$|\mathrm{Cov}(I_{12}, I_{13})| = |EI_{12}I_{13} - (EI_{12})^2|$$

$$\leqslant EI_{12} + (EI_{12})^2.$$

But $EI_{12} = F^*(b/n^{1/2}) - F^*(0) \to 0$ and so $\mathrm{Var}\, U_n \to 0$. By Theorem A5, $U_n(b) \xrightarrow{P} -b\int f^2(x)\,dx$ and $V_n(b) \to 0$. Since $U_n(b)$ is nondecreasing in b we can apply Theorem 2.7.1 and the proof is complete.

Now, (2.7.3) can be expanded and rewritten in various ways. Equations (2.7.4)–(2.7.6) provide three forms that are useful. First we have

$$\sqrt{n}\,\overline{T}(b/\sqrt{n}) - \sqrt{n}\,\overline{T}(a/\sqrt{n}) = -(b-a)\int f^2(x)\,dx + o_p(1) \quad (2.7.4)$$

uniformly for a, b such that $-A \leqslant a < b \leqslant A$. Next, if we take $\theta = b/n^{1/2}$ and $\theta_0 = a/n^{1/2}$ so that $n^{1/2}|\theta - \theta_0| \leqslant A$ then

$$\sqrt{n}\,\overline{T}(\theta) \doteq \sqrt{n}\,\overline{T}(\theta_0) - \sqrt{n}\,(\theta - \theta_0)\int f^2(x)\,dx \quad (2.7.5)$$

which provides a kind of Taylor series approximation to $n^{1/2}\overline{T}(\theta)$ about θ_0. If θ_0 denotes the true value of θ so that $E_{\theta_0}\overline{T}(\theta_0) \sim \frac{1}{4}$, and $\mathrm{Var}_{\theta_0}\overline{T}(\theta_0) \sim 1/12n$, then we have, for $n^{1/2}|\theta - \theta_0| \leqslant A$,

$$\frac{\sqrt{n}\,[\overline{T}(\theta) - 1/4]}{\sqrt{1/12}} \doteq \frac{\sqrt{n}\,[\overline{T}(\theta_0) - 1/4]}{\sqrt{1/12}} - \sqrt{n}\,(\theta - \theta_0)\sqrt{12}\int f^2(x)\,dx.$$

$$(2.7.6)$$

From (2.7.6) we see that $\bar{T}(\theta)$, when standardized, has approximate slope $n^{1/2}c_T$, the efficacy of T given in (2.6.13). This ties together the properties of tests and the estimators derived from them.

Our first application of (2.7.6) is a derivation of the asymptotic length of the confidence interval, (2.3.3), derived from the Wilcoxon signed rank test T. Recall that if $P(T \leqslant C_1) = P(T \geqslant N - C_1) = \alpha/2$ (let $C_2 = N - C_1$ denote the upper critical point) then $\hat{\theta}_L$ is the $(C_1 + 1)$st or $(N - C_2 + 1)$st ordered Walsh average. From the discussion following Definition 1.5.1, $T(\hat{\theta}_L) = C_2 - 1$. Since the result is an asymptotic one, we will transfer the argument to $\bar{T}$ with $E\bar{T} = 1/4$ and $\operatorname{Var}\bar{T} \sim 1/12n$. Since $\bar{T}$ has an approximate normal distribution we can write

$$\frac{\sqrt{n}\left[\bar{T}(\hat{\theta}_L) - 1/4\right]}{\sqrt{1/12}} \doteq Z_{\alpha/2} \qquad (2.7.7)$$

where $Z_{\alpha/2}$ is the upper $\alpha/2$ percentile of the standard normal distribution. From Exercise 2.10.17, $n^{1/2}(\hat{\theta}_L - \theta_0)$ has an asymptotic normal distribution. Hence, for $\epsilon > 0$, there exists a positive integer N and a positve number A such that when $n \geqslant N$, $P(n^{1/2}|\hat{\theta}_L - \theta_0| \leqslant A) \geqslant 1 - \epsilon$, and we will say $n^{1/2}(\hat{\theta}_L - \theta_0)$ is bounded in probability.

The boundedness in probability of $n^{1/2}(\hat{\theta}_L - \theta_0)$ allows us to combine (2.7.7) with (2.7.6) for large n with high probability and write

$$Z_{\alpha/2} \doteq -\sqrt{n}\,(\hat{\theta}_L - \theta_0)\sqrt{12}\int f^2(x)\,dx.$$

Similarly,

$$-Z_{\alpha/2} \doteq -\sqrt{n}\,(\hat{\theta}_U - \theta_0)\sqrt{12}\int f^2(x)\,dx.$$

Subtracting and rearranging we find

$$\frac{\sqrt{n}\,(\hat{\theta}_U - \hat{\theta}_L)}{2Z_{\alpha/2}} \doteq \frac{1}{\sqrt{12}\int f^2(x)\,dx}, \qquad (2.7.8)$$

where the approximation means for large n with high probability. More formally we see that the standardized length of the Wilcoxon confidence interval $n^{1/2}(\hat{\theta}_U - \hat{\theta}_L)/2Z_{\alpha/2}$ converges in probability to $1/(12)^{1/2}\int f^2(x)\,dx = 1/c_T$, the reciprocal of the efficacy. Note that we need the uniform convergence to replace θ in (2.7.6) by $\hat{\theta}_L$ and $\hat{\theta}_U$.

Later we will see this is a general property of confidence intervals derived from tests. In the exercises, you are asked to verify this for the sign and t tests. The ratio of squared lengths of two intervals is a measure of the relative efficiency of the intervals, and this ratio converges in probability to the Pitman efficiency of the respective tests. In this sense the test, point estimate, and confidence interval all share common efficiency properties. The results of Examples 2.6.3 and 2.6.5 on the sign, Wilcoxon, and t procedures now extend to the confidence intervals.

Our next application of (2.7.6) is a heuristic derivation of the asymptotic normality of $n^{1/2}(\hat{\theta} - \theta_0)$ where θ_0 is the true parameter value and $\hat{\theta} = \text{med}_{i \leqslant j}(X_i + X_j)/2$; see Example 2.6.5. The estimate $\hat{\theta}$ can be defined by

$$\frac{\sqrt{n}\left[\bar{T}(\hat{\theta}) - 1/4\right]}{\sqrt{1/12}} \doteq 0$$

$$\doteq \frac{\sqrt{n}\left[\bar{T}(\theta_0) - 1/4\right]}{\sqrt{1/12}} - \sqrt{n}\,(\hat{\theta} - \theta_0)\sqrt{12} \int f^2(x)\,dx.$$

Hence

$$\sqrt{n}\,(\hat{\theta} - \theta_0) \doteq \frac{1}{\sqrt{12}\int f^2(x)\,dx} \frac{\sqrt{n}\left[\bar{T}(\theta_0) - 1/4\right]}{\sqrt{1/12}}.$$

The second factor is the standardized Wilcoxon signed rank test which is asymptotically standard normal. Hence the right side is asymptotically normal with mean 0 and variance $1/12(\int f^2(x)\,dx)^2$. This heuristic derivation shows how the asymptotic distributions of the test and estimate are related through the approximate linearity with the efficacy playing an important part. A rigorous argument would require that we first show $n^{1/2}(\hat{\theta} - \theta_0)$ is bounded in probability (Theorem 2.6.4).

The final application of (2.7.6) is a heuristic derivation of the local asymptotic power; see Theorem 2.6.1. Let $\theta_n = \theta/n^{1/2}$. The asymptotic size α critical region for T is determined by

$$\frac{\sqrt{n}\left[\bar{T}(0) - 1/4\right]}{\sqrt{1/12}} \geqslant Z_\alpha,$$

where Z_α is the upper α percentile of the standard normal distribution. The power is then

$$
P_{\theta_n} \left\{ \frac{\sqrt{n} \left[\overline{T}(0) - 1/4 \right]}{\sqrt{1/12}} \geq Z_\alpha \right\}
$$

$$
= P_0 \left\{ \frac{\sqrt{n} \left[\overline{T}(-\theta/\sqrt{n}) - 1/4 \right]}{\sqrt{1/12}} \geq Z_\alpha \right\}
$$

$$
\doteq P_0 \left\{ \frac{\sqrt{n} \left[\overline{T}(0) - 1/4 \right]}{\sqrt{1/12}} - \sqrt{n} \left(-\theta/\sqrt{n} \right)\sqrt{12} \int f^2(x)\,dx \geq Z_\alpha \right\}
$$

$$
= P_0 \left\{ \frac{\sqrt{n} \left[\overline{T}(0) - 1/4 \right]}{\sqrt{1/12}} \geq Z_\alpha - \theta\sqrt{12} \int f^2(x)\,dx \right\}
$$

$$
\doteq 1 - \Phi\left(Z_\alpha - \theta\sqrt{12} \int f^2(x)\,dx \right).
$$

We now summarize the properties of the Wilcoxon procedures. We suppose that $X_1, \ldots, X_n$ are i.i.d. $F(x - \theta)$, $F \in \Omega_s$; hence we have the symmetric location model. The Wilcoxon signed rank statistic T is distribution free under $H_0 : \theta = 0$. The test is unbiased (apply Exercise 1.8.4), consistent, and has positive tolerance (asymptotically 0.29) to both acceptance and rejection. The symmetric distribution of T, under H_0, is easy to table and the asymptotic distribution is normal under both null and alternative hypotheses. The Pitman efficacy is given by $c_T = 12^{1/2}\int f^2(x)\,dx$ and the Pitman asymptotic efficiency relative to the t test is 0.955 for an underlying normal distribution, 1.19 for a 5% contaminated normal distribution, and is never less than 0.864. For heavy-tailed distributions around a double exponential distribution the sign test is more efficient. Furthermore, the Pitman efficiency reflects quite well the small sample and nonlocal alternative properties of T. In Section 6.2, the Wilcoxon signed rank test is extended to the one-sample multivariate location model.

The Hodges–Lehmann estimate of θ is $\hat{\theta} = \mathrm{med}(X_i + X_j)/2$, $i \leq j$. The estimate is unbiased and symmetrically distributed about θ. It has positive tolerance (asymptotically .29) and a bounded, continuous influence curve. Hence it is a robust estimate. Moreover, $n^{1/2}(\hat{\theta} - \theta)$ is asymptotically normally distributed with asymptotic variance $1/c_T^2 = 1/12(\int f^2(x)\,dx)^2$. Hence, $\hat{\theta}$ inherits the efficiency properties of T.

The confidence interval generated from T is distribution free and $n^{1/2}$ (length)/$2Z_{\alpha/2}$ converges in probability to $1/c_T$. Hence, when efficiency of confidence intervals is defined in terms of their lengths, the Wilcoxon confidence interval inherits its efficiency properties directly from T. Finally, $T(\theta)$, as a function of θ, is a nonincreasing step function with steps at the Walsh averages and is approximately linear in θ for large n. The slope of the linear approximation is proportional to c_T. The approximate linearity can be used to show how c_T determines the properties of the efficiency of the test (through local asymptotic power), the asymptotic variance of the estimate, and the asymptotic length of the confidence interval.

We now consider the simple model of serial correlation introduced at the end of Section 1.7. The model specifies that (X_i, X_{i+1}), $i = 1, 2, \ldots$, has a bivariate normal distribution with means 0, variances 1, and correlation ρ. The results of Gastwirth and Rubin (1975), which will not be derived here, show that, for the serial correlation model, the projection still determines the limiting distribution of the Wilcoxon signed rank statistic. (See their equation 3.25.) In our model, the marginal distributions are standard normal, denoted $\Phi(\cdot)$. Hence, from (2.5.18) in Example 2.5.5, $n^{-1/2}V_p'$ $= n^{-1/2}\sum(\Phi(X_i) - 1/2)$ determines the limiting distribution of T under $H_0: \theta = 0$. The sequence $\Phi(X_1), \Phi(X_2), \ldots$ is a 1-dependent sequence, and Theorem A16 implies that $n^{-1/2}V_p'$ is asymptotically $n(0, \sigma^2)$ with σ^2 $= \operatorname{Var}\Phi(X_1) + 2\operatorname{Cov}(\Phi(X_1), \Phi(X_2))$.

Hence, from the discussion in Example 2.5.5, we have $n^{-3/2}(T - ET)$ or $n^{-1/2}(\overline{T} - 1/4)$, under H_0, have the same limiting distribution as $n^{-1/2}V_p'$. The random variable $\Phi(X_1)$ has a uniform distribution on $(0, 1)$, and so $\operatorname{Var}\Phi(X_1) = 1/12$. Next, we consider:

$$\operatorname{Cov}(\Phi(X_1), \Phi(X_2)) = E\big[\Phi(X_1)\Phi(X_2)\big] - E\Phi(X_1)E\Phi(X_2). \quad (2.7.9)$$

The expectation is taken with respect to the bivariate normal distribution, denoted $\Phi(x, y)$. Let U and V be independent $n(0, 1)$ variables, also independent of (X_1, X_2), then

$$E\Phi(X_1)\Phi(X_2) = \int\int P(U \leqslant x_1)P(V \leqslant x_2)\,d\Phi(x_1, x_2)$$

$$= P(U \leqslant X_1, V \leqslant X_2)$$

$$= P(U - X_1 \leqslant 0, V - X_2 \leqslant 0). \quad (2.7.10)$$

Note that $E(U - X_1)(V - X_2) = EX_1X_2 = \rho$ and $\operatorname{Var}(U - X_1) = \operatorname{Var}(V - X_2) = 2$; hence the correlation between $U - X_1$ and $V - X_2$ is $\rho/2$. In fact, $(U - X_1, V - X_2)$ has a bivariate normal distribution with means 0, variances 2, and correlation $\rho/2$. The $P(U - X_1 \leqslant 0, V - X_2 \leqslant 0)$ does not depend on the variances (divide the inequalities by $2^{1/2}$); hence Exercise 1.8.14 shows the probability to be equal to $1/4 + (1/2\pi)\sin^{-1}(\rho/2)$. Since $E\Phi(X_1)E\Phi(X_2) = 1/4$, (2.7.9) becomes $(1/2\pi)\sin^{-1}(\rho/2)$ and

$$\sigma^2 = \frac{1}{12} + \frac{1}{\pi}\sin^{-1}(\rho/2). \tag{2.7.11}$$

Thus, $n^{1/2}(\overline{T} - 1/4)$ is asymptotically $n(0, \sigma^2)$, and, if we suppose that we are sampling from an i.i.d. sequence, then a nominal 5% test, for large n, is given by $\overline{T} \geqslant 1/4 + 1.645/(12n)^{1/2}$. However, the true level is

$$\alpha_T = P\left(\overline{T} \geqslant 1/4 + 1.645/\sqrt{12n}\right)$$

$$= P\left(\sqrt{n}\left(\overline{T} - 1/4\right) \geqslant 1.645/\sqrt{12}\right)$$

$$= P\left[\frac{\sqrt{n}\left(\overline{T} - 1/4\right)}{\sigma} \geqslant \frac{1.645}{\sqrt{1 + \dfrac{12}{\pi}\sin^{-1}(\rho/2)}}\right]$$

$$\doteq 1 - \Phi\left[\frac{1.645}{\sqrt{1 + \dfrac{12}{\pi}\sin^{-1}(\rho/2)}}\right]. \tag{2.7.12}$$

Another line can now be added to Table 1.4 reflecting the true level of $\overline{T}$. However, the line for $\overline{T}$ is almost identical to that of $\overline{X}$. This is not surprising since $\sin^{-1}t = t + t^3/6 + \cdots$, and so $1 + (12/\pi)\sin^{-1}(\rho/2) \doteq 1 + (12/2\pi)\rho \doteq 1 + 2\rho$, the asymptotic variance corresponding to $\overline{X}$. Hence, with this simple model of dependence in the data, the superior stability of level of T over t completely vanishes.

2.8. GENERAL SCORES STATISTICS

In the first seven sections of this chapter we have developed methods for testing and estimation based on the Wilcoxon signed rank statistic. The necessary distribution theory, both finite and asymptotic, for constructing tests and confidence intervals is given in Sections 2.1–2.3. In the later sections we developed the local asymptotic power and efficiency of T. In particular, we find that if the underlying population is symmetric, then T often provides a more efficient set of inference methods than the sign or t tests. However, T is not uniformly best, and no optimality theory similar to that of the sign test in Chapter 1 is possible.

In this section we generalize from T, which is based on the ranks of the absolute values of the observations, to statistics based on functions (called scores) of the ranks of the absolute values. We provide the asymptotic theory necessary to construct approximate tests and confidence intervals. We give a heuristic development of the efficiency properties in the next section and consider the problem of determining the most efficient rank score statistic for a given underlying distribution.

Definition 2.8.1. Let $0 = a(0) \leqslant a(1) \leqslant \cdots \leqslant a(n)$ be a nonconstant sequence and define

$$V = \sum_{j=1}^{n} a(R_j) s(X_j)$$

$$= \sum_{j=1}^{n} a_j s(X_{D_j})$$

$$= \sum_{j=1}^{n} a_j W_j \tag{2.8.1}$$

where R_j is the rank of $|X_j|$ among $|X_1|, \ldots, |X_n|$, W_j is given by (2.2.1), and D_j is the antirank in Definition 2.2.1. Then V is called a signed rank score statistic. Note that if $a_j = 1$, $j = 1, \ldots, n$, then $V = S$, and if $a_j = j$, $j = 1, \ldots, n$, then $V = T$.

In Exercise 2.10.4 you are asked to show that if $X_1, \ldots, X_n$ are i.i.d. $F(x)$, $F \in \Omega_s$, then

$$EV_1 = \frac{1}{2} \sum_{j=1}^{n} a_j$$

$$\operatorname{Var} V_1 = \frac{1}{4} \sum_{j=1}^{n} a_j^2 \tag{2.8.2}$$

and

$$\operatorname{Cov}(V_1, V_2) = \frac{1}{4} \sum_{j=1}^{n} a_j b_j$$

where $V_1 = \sum a_j W_j$ and $V_2 = \sum b_j W_j$. It follows at once from Theorem A10 that $(V - EV)/(\text{Var } V)^{1/2}$ has an asymptotic standard normal distribution, provided

$$\lim_{n \to \infty} \frac{\max\limits_{1 \le i \le n} a_i}{\sqrt{\sum\limits_{1}^{n} a_j^2}} = 0. \tag{2.8.3}$$

Hence using (2.8.2) and (2.8.3) we can easily determine when a normal approximation is possible for a signed rank score statistic V. Often the scores are given by a score generating function.

Definition 2.8.2.   Suppose $\phi(u)$, $0 < u < 1$, is nonnegative and nondecreasing. Suppose further that $\int_0^1 \phi(u)\,du < \infty$ and $0 < \int_0^1 \phi^2(u)\,du < \infty$. Define $a(i) = \phi[i/(n+1)]$, then

$$\overline{V} = \frac{1}{n} \sum_{j=1}^{n} \phi\left(\frac{R_j}{n+1} \right) s(X_j)$$

is the statistic generated by the score generating function $\phi(\cdot)$. Note that $\phi(u) = 1$, $0 < u < 1$, produces $\overline{S}$ and $\phi(u) = u$ produces $\overline{T}$.

Theorem 2.8.1.   If $\phi(\cdot)$ generates $\overline{V}$, then

$$E\overline{V} = \frac{1}{2n} \sum \phi\left(\frac{j}{n+1} \right) \to \frac{1}{2} \int_0^1 \phi(u)\,du \qquad \text{as } n \to \infty$$

$$n \text{ Var } \overline{V} = \frac{1}{4n} \sum \phi^2\left(\frac{j}{n+1} \right) \to \frac{1}{4} \int_0^1 \phi^2(u)\,du \qquad \text{as } n \to \infty$$

and

$$\frac{\overline{V} - E\overline{V}}{\sqrt{\text{Var } \overline{V}}} \qquad \text{or} \qquad \frac{\sqrt{n}\left(\overline{V} - \frac{1}{2}\int \phi(u)\,du \right)}{\sqrt{\frac{1}{4}\int \phi^2(u)\,du}}$$

has an asymptotically standard normal distribution.

Proof.   The moments follow directly from (2.8.2) and the definition of the Riemann integral. To establish asymptotic normality, from (2.8.3) we must show

$$\frac{\max \phi\left(\dfrac{j}{n+1}\right)}{\sqrt{\Sigma \phi^2\left(\dfrac{j}{n+1}\right)}} \to 0 \qquad \text{as} \quad n \to \infty.$$

We write the square of the left side as

$$\frac{\left(\dfrac{1}{n+1}\right)\phi^2\left(\dfrac{n}{n+1}\right)}{\left(\dfrac{1}{n+1}\right)\Sigma \phi^2\left(\dfrac{j}{n+1}\right)}.$$

We now need only show the numerator converges to 0 since the denominator converges to $0 < \int \phi^2(u)\,du < \infty$. Since $\phi(u)$ is nondecreasing, we have

$$0 \leqslant \left(\frac{1}{n+1}\right)\phi^2\left(\frac{n}{n+1}\right) \leqslant \int_{n/(n+1)}^{1} \phi^2(u)\,du.$$

Since $0 < \int \phi^2(u)\,du < \infty$, as $n \to \infty$ the right side tends to 0.

This theorem shows that for a large class of possible test statistics, we can use a normal approximation to construct critical values for the tests. In Exercise 2.10.20 it is shown that $(\overline{V}_1, \overline{V}_2)$, when properly standardized, has an asymptotic bivariate normal distribution.

Bickel (1974) provides an Edgeworth approximation for the general scores statistic. Provided the regularity conditions discussed by Bickel are satisfied, we have

$$P\left(\frac{\overline{V} - E\overline{V}}{\sqrt{\operatorname{Var} \overline{V}}} \leqslant t\right) \doteq \Phi(t) + \frac{\int_0^1 \phi^4(u)\,du}{12n\left(\int_0^1 \phi^2(u)\,du\right)^2} \psi(t)\left[t^3 - 3t\right] \quad (2.8.4)$$

where $\psi(x)$ is the $n(0,1)$ pdf.

A heuristic development of the asymptotic testing tolerance, Definition 1.6.2, can be given for the general scores test. In accordance with the definition of tolerance to acceptance, for given values of $x_{a+2}, \ldots, x_n$, we will take $x_1, \ldots, x_{a+1}$ to be negative with large absolute values. Then we have

$$\overline{V} \leqslant \frac{1}{n} \sum_{i=1}^{n-a-1} \phi\left(\frac{i}{n+1}\right)$$

and we fail to reject if

$$\overline{V} < \frac{1}{2}\int_0^1 \phi(u)\,du + Z_{\alpha/2}\sqrt{\frac{1}{4n}\int_0^1 \phi^2(u)\,du} \ .$$

We have used the asymptotic mean and standard deviation to define the approximate critical value. Tolerance to acceptance is the smallest such a, and in this heuristic development we will suppose that for large n, $a \doteq \epsilon n$. Hence ϵ is the asymptotic tolerance to acceptance. Then for large n, $\overline{V} \sim \int_0^{1-\epsilon} \phi(u)\,du$ and ϵ is the solution of

$$\int_0^{1-\epsilon} \phi(u)\,du = \frac{1}{2}\int_0^1 \phi(u)\,du \tag{2.8.5}$$

since the second term in the definition of the critical value tends to 0. In the case of tolerance to rejection, fix $x_{b+2}, \ldots, x_n$ and choose $x_1, \ldots, x_{b+1}$ positive with large absolute value. Then

$$\overline{V} \geqslant \frac{1}{n} \sum_{i=n-b}^{n} \phi\left(\frac{i}{n+1}\right)$$

and we reject if

$$\overline{V} > \frac{1}{2}\int_0^1 \phi(u)\,du + Z_{\alpha/2}\sqrt{\frac{1}{4n}\int_0^1 \phi^2(u)\,du} \ .$$

Again, if we suppose that $b \doteq \delta n$ for large n, then the asymptotic tolerance to rejection is δ, defined by

$$\int_{1-\delta}^1 \phi(u)\,du = \frac{1}{2}\int_0^1 \phi(u)\,du.$$

In typical cases, $\delta = \epsilon$, and this asymptotic testing tolerance is the same as that defined in a much more abstract setting by Rieder (1982). It is also the same as the breakdown point of an estimator derived from a rank statistic; see Huber (1981, p. 67).

Example 2.8.1. Normal Scores Statistic. Let $\Phi(\cdot)$ denote the standard normal cdf and define $\Phi_+(x) = 2\Phi(x) - 1 = P(|X| \leqslant x)$. Let $\phi(u) = \Phi_+^{-1}(u)$ and the statistic $\overline{V}$ is then

$$\overline{V} = \frac{1}{n} \sum \Phi_+^{-1}\left(\frac{R_j}{n+1}\right) s(X_j),$$

proposed by Fraser (1957) as a one-sample normal scores statistic. Note that $\Phi_+^{-1}(i/(n+1)) = \Phi^{-1}(1/2 + i/2(n+1))$ is roughly equal to $E|X|_{(i)}$. In Exercise 2.10.21 you are asked to show $\phi(u) = \Phi_+^{-1}(u)$ satisfies Definition 2.8.2 and hence is a scores-generating function. Approximate normality then follows from Theorem 2.8.1. If $E|X|_{(i)}$ rather than $\Phi_+^{-1}(i/(n+1))$ is used in the computation of $\overline{V}$, then tables of the values of $E|X|_{(i)}$ are available in Govindarajulu and Eisenstat (1965) and Klotz (1963). A simple approximation to $\Phi^{-1}(u)$ is given by 4.91 $[u^{0.14} - (1 - u)^{0.14}]$. This approximation is based on Tukey's λ-distribution and its accuracy is discussed by Joiner and Rosenblatt (1971).

Since $\Phi_+^{-1}(u) = \Phi^{-1}[(u+1)/2]$, the asymptotic tolerance to acceptance is defined by

$$\int_0^{1-\epsilon} \Phi^{-1}\left(\frac{u+1}{2}\right) du = \frac{1}{2}\int_0^1 \Phi^{-1}\left(\frac{u+1}{2}\right) du.$$

Exercise 2.10.21 can be used to reduce this to

$$1 - \exp\left\{-\frac{1}{2}\left[\Phi^{-1}\left(1 - \frac{\epsilon}{2}\right)\right]^2\right\} = \frac{1}{2}$$

and $\epsilon = 2(1 - \Phi(\sqrt{\log 4})) \doteq .239$. A similar computation shows that the asymptotic tolerance to rejection is also .239. This is a bit less than .293, the asymptotic tolerance of the Wilcoxon signed rank test (Exercise 2.10.5); however, the normal scores test is still much more resistant to outliers than the t test.

Example 2.8.2. Winsorized Signed Rank Statistics. Let $\phi(u) = \min(u, 1 - \gamma)$, $0 < u < 1$; see Fig. 2.5. The statistic $\overline{V} = n^{-1}\sum \phi(R_i/(n+1))s(X_i)$ assigns the rank to the $(1 - \gamma)$ 100% observations smallest in absolute value and $(1 - \gamma)$ times $s(x)$ to the (γ) 100% observations with largest absolute value. Hence $\overline{V}$ represents a mixture of $(1 - \gamma)$ 100% Wilcoxon and (γ) 100% sign scores. Winsorization is a term coined by John Tukey to denote

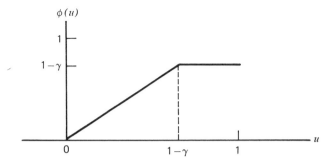

Figure 2.5. Winsorized signed rank score function.

the process of replacing values beyond a point with the value at that point, for example, the Winsorized mean. We see from Exercise 2.10.22 that the asymptotic parameters in Theorem 2.8.1 are $(1 - \gamma^2)/4$ and $(1 - \gamma)^2(1 + 2\gamma)/12$, and hence $n^{1/2}(\overline{V} - (1 - \gamma^2)/4)/[(1 - \gamma)^2(1 + 2\gamma)/12]^{1/2}$ has an approximate standard normal distribution. A comparison of the exact and asymptotic parameters is discussed in Exercise 2.10.23.

The asymptotic tolerance to acceptance is ϵ, defined by

$$\int_0^{1-\epsilon} \min(u, 1 - \gamma)\, du = \frac{1 - \gamma^2}{4}.$$

If $1 - \epsilon < 1 - \gamma$, then $(1 - \epsilon)^2/2 = (1 - \gamma^2)/4$ and $\epsilon = 1 - [(1 - \gamma^2)/2]^{1/2}$. If $1 - \epsilon \geqslant 1 - \gamma$, then $(1 - \gamma)^2/2 + (1 - \gamma)(\gamma - \epsilon) = (1 - \gamma^2)/4$ and $\epsilon = (1 + \gamma)/4$. Note that in the first case $1 - \epsilon = [(1 - \gamma^2)/2]^{1/2} < 1 - \gamma$. This requires $\gamma < 1/3$. Likewise, in the second case $\gamma \geqslant \frac{1}{3}$. Hence the asymptotic tolerance to acceptance is

$$\epsilon = \begin{cases} 1 - \sqrt{(1 - \gamma^2)/2} & \gamma < \frac{1}{3} \\ (1 + \gamma)/4 & \gamma \geqslant \frac{1}{3}. \end{cases}$$

It is easy to check that the asymptotic tolerance to rejection is the same. Estimation tolerance is discussed at the end of Example 2.8.4.

By letting γ range from 0 to 1, the statistics range from the Wilcoxon signed rank test to the sign test. Some prior information concerning the tail weight of the underlying population may be used to select a highly efficient test; something around a normal distribution suggests γ around 0, and something closer to a double exponential suggests γ around 1. Rieder (1981, 1982) has developed an asymptotic robustness theory for rank tests. His work indicates that, in any case, some Winsorization may be desirable to achieve the stability properties discussed in his work. For a further discussion of Winsorized signed rank statistics, see Policello and Hettmansperger (1976) and Hettmansperger and Utts (1977).

Example 2.8.3. Modified Sign Statistics. Let $\phi(u) = 0$ if $0 < u \leqslant 1 - \gamma$ and 1 if $1 - \gamma < u < 1$. Then $\overline{V} = n^{-1}\sum\phi(R_j/(n + 1))s(X_j)$. Let $V = n\overline{V}$, then

> V = number of positive observations for which the ranks of
> their absolute values are greater than $(1 - \gamma)(n + 1)$.

When $\gamma = 1$, $V = S$. Note that

$$V = \sum_{j=[(1-\gamma)(n+1)]+1}^{n} W_j$$

where $[\cdot]$ is the greater integer function and W_j is defined in (2.2.3). It now follows from Theorem 2.2.1 that, under the null hypothesis, V has a binomial distribution with parameters $p = 1/2$ and $n - [(1 - \gamma)(n + 1)]$, so that critical values are easily determined.

By varying γ, we generate the family of modified sign statistics. We will see later that the efficiency can be increased over that of the sign test by judicious choice of γ. Unlike the sign statistic, under alternative hypotheses the distribution is no longer binomial; see Example 2.4.2. The asymptotic distribution under the null hypothesis is discussed in Exercise 2.10.24. For a further discussion of these statistics see Noether (1973) and Markowski and Hettmansperger (1982).

We now return to the general signed rank score statistic $V = \sum a(R_j) s(X_j)$ given in (2.8.1) and consider estimates derived from V. Define, as usual,

$$V(\theta) = \sum a(R_j(\theta))s(X_j - \theta) \tag{2.8.6}$$

where $R_j(\theta)$ is the rank of $|X_j - \theta|$ among $|X_1 - \theta|, \ldots, |X_n - \theta|$. The order statistics of the sample are given by $X_{(1)} \leqslant \cdots \leqslant X_{(n)}$. We now present a theorem due to Bauer (1972) that characterizes $V(\theta)$ in terms of its behavior at the Walsh averages, which we will write as $(X_{(i)} + X_{(j)})/2$, $1 \leqslant i \leqslant j \leqslant n$.

Theorem 2.8.2. As a function of θ, $V(\theta)$ is a nonincreasing step function such that

$V(\theta)$ decreases by the amount a_1 at each $X_{(j)}$

$V(\theta)$ decreases by the amount $a_{j-i+1} - a_{j-i}$ at $(X_{(i)} + X_{(j)})/2$.

Proof. First consider values of θ just to the left of $X_{(j)}$. The value $X_{(j)} - \theta$ is positive, small, and has rank 1 among the absolute values. Hence a_1 is included in $V(\theta)$. However, as soon as we consider values of θ just to the right of $X_{(j)}$, we see that $X_{(j)} - \theta$ is negative but its absolute value still has rank 1. Hence a_1 is not included in $V(\theta)$, and so $V(\theta)$ must decrease by the amount a_1 at $X_{(j)}$ for each $j = 1, \ldots, n$.

Next we consider $X_{(i)}$ and $X_{(j)}$ for some pair $i < j$. Consider values of θ just to the left of $(X_{(i)} + X_{(j)})/2$. The value of $X_{(j)} - \theta$ is a bit larger than $|X_{(i)} - \theta|$ and also larger than $|X_{(k)} - \theta|$, $i < k < j$, for those order statistics between $X_{(i)}$ and $X_{(j)}$. Hence the rank of $X_{(j)} - \theta$ is seen to be $j - (i - 1) = j - i + 1$; and since $X_{(j)} - \theta > 0$, a_{j-i+1} is included in the computation of $V(\theta)$. For values of θ just to the right of $(X_{(i)} + X_{(j)})/2$ the situation is reversed. Now $|X_{(i)} - \theta|$ is slightly larger than $X_{(j)} - \theta$, and the rank of $X_{(j)} - \theta$ has decreased to $j - i$. Hence, at $(X_{(i)} + X_{(j)})/2$, the value of $V(\theta)$ must decrease in a jump of size $a_{j-i+1} - a_{j-i}$.

Define $T_{ij}(\theta) = 1$ if $(X_{(i)} + X_{(j)})/2 > \theta$ and 0 otherwise, then we can construct the counting form of $V(\theta)$. Since $V(\theta)$ is a step function, its value (height) is the sum of accumulated steps at Walsh averages to the right of θ. Hence

$$V(\theta) = \sum_{i \leq j} \sum (a_{j-i+1} - a_{j-i}) T_{ij}(\theta). \tag{2.8.7}$$

From Exercise 2.10.2, under the model that we are sampling from a symmetric distribution $F(x - \theta)$, $F \in \Omega_s$, we see that $V(\theta)$ is symmetrically distributed about $\sum a_i/2$. Generally a Hodges–Lehmann estimate, Definition 1.5.1, can be determined from $V(\hat{\theta}) = \sum a_i/2$.

If we restrict attention to scores such that

$$a_{j-i+1} - a_{j-i} = \frac{1}{n+1} \quad \text{or } 0, \tag{2.8.8}$$

then the steps are constant and occur at selected Walsh averages; see (2.4.1) in Section 2.4. Let the set B be defined by

$$B = \{(i, j) : 1 \leq i \leq j \leq n, a_{j-i+1} - a_{j-i} = (n+1)^{-1}\}. \tag{2.8.9}$$

Then B specifies the set of Walsh averages where steps occur. We can now write $V(\theta)$ as follows:

$$\overline{V}(\theta) = \frac{1}{n} \sum\sum_{i \leqslant j} (a_{j-i+1} - a_{j-i}) T_{ij}(\theta)$$

$$= \frac{1}{n(n+1)} \sum\sum_{(i,j) \in B} T_{ij}(\theta)$$

$$= \frac{1}{n(n+1)} \# \left\{ \frac{X_{(i)} + X_{(j)}}{2} > \theta, (i,j) \in B \right\}. \quad (2.8.10)$$

The restriction in (2.8.8) reduces the statistic to counting Walsh averages.

In the next example we discuss the point and interval estimates of θ derived from the Winsorized Wilcoxon statistic of Example 2.8.2. We also derive the estimation tolerance and compare it to that of the trimmed mean.

Example 2.8.4. The Winsorized signed rank statistics in Example 2.8.2 provide an example of (2.8.10) with $a_i = \phi(i/(n+1)) = \min(i/(n+1), 1 - \gamma)$. Hence, if $[\cdot]$ denotes the greatest integer function,

$$a_i = \begin{cases} \dfrac{i}{n+1}, & 0 < \dfrac{i}{n+1} \leqslant 1 - \gamma \text{ i.e., } 0 < i \leqslant \left[(1-\gamma)(n+1)\right] \\[3mm] 1 - \gamma, & 1 - \gamma < \dfrac{i}{n+1} < 1 \text{ i.e., } \left[(1-\gamma)(n+1)\right] < i < n + 1, \end{cases}$$

and

$$a_{j-i+1} - a_{j-i} = \frac{1}{n+1} \quad \text{if } j - i \leqslant \left[(1-\gamma)(n+1)\right]. \quad (2.8.11)$$

[If $(1 - \gamma)(n + 1)$ is an integer, it must be reduced by 1 in (2.8.11).] Thus we see that Winsorization of the ranks results in the restriction to Walsh averages that are formed from order statistics that are not too far apart in the sample. For example, $(X_{(1)} + X_{(n)})/2$, the midrange, would be the first to be excluded.

The Hodges–Lehmann estimate based on the Winsorized signed rank statistic is

$$\operatorname*{med}_{i,j \in B} \left\{ (X_{(i)} + X_{(j)})/2 \right\}$$

where $B = \{(i, j) : j - i \leqslant [(1 - \gamma)(n + 1)]\}$. Theorem 2.4.1 can be applied to show that $\hat{\theta}$ is symmetrically distributed about θ.

Now let $N_\gamma = [(1 - \gamma)(n + 1)]$, then there are $N^* = \{n(n + 1) - (n - N_\gamma - 1)(n - N_\gamma)\}/2$ pairs (i, j) such that $j - i \leqslant N_\gamma$. Let $V = n(n + 1)\bar{V}$ in (2.8.10), so $V = \#(X_{(i)} + X_{(j)})/2 > 0$, $(i, j) \in B$, and V is the number of positive Walsh averages subject to the restriction imposed by B. From the asymptotic normality of $\bar{V}$ we can write (from Example 2.8.2)

$$P(V \leqslant c) = P\left(\bar{V} \leqslant c/n(n + 1)\right)$$

$$\doteq \Phi\left[\frac{\sqrt{n}\left[c/n(n + 1) - (1 - \gamma^2)/4\right]}{\sqrt{(1 - \gamma)^2(1 + 2\gamma)/12}}\right]$$

$$= \alpha/2$$

and equating the standardized critical value with $-Z_{\alpha/2}$, the lower standard normal $\alpha/2$ percentile,

$$c \doteq (1 - \gamma^2)\frac{n(n + 1)}{4} - Z_{\alpha/2}\sqrt{\frac{(1 - \gamma)^2(1 + 2\gamma)n(n + 1)^2}{12}}.$$

The $(1 - \alpha)$ 100% approximate confidence interval for θ, based on the Winsorized signed rank statistic, is then $[W^*_{(c+1)}, W^*_{(N^*-c)})$ where $W^*_{(1)} \leqslant \cdots \leqslant W^*_{(N^*)}$ are the ordered Walsh averages determined by B; compare this to (2.3.3).

We now illustrate the calculations in a simple example using the first six observed differences in Table 2.2. Take $\gamma = 1/3$ so $N_\gamma = [14/3] = 4$ and $N^* = 20$. Hence we need the Walsh averages $(X_{(i)} + X_{(j)})/2$ such that $j - i \leqslant 4$. Arrange the observed differences in order and form Walsh averages of pairs at intersections of diagonals as follows:

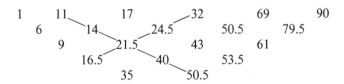

For example, $21.5 = (11 + 32)/2$. The restriction $j - i \leqslant 4$ means we need the five rows displayed. The estimate $\hat{\theta}$ is $(W^*_{(10)} + W^*_{(11)})/2 = (32 + 35)/2$

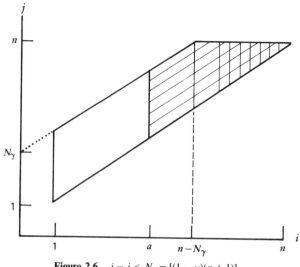

Figure 2.6. $j - i \leqslant N_\gamma = [(1 - \gamma)(n + 1)]$.

$= 33.5$. For an approximate 90% confidence interval take $Z_{\alpha/2} = 1.645$ and compute $c = 2.33$; so, to be conservative, we use $c = 2$. The interval is then $[W^*_{(3)}, W^*_{(18)}) = [9, 69)$.

We now consider the asymptotic tolerance of $\hat{\theta}$, the estimate derived from the Winsorized signed rank statistic. Figure 2.6 helps identify the (i, j) pairs in B.

The number of lattice points in the trapezoidal region is $\# B$, needed in Theorem 2.4.2. Further, $\# B = N^* = \{n(n + 1) - (n - N_\gamma - 1)(n - N_\gamma)\}/2$ where $N_\gamma = [(1 - \gamma)(n + 1)]$. We are interested in the asymptotic tolerance, so in the following argument we replace $n + 1$ by n and $N_\gamma = [(1 - \gamma)(n + 1)]$ by $(1 - \gamma)n$. Then $\# B \doteq \{n^2 - (n - N_\gamma)^2\}/2 = (1 - \gamma^2) n^2/2$. Next we need $\# K_{a+1}$ and again we consider $\# K_a$ since the 1 will not matter much for large n. The hatched region in Fig. 2.6 contains the lattice points defined by K_a when $a < n - N_\gamma = n\gamma$, and the cross-hatched region corresponds to K_a when $a \geqslant n\gamma$. The area of these two regions approximates $\# K_a$ as

$$\# K_a \doteq \begin{cases} (n - a)^2/2 & \text{if } a > n\gamma \\ (1 - \gamma^2) n^2/2 - a(1 - \gamma)n & \text{if } a \leqslant n\gamma. \end{cases}$$

The case $a \leqslant n\gamma$ is computed from $N_\gamma^2/2$ for the triangle and $N_\gamma(n - N_\gamma - $

a) for the parallelogram. We are now ready to solve for a such that

$$\frac{1}{2} \# B = \# K_a ,$$

or

$$\frac{(1 - \gamma^2)n^2}{4} = \begin{cases} (n - a)^2/2 & \text{if } a > n\gamma \\ (1 - \gamma^2)n^2/2 - a(1 - \gamma)n & \text{if } a \leqslant n\gamma. \end{cases}$$

Under the condition $a > n\gamma$, we have a quadratic equation in a with (admissible) solution:

$$\frac{a}{n} = 1 - \sqrt{(1 - \gamma^2)/2} .$$

This is an approximation to the tolerance provided $a/n > \gamma$, that is, provided $1 - [(1 - \gamma^2)/2]^{1/2} > \gamma$. This holds provided $\gamma < 1/3$. The second condition, $a \leqslant n\gamma$, yields

$$\frac{a}{n} = \frac{(1 + \gamma)}{4}$$

and is compatible with $\gamma \geqslant 1/3$. Hence the asymptotic tolerance of $\hat{\theta}$ is

$$\tau = \begin{cases} 1 - \sqrt{(1 - \gamma)^2/2} & \text{if } \gamma < 1/3 \\ (1 + \gamma)/4 & \text{if } \gamma \geqslant 1/3. \end{cases}$$

Note that if $\gamma = 0$ then $\tau = 1 - (1/2)^{1/2}$, the tolerance of $\hat{\theta} = \text{med}(X_i + X_j)/2$, $i \leqslant j$, and if $\gamma = 1$ then $\tau = 1/2$, the tolerance of $\hat{\theta} = \text{med} X_i$. The result also matches the asymptotic testing tolerance in Example 2.8.2.

Recall from Example 1.6.1 that the trimmed mean based on the middle $n - 2[n\alpha]$ observations has tolerance $\tau(\overline{X}_\alpha) = \alpha$. If we compare this tolerance to that of $\hat{\theta}_{2\alpha}$, the estimate derived from a Winsorized Wilcoxon statistic using (2α) 100% sign scores and $(1 - 2\alpha)$ 100% Wilcoxon scores, we find $\tau(\hat{\theta}_{2\alpha}) > \tau(\overline{X}_\alpha)$ for every α, $0 < \alpha < 1/2$. Hence these estimates are more resistant to outliers than the trimmed means. See Hettmansperger and Utts (1977) for further discussion.

A similar analysis of estimates derived from the modified sign statistics of Example 2.8.3 is outlined in Exercise 2.10.25. The structure of these estimates is simpler than that of the estimates derived from the Winsorized signed rank statistics.

For convenience in computing the influence curve for an estimate derived from V, we will put the statistic into a more symmetric form. Let $\mathrm{sgn}(x) = -1$ if $x < 0$, 0 if $x = 0$, and $+1$ if $x > 0$. Then define

$$\overline{V}^*(\theta) = \frac{1}{n} \sum_{i=1}^{n} \phi\left(\frac{R_i(\theta)}{n+1} \right) \mathrm{sgn}(X_i - \theta). \qquad (2.8.12)$$

Note that $\overline{V}^*(\theta) = 2\overline{V}(\theta) - n^{-1}\sum\phi(i/(n+1))$, so the Hodges–Lehmann estimate mentioned under (2.8.7) can be defined equivalently by $\overline{V}^*(\hat{\theta}) \doteq 0$.

Let $H(x) = F(x - \theta)$ be the underlying cdf and $H_n(x)$ denote the empirical cdf. As in Example 2.4.2, we define the functional $\theta = T(H)$ implicitly. We begin by expressing $\overline{V}^*(\theta)$ in terms of the empirical cdf. First extend the definition of ϕ in Definition 2.8.2 from $(0, 1)$ to $(-1, 1)$ by $\phi(-u) = -\phi(u)$.

Theorem 2.8.3. The equation $\overline{V}^*(\theta) = 0$ is asymptotically equivalent to

$$\int_{-\infty}^{\infty} \phi\left(\frac{n}{n+1} \left[H_n(x) - H_n(-x + 2\theta)_- \right] \right) dH_n(x) = 0 \quad (2.8.13)$$

and the functional $\theta = T(H)$ is defined by

$$\int_{-\infty}^{\infty} \phi(H(x) - H(-x + 2\theta)) \, dH(x) = 0. \qquad (2.8.14)$$

Proof. The rank $R_i(\theta)$ is equal to the number of x_j's such that $|x_j - \theta| \leqslant |x_i - \theta|$. We can then write (2.8.12) as

$$\sum_{i\,:\,\theta \leqslant x_i} \phi\left[\frac{1}{n+1} (\# x_j \ni -x_i + 2\theta \leqslant x_j < x_i) \right]$$

$$- \sum_{i\,:\,\theta > x_i} \phi\left[\frac{1}{n+1} (\# x_j \ni +x_i \leqslant x_j \leqslant -x_i + 2\theta) \right].$$

Using the definition of the empirical cdf, we have

$$\sum_{i\,:\,\theta \leqslant x_i} \phi\left\{ \frac{n}{n+1} \left[H_n(x_i) - H_n(-x_i + 2\theta)_- \right] \right\}$$

$$- \sum_{i\,:\,\theta > x_i} \phi\left\{ \frac{n}{n+1} \left[H_n(-x_i + 2\theta) - H_n(x_i)_- \right] \right\}.$$

Since $\phi(-u) = -\phi(u)$, the two sums can be combined, and the first equation follows by representing the sum as a Stiltjes integral. The second follows by letting $n \to \infty$.

Hence the location functional $\theta = T(H)$ is implicitily defined by (2.8.14), and the Hodges–Lehmann estimate $\hat{\theta}$ is (asymptotically) equivalent to the solution of (2.8.13). We now derive the influence curve for the functional $\theta = T(H)$. Typical scores functions used for the symmetric location model satisfy

$$\phi(2u - 1) + \phi(1 - 2u) = 0. \tag{2.8.15}$$

In the next theorem we derive the influence curve under this condition. All examples in this chapter satisfy (2.8.15).

Theorem 2.8.4.   Suppose $F \in \Omega_s$ and ϕ satisfies the Definition 2.8.2, is differentiable, and satisfies (2.8.15). Then the functional $T(F)$ has influence curve

$$\Omega(y) = \frac{\phi(2F(y) - 1)}{2\int_{-\infty}^{\infty} \phi'(2F(x) - 1)f^2(x)\,dx}.$$

Proof.   We begin with $H(x) = F(x - \theta)$ and define $T(H)$ through (2.8.14) by

$$\int_{-\infty}^{\infty} \phi[H(x) - H(-x + 2T(H))]\,dH(x) = 0. \tag{2.8.16}$$

Define $H_t(x) = H(x) + t(\delta_y(x) - H(x))$ and apply Definition 1.6.4. Substituting $H_t(x)$ into (2.8.16), we get

$$\int_{-\infty}^{\infty} \phi[H_t(x) - H_t(-x + 2T(H_t))]\,dH(x)$$

$$+ t\int_{-\infty}^{\infty} \phi[H_t(x) - H_t(-x + 2T(H_t))]\,d[\delta_y(x) - H(x)] = 0.$$

We now differentiate with respect to t and then set $t = 0$. Further, the derivative simplifies if we take $\theta = 0$, without loss of generality. Hence the derivative is presented with $t = 0$ and $\theta = 0$, so $H_t(x)$ is replaced by $F(x)$.

Finally, recall $\Omega(y)$ is the derivative of $T(H_t)$ evaluated at $t = 0$. Hence

$$\int_{-\infty}^{\infty} \phi'\big[F(x) - F(-x)\big]\{\delta_y(x) - F(x) - f(-x)2\Omega(y)$$

$$- \delta_y(-x) + F(-x)\} \, dF(x)$$

$$+ \int_{-\infty}^{\infty} \phi\big[F(x) - F(-x)\big] \, d\delta_y(x) - \int_{-\infty}^{\infty} \phi\big[F(x) - F(-x)\big] \, dF(x) = 0.$$

Now, $F(x) - F(-x) = 2F(x) - 1$, $f(-x) = f(x)$; hence

$$- 2\Omega(y)\int_{-\infty}^{\infty} \phi'(2F(x) - 1)f^2(x) \, dx + \int_{-\infty}^{\infty} \phi'(2F(x) - 1)\delta_y(x) \, dF(x)$$

$$- \int_{-\infty}^{\infty} \phi'(2F(x) - 1)(2F(x) - 1) \, dF(x)$$

$$- \int_{-\infty}^{\infty} \phi'(2F(x) - 1)\delta_y(-x) \, dF(x)$$

$$+ \int_{-\infty}^{\infty} \phi(2F(x) - 1) \, d\delta_y(x) - \int_{-\infty}^{\infty} \phi(2F(x) - 1) \, dF(x) = 0.$$

Make a change of variable $t = -x$ in the fourth integtral, then $\phi(2u - 1)$ $= -\phi(1 - 2u)$ and $\phi'(2u - 1) = \phi'(1 - 2u)$ imply that the second and fourth integrals cancel. The third integral is $2^{-1}\int_{-1}^{1} u\phi'(u) \, du = 0$, after using an integration by parts. The last integral is 0 also, because $\phi(u) = -\phi(-u)$. Now solve for $\Omega(y)$ to get

$$\Omega(y) = \frac{\int_{-\infty}^{\infty} \phi(2F(x) - 1) \, d\delta_y(x)}{2\int_{-\infty}^{\infty} \phi'(2F(x) - 1)f^2(x) \, dx}$$

$$= \frac{\phi(2F(y) - 1)}{2\int_{-\infty}^{\infty} \phi'(2F(x) - 1)f^2(x) \, dx} \, .$$

Huber (1981, p. 64) gives the influence curve in the general case of asymmetric F. He points out that asymmetric F may result in arbitrarily large values of the influence curve. This implies that rank estimates, such as the median of the Walsh averages, may lose their superior robustness in the face of asymmetric contamination.

The influence function suggests the asymptotic distribution of $n^{1/2}(\hat{\theta} - \theta)$, recall (1.6.2). We need

$$\int \Omega^2(y)\,dF(y) = \frac{\int_{-\infty}^{\infty}\phi^2(2F(y)-1)\,dF(y)}{4\left\{\int_{-\infty}^{\infty}\phi'(2F(x)-1)f^2(x)\,dx\right\}^2}.$$

Note that

$$\int_{-\infty}^{\infty}\phi^2(2F(y)-1)\,dF(y) = \frac{1}{2}\int_{-1}^{1}\phi^2(u)\,du = \int_{0}^{1}\phi^2(u)\,du,$$

and hence

$$\int \Omega^2(y)\,dF(y) = \frac{\frac{1}{4}\int_{0}^{1}\phi^2(u)\,du}{\left\{\int_{-\infty}^{\infty}\phi'(2F(y)-1)f^2(y)\,dy\right\}^2}. \qquad (2.8.17)$$

Then $n^{1/2}(\hat{\theta} - \theta) \xrightarrow{D} Z \sim n(0, \int\Omega^2(y)\,dF(y))$.

Example 2.8.5. The Winsorized Wilcoxon estimate in Example 2.8.4 is generated by $\phi(u) = \min(u, 1 - \gamma)$, $0 < u < 1$. Let k be defined by $F(-k) = \gamma/2$. Then $\phi(2F(x)-1) = \min(2F(x)-1, 1-\gamma) = 2F(x)-1$ if $0 < 2F(x)-1 < 1-\gamma$ or $0 < x < F^{-1}(1-\gamma/2) = k$, and $1-\gamma$ if $x > k$. Extend $\phi(u)$ to $(-1, 1)$ by $\phi(-u) = -\phi(u)$ then

$$\Omega(y) = \begin{cases} (1-\gamma)\mathrm{sgn}(y)/2\int_{-k}^{k}f^2(x)\,dx, & |y| > k \\ (2F(y)-1)/2\int_{-k}^{k}f^2(x)\,dx, & |y| \leqslant k. \end{cases}$$

Hence the influence is not only bounded, it is Winsorized. Sensitivity curves and further discussion can be found in Hettmansperger and Utts (1977). The results of Jaeckel (1971) indicate that estimates with Winsorized influence curves have minimax asymptotic variance for contamination models; see Example 2.9.5. See also Huber (1981, p. 97–99) and Rieder (1981).

Example 2.8.6. The normal scores estimate is determined by $\bar{V}(\hat{\theta}) = \frac{1}{2}\sum\Phi_{+}^{-1}(i/(n+1))$. Recall, from Example 2.8.1, the score function can be written as $\phi(u) = \Phi_{+}^{-1}(u) = \Phi^{-1}[(u+1)/2]$. To compute the influence curve we need

$$\phi'(u) = \frac{d}{dx}\Phi^{-1}\left(\frac{u+1}{2}\right)$$

$$= \frac{1}{2\Psi\left(\Phi^{-1}[(u+1)/2]\right)}$$

where $\Psi(x) = \Phi'(x)$, the $n(0,1)$ pdf. Hence

$$\Omega(y) = \Phi^{-1}(F(y))\Big/\left\{\int_{-\infty}^{\infty}\left[f^2(x)/\Psi(\Phi^{-1}(F(x)))\right]dx\right\}.$$

Now suppose the underlying distribution is normal so that $F = \Phi$. Then the denominator of $\Omega(y)$ is 1, and so $\Omega(y) = y$, the same unbounded influence curve as $\bar{X}$! The unusual feature of the normal scores estimate is that it has positive asymptotic estimation tolerance of .239 (see Huber, 1981, p. 67) and yet has unbounded influence. In the ϵ-contaminated normal model the asymptotically minimax rank estimate is the Hodges–Lehmann estimate generated from a Winsorized normal scores function. Winsorization results in bounded influence. For further details see Huber (1981, Section 4.7).

2.9. EFFICIENCY OF GENERAL SCORES STATISTICS

The previous section provided a rigorous development of the asymptotic distribution theory needed to develop tests and estimates from a general scores statistic. In this section we discuss, without proof, Theorem 2.6.1 and then apply it to find the efficacy of a general scores test. We then establish an upper bound on the efficacy and find the score function that generates the maximum efficacy for a given distribution. We also find the greatest lower bound on the efficiency of the normal scores tests in Example 2.8.1 relative to the t test, a result similar to the bound on the efficiency of the T relative to t given in Theorem 2.6.3. We introduce an ordering on the tailweight of distributions and use it to compare Wilcoxon to normal scores procedures. Finally, we develop asymptotic maximin tests and minimax asymptotic variance estimates in the contamination model.

Definition 2.9.1. The expression

$$I(f) = E\left[\frac{\partial}{\partial\theta}\log f(x-\theta)\right]^2 = \int_{-\infty}^{\infty}\frac{\left[f'(x)\right]^2}{f(x)}dx$$

is called the Fisher information in f, provicded f is absolutely continuous and $I(f) < \infty$. See Huber (1981, p. 77) for an extension to the case $I(f) = \infty$.

Hajek and Sidak (1967, p. 220) show that if a score-generating function $\phi(u)$ satisfies the conditions in Definition 2.8.2, then Theorem 2.6.1 holds for the statistic generated by $\phi(u)$, provided, in addition, that the sampled distribution has finite Fisher information; see Definition 2.9.1. Namely, if $\theta_n = \theta/n^{1/2}$, then

$$\frac{\sqrt{n}\left(\overline{V} - \int \phi(u)\,du/2\right)}{\sqrt{\frac{1}{4}\int \phi^2(u)\,du}} \xrightarrow{D_{\theta_n}} Z \sim n(c\theta, 1),$$

where c is the efficacy defined in condition 5 of the Pitman regularity conditions in Section 2.6. To compute c we need the asymptotic mean under a fixed alternative.

Let $X_1, \ldots, X_n$ be a random sample from $G(x) = F(x - \theta)$, $F \in \Omega_s$, and let $G_n(x) = (\# X_i \leqslant x)/n$, the empirical distribution function. Then $H(x) = P(|X| \leqslant x) = G(x) - G(-x)$ and $H_n(x) = (\#|X_i| \leqslant x)/n$. If $X_j > 0$ then the rank among the absolute values can be written as $R_j = nH_n(X_j)$. The general scores statistic, (2.8.4), can be written

$$\overline{V} = \frac{1}{n}\sum_{j=1}^{n}\phi\left(\frac{R_j}{n+1}\right)s(X_j)$$

$$= \int_0^{\infty}\phi\left(\frac{n}{n+1}H_n(x)\right)dG_n(x). \tag{2.9.1}$$

Since $H_n(x)$ and $G_n(x)$ converge in probability to $H(x)$ and $G(x)$, it can be shown under regularity conditions discussed later that

$$\overline{V} \xrightarrow{P} \int_0^{\infty}\phi(H(x))\,dG(x).$$

From the definitions of $H(x)$ and $G(x)$, it can be further shown that the asymptotic mean $\mu(\theta)$, suggested by the stochastic limit of $\overline{V}$, is given by

$$\mu(\theta) = \int_0^{\infty}\phi\big[F(x-\theta) - F(-x-\theta)\big]f(x-\theta)\,dx$$

$$= \int_{-\theta}^{\infty}\phi\big[F(x) - F(-x-2\theta)\big]f(x)\,dx. \tag{2.9.2}$$

To compute the efficacy we need $\mu'(0)$ given by

$$\mu'(0) = 2 \int_0^\infty \phi'(2F(x) - 1)f^2(x)\,dx,$$

since $F \in \Omega_s$. Furthermore, $\sigma^2(0) = \int \phi^2(u)\,du/4$, hence

$$c = \frac{2 \int_0^\infty \phi'(2F(x) - 1)f^2(x)\,dx}{\sqrt{\dfrac{1}{4} \int_0^1 \phi^2(u)\,du}}. \tag{2.9.3}$$

The derivation which leads to (2.9.3) is made rigorous by Puri and Sen (1971, Section 3.6). They place more restrictive conditions on $\phi(u)$ than are needed by Hajek and Sidak (1967); however, they do not need finite Fisher information. The main assumption made by Puri and Sen is that $|\phi(u)| \leq K[u(1 - u)]^{\delta - 1/2}$ and $|\phi'(u)| \leq K[u(1 - u)]^{\delta - 3/2}$ for some $\delta > 0$.

We next derive an alternative form of c. Let $I = \int_0^\infty \phi'(2F(y) - 1)f^2(y)\,dy$ and make the change of variable $x = 2F(y) - 1$. Then

$$I = \int_0^1 \phi'(x)f\left(F^{-1}\left(\frac{x+1}{2}\right)\right)\frac{1}{2}\,dx.$$

Now $F(F^{-1}(x)) = x$, so differentiating both sides shows that

$$\frac{d}{dx}\left(F^{-1}(x)\right) = \frac{1}{f\left(F^{-1}(x)\right)}.$$

Let $u = f(F^{-1}((x + 1)/2))$ and $dv = \frac{1}{2}\phi'(x)\,dx$, so $du = f'(F^{-1}((x + 1)/2))/2f(F^{-1}((x + 1)/2))\,dx$ and $v = \phi(x)/2$. Finally, if $\phi(x)f(F^{-1}((x + 1)/2)) \to 0$ as $x \to 0$ or 1, then integrating by parts yields

$$I = -\frac{1}{4} \int_0^1 \phi(x) \frac{f'\left(F^{-1}\left(\dfrac{x+1}{2}\right)\right)}{f\left(F^{-1}\left(\dfrac{x+1}{2}\right)\right)}\,dx.$$

Now c in (2.9.3) can be written as

$$c = \frac{\int_0^1 \phi(u)\phi_f(u)\,du}{\sqrt{\int_0^1 \phi^2(u)\,du}} \tag{2.9.4}$$

where

$$\phi_f(u) = -\frac{f'\left(F^{-1}\left(\frac{u+1}{2}\right)\right)}{f\left(F^{-1}\left(\frac{u+1}{2}\right)\right)}. \tag{2.9.5}$$

Note that this form of c does not require differentiability of $\phi(u)$. This is the form of c derived by Hajek and Sidak (1967) under the restriction of finite Fisher information.

The estimates $\hat{\theta}$ discussed in Section 2.8 and derived from the general scores statistic have asymptotic normal distributions. In fact, from Theorem 2.6.5 we have $n^{1/2}(\hat{\theta} - \theta)$ is approximately normal with mean 0 and variance $1/c^2$ with c given by (2.9.3) or (2.9.4). This agrees with (2.8.17) derived from the influence curve.

There is a result on the approximate linearity of $\overline{V}(\theta)$ corresponding to Theorem 2.7.1. The version that corresponds to (2.7.6) states

$$\frac{\sqrt{n}\left[\overline{V}(\theta) - \int_0^1 \phi(u)\,du/2\right]}{\sqrt{\frac{1}{4}\int_0^1 \phi^2(u)\,du}} \doteq \frac{\sqrt{n}\left[\overline{V}(\theta_0) - \int_0^1 \phi(u)\,du/2\right]}{\sqrt{\frac{1}{4}\int_0^1 \phi^2(u)\,du}} - \sqrt{n}\,(\theta - \theta_0)c$$

$$\tag{2.9.6}$$

with c given by (2.9.3). The approximation is uniform in probability in the same sense as in Theorem 2.7.2. See van Eeden (1972).

If $[\hat{\theta}_L, \hat{\theta}_U]$ denotes the confidence interval derived from $\overline{V}$, then the approximate linearity can be used to show that

$$\frac{\sqrt{n}\,(\hat{\theta}_U - \hat{\theta}_L)}{2Z_{\alpha/2}} \xrightarrow{P} 1/c, \tag{2.9.7}$$

similar to (2.7.8) in the Wilcoxon case. See Sen (1966) also.

Example 2.9.1. From Example 2.8.2 the score function for a Winsorized signed rank statistic is $\phi(u) = \min(u, 1 - \gamma)$. Then $\phi'(u) = 1$ if $0 < u \leqslant 1 - \gamma$ and 0 otherwise, and hence

$$\phi'(2F(x) - 1) = \begin{array}{ll} 1 & 0 < 2F(x) - 1 \leqslant 1 - \gamma \\ 0 & 1 - \gamma < 2F(x) - 1 < 1. \end{array} \tag{2.9.8}$$

If we let $x(\alpha) = F^{-1}(\alpha)$, the α percentile of the underlying distribution, then (2.9.8) becomes

$$\phi'(2F(x) - 1) = 1 \qquad 0 < x \leqslant x(1 - \gamma/2)$$
$$0 \qquad x(1 - \gamma/2) < x < 1,$$

and the efficacy becomes

$$c = \frac{2\int_0^{x(1-\gamma/2)} f^2(x)\,dx}{\sqrt{(1 - \gamma)^2(1 + 2\gamma)/12}}$$

$$= \frac{\sqrt{12} \int_{x(\gamma/2)}^{x(1-\gamma/2)} f^2(x)\,dx}{\sqrt{(1 - \gamma)^2(1 + 2\gamma)}}.$$

The efficiency of the Winsorized Wilcoxon signed rank test, estimate, or confidence interval relative to the ordinary Wilcoxon procedures is then

$$e = \frac{\left[\int_{x(\gamma/2)}^{x(1-\gamma/2)} f^2(x)\,dx\right]^2}{(1 - \gamma)^2(1 + 2\gamma)\left[\int_{-\infty}^{\infty} f^2(x)\,dx\right]^2}.$$

Table 2.5 gives the values of e for various values of γ and for normal, double exponential (DE), and Cauchy distributions. Note that the optimal Winsorization for the Cauchy distribution is $\gamma = .75$, not $\gamma = 1$ which you might expect for such a heavy-tailed distribution.

Now, provided that $\phi_f(u)$ satisfies Definition 2.8.2, we next show that $\overline{V}_f$, the statistic generated by $\phi_f(u)$, has maximum efficacy. Hence the test and estimates derived from $\overline{V}_f$ are optimal in the sense of asymptotic efficiency.

Theorem 2.9.1. Suppose $X_1, \ldots, X_n$ is a random sample from $F(x - \theta)$, $F \in \Omega_s$, which satisfies Definition 2.9.1. Suppose $\phi_f(u)$, defined by (2.9.5),

Table 2.5. Efficiency of Winsorized Wilcoxon Relative to T

Distribution	γ						
	.1	.2	.5	.7	.8	.9	.98
Normal	0.99	0.94	0.92	0.81	0.75	0.70	0.66
DE	1.01	1.03	1.13	1.20	1.25	1.29	1.33
Cauchy	1.03	1.09	1.34	1.43	1.44	1.40	1.35

satisfies Definition 2.8.2 and generates $\overline{V}_f$. If $\overline{V}$ is any other statistic generated by a generating function $\phi(u)$, then

$$c \leqslant \sqrt{I(f)}$$

and

$$c_f = \sqrt{I(f)} \, ,$$

where c and c_f are the efficacies of $\overline{V}$ and $\overline{V}_f$, respectively.

Proof.   From (2.9.4) and the Cauchy–Schwarz inequality we have

$$c = \frac{\int_0^1 \phi(u)\phi_f(u)\,du}{\sqrt{\int_0^1 \phi^2(u)\,du}}$$

$$\leqslant \frac{\sqrt{\int_0^1 \phi^2(u)\,du \int_0^1 \phi_f^2(u)\,du}}{\sqrt{\int_0^1 \phi^2(u)\,du}}$$

$$= \sqrt{\int_0^1 \phi_f^2(u)\,du} \, .$$

Next, making the change of variable $y = F^{-1}[(u + 1)/2]$, we have

$$\int_0^1 \phi_f^2(u)\,du = \int_0^1 \left[\frac{f'\!\left(F^{-1}\!\left(\frac{u+1}{2}\right)\right)}{f\!\left(F^{-1}\!\left(\frac{u+1}{2}\right)\right)} \right]^2 du$$

$$= \int_0^\infty \left(\frac{f'(y)}{f(y)} \right)^2 2f(y)\,dy$$

$$= \int_{-\infty}^\infty \frac{[f'(y)]^2}{f(y)}\,dy = I(f).$$

Now, from (2.9.4), $c_f = \int_0^1 \phi_f^2(u)\,du / (\int_0^1 \phi_f^2(u)\,du)^{1/2} = (I(f))^{1/2}$.

Definition 2.9.2. Suppose $F \in \Omega_s$ and $-\log f(x)$ is convex. Then we say f is strongly unimodal.

Note that if f is strongly unimodal and if $0 < I(f) < \infty$, then $\phi_f(u)$ satisfies Definition 2.8.2 and is a scores-generating function. In fact, strong unimodality is necessary.

From Theorem 2.6.1, the asymptotic local power of the test based on $\overline{V}_f$ is $1 - \Phi(Z_\alpha - \theta c_f)$. When this is compared to $1 - \Phi(Z_\alpha - \theta c)$ for some other statistic $\overline{V}$, we have $1 - \Phi(Z_\alpha - \theta c_f) \geqslant 1 - \Phi(Z_\alpha - \theta c)$ since $c_f \geqslant c$. Hence we refer to $\overline{V}_f$ as the asymptotically most powerful rank test (AMPRT). The asymptotic testing tolerance of $\overline{V}_f$ is given in Exercise 2.10.30. A different type of optimal test, called a locally most powerful rank test (LMPRT), is developed for the two-sample location model in the next chapter. At the end of Section 3.5, we show how the corresponding one-sample LMPRT can be constructed. See (3.5.15) and (3.5.16).

Note also that the estimate $\hat{\theta}_f$ derived from $\overline{V}_f$ has the property that $n^{1/2}(\hat{\theta}_f - \theta)$ has an asymptotic normal distribution with mean 0 and variance $1/c_f^2 = 1/I(f)$, the Rao–Cramér lower bound. Hence $\hat{\theta}_f$ has the same asymptotic variance as the maximum likelihood estimate. In this sense $\hat{\theta}_f$ is an asymptotically efficient estimate. Bickel and Doksum (1977, Section 4.4.c) briefly discuss the asymptotic efficiency of the maximum likelihood estimate and give further references. See also Kendall and Stuart (1973, Chapter 18). The influence curve for $\hat{\theta}_f$ is given in Exercise 2.10.31.

Example 2.9.2. Take $\psi(x)$ to be the standard normal density function. Then $-\psi'(x)/\psi(x) = x$ and we have

$$\phi_\psi(u) = \Phi^{-1}\left(\frac{u+1}{2}\right)$$

where $\Phi(\cdot)$ is the standard normal distribution function. Recall from Example 2.8.1 that $\Phi_+(x) = P(|X| \leqslant x) = 2\Phi(x) - 1$. Hence $\Phi(x) = [\Phi_+(x) + 1]/2$. If $t = \Phi(x)$, then $x = \Phi^{-1}(t)$, and likewise $t = [\Phi_+(x) + 1]/2$ implies $x = \Phi_+^{-1}(2t - 1)$. Thus,

$$\phi_\psi(u) = \Phi^{-1}\left(\frac{u+1}{2}\right)$$

$$= \Phi_+^{-1}(u)$$

and the optimal rank statistic $\overline{V}_\psi$ is the one-sample normal scores statistic of Example 2.8.1. Next note that

$$\phi'_\psi(u) = \frac{d}{du}\,\Phi^{-1}\!\left(\frac{u+1}{2}\right)$$

$$= \frac{1}{2\psi\!\left(\Phi^{-1}\!\left(\frac{u+1}{2}\right)\right)}.$$

Hence, if the underlying distribution is F, then

$$\phi'_\psi(2F(x)-1) = \frac{1}{2\psi\big(\Phi^{-1}(F(x))\big)}.$$

The calculations in Exercise 2.10.21 show $\int \phi^2_\psi(u)\,du = 1$, and from (2.9.3) the efficacy is

$$c_\psi = \int_{-\infty}^{\infty} \frac{f^2(x)}{\psi\big(\Phi^{-1}(F(x))\big)}\,dx. \qquad (2.9.9)$$

The efficiency of the normal score procedures (test or estimates) relative to the t procedures when the underlying distribution is F, symmetric, is $e(\overline{V}_\psi, t) = \sigma_f^2 c_\psi^2$, where σ_f^2 is the variance of F. From the scale invariance of the efficiency, we can let $\sigma_f^2 = 1$ without loss of generality, and write

$$e(\overline{V}_\psi, t) = \left[\int_{-\infty}^{\infty} \frac{f^2(x)}{\psi\big(\Phi^{-1}(F(x))\big)}\,dx\right]^2. \qquad (2.9.10)$$

Note if we are sampling from a normal distribution, where t is optimal, then $F(x) = \Phi(x)$ and $f(x) = \psi(x)$, and (2.9.10) becomes $e(\overline{V}_\psi, t) = 1$. Hence, at the normal distribution, the normal scores procedures are fully efficient in the sense of Pitman efficiency. In the next theorem we consider what happens to the efficiency when $F(x)$ is not normal. The result was first proved by Chernoff and Savage (1958). The proof given here is due to Gastwirth and Wolff (1968).

Theorem 2.9.2. Let $X_1, \ldots, X_n$ be a random sample from $F(x - \theta)$, $F \in \Omega_s$, then

$$\inf_{\Omega_s} e(\overline{V}_\psi, t) = 1.$$

Hence the efficiency of normal scores procedures is never less than 1 when sampling from a symmetric distribution.

Proof. If $\sigma_f^2 = \infty$ then $e(\overline{V}_\psi, t) > 1$, hence we suppose $\sigma_f^2 = 1$. From (2.9.10) we can write

$$\sqrt{e} = E\left[\frac{f(X)}{\psi(\Phi^{-1}(F(X)))} \right]$$

$$= E\left[\frac{1}{\psi(\Phi^{-1}(F(X)))/f(X)} \right].$$

Applying Jensen's inequality (Theorem A20 in the Appendix) to the convex function $h(x) = 1/x$, we have

$$\sqrt{e} \geqslant \frac{1}{E\left[\psi(\Phi^{-1}(F(X)))/f(X) \right]} .$$

Hence

$$\frac{1}{\sqrt{e}} \leqslant E\left[\frac{\psi(\Phi^{-1}(F(X)))}{f(X)} \right]$$

$$= \int \psi(\Phi^{-1}(F(x)))\, dx.$$

We now integrate by parts, using $u = \psi(\Phi^{-1}(F(x)))$, $du = \psi'(\Phi^{-1}(F(x)))$ $f(x)\,dx/\psi(\Phi^{-1}(F(x))) = -\Phi^{-1}(F(x))f(x)\,dx$ since $\psi'(x)/\psi(x) = -x$. Hence, with $dv = dx$, we have

$$\int_{-\infty}^{\infty} \psi(\Phi^{-1}(F(x)))\, dx = x\psi(\Phi^{-1}(F(x)))\big|_{-\infty}^{\infty}$$

$$+ \int_{-\infty}^{\infty} x\Phi^{-1}(F(x))f(x)\,dx. \qquad (2.9.11)$$

Now transform $x\psi(\Phi^{-1}(F(x)))$ into $F^{-1}(\Phi(w))\psi(w)$ by first letting $t = F(x)$ and then $w = \Phi^{-1}(t)$. The integral $\int F^{-1}(\Phi(w))\psi(w)\,dw = \int xf(x)\,dx < \infty$, hence the limit of the integrand must be 0 as $x \to \pm\infty$. This implies that the first term on the right side of (2.9.11) is 0. Hence applying the

Cauchy–Schwarz inequality,

$$\frac{1}{\sqrt{e}} \leqslant \int_{-\infty}^{\infty} x \Phi^{-1}(F(x)) f(x)\, dx$$

$$= \int_{-\infty}^{\infty} x \sqrt{f(x)}\ \Phi^{-1}(F(x)) \sqrt{f(x)}\, dx$$

$$\leqslant \left[\int_{-\infty}^{\infty} x^2 f(x)\, dx \int_{-\infty}^{\infty} \left\{ \Phi^{-1}(F(x)) \right\}^2 f(x)\, dx \right]^{1/2}$$

$$= 1$$

since $\int x^2 f(x)\, dx = 1$ and $\int x^2 \psi(x)\, dx = 1$. Hence $e^{1/2} \geqslant 1$ and $e \geqslant 1$, which completes the proof. It should be noted that the inequality is strict except at the normal distribution. Hence the normal scores procedures are strictly more efficient than the t procedures except at the normal model where the asymptotic relative efficiency is 1.

The implication of many of the results in this section is that as the tailweight of the underlying distribution increases we should use statistical methods that emphasize the extreme sample values less. Resistant methods provide such protection and also have excellent efficiency properties. Examples are the regular or Winsorized Wilcoxon and normal scores procedures relative to the t procedures. Provided the tails are sufficiently heavy, the sign procedures are also quite good.

We now provide a formal definition of tail ordering for distributions and use this concept to compare the Wilcoxon and normal scores methods.

Definition 2.9.3. Suppose F and G are in Ω_s. We say F has lighter tails than G (or G has heavier tails than F), denoted $F < G$, if $G^{-1}(F(x))$ is convex for $x \geqslant 0$.

This definition was introduced by van Zwet (1970) in his study of convex transformations of random variables. A survey of other definitions of tail ordering is given by Hettmansperger and Keenan (1975). See also Gastwirth (1970b).

Note that (1) $F < F$ and (2) $F < G$ and $G < H$ imply $F < H$. Hence $<$ is a weak ordering. If $F < G$ and $G < F$ we say F and G are equivalent. Let $F(x) = G(ax)$ for $a > 0$, then $G^{-1}(F(x)) = ax$ so $F < G$; also $F^{-1}(G(x)) = x/a$ so $G < F$. Hence distributions that differ only in scale are equivalent. This means that tail ordering is a family property and does not depend upon scale.

Suppose $F, G \in \Omega_s$ with positive densities at 0. Further, without loss of generality take $f(0) = g(0)$ since this can be achieved by rescaling the distribution functions: $F(x) = F_1(x/\sigma)$ with $\sigma = f(0)/g(0)$. Now suppose $F < G$, not equivalent. Then $q(x) = G^{-1}(F(x))$ is strictly convex for some x and $q'(x) = f(x)/g(G^{-1}(F(x)))$ is strictly increasing for some x. Since $q'(0) = f(0)/g(0) = 1$, $q'(x) > 1$ for some x, and eventually $G^{-1}(F(x)) > x$. Hence $F(x) > G(x)$ eventually, and $1 - G(x) > 1 - F(x)$; so there is more probability in the tail of G. In the next example we describe the ordering of several distributions often used in efficiency calculations.

Example 2.9.3. Uniform $<$ normal $<$ logistic $<$ double exponential $<$ Cauchy. We will verify some of these and leave others as exercises. First suppose that $F \in \Omega_s$ and $f(x)$ is nonincreasing for $x \geqslant 0$. Let $U(x) = 0$ if $x < -1$, $(x + 1)/2$ if $-1 \leqslant x < 1$, and 1 if $1 \leqslant x$, the uniform cdf on $(-1, 1)$. Then if $q(x) = F^{-1}(U(x)) = F^{-1}[(x + 1)/2]$ we have $q'(x) = 1/[2f(F^{-1}[(x + 1)/2])]$ is nondecreasing for $x \geqslant 0$, and so $U < F$.

Let $L(x) = 1/(1 + e^{-x})$, $-\infty < x < \infty$, denote the logistic cdf. We now show $\Phi < L$. The inverse $L^{-1}(y) = \log y - \log(1 - y)$. Let $\psi(x) = \Phi'(x)$, the $n(0, 1)$ density function, and note that $\psi'(x) = -x\psi(x)$, then two differentiations show that

$$\frac{d^2}{dx^2} L^{-1}(\Phi(x))$$

$$= \frac{\psi(x)}{\left[\Phi(x)(1 - \Phi(x))\right]^2} \{\psi(x)(2\Phi(x) - 1) - x\Phi(x)(1 - \Phi(x))\}.$$

Now $L^{-1}(\Phi(x))$ is convex if the function in the brackets, denoted $r(x)$, satisfies $r(x) \geqslant 0$. Since $r(0) = 0$ and $r(x) \to 0$ as $x \to \infty$, it is sufficient to show $r''(x)$ changes from negative to positive. Now

$$r''(x) = -\psi(x) + 2\psi(x)\Phi(x) - 4x\psi^2(x)$$

$$= \psi(x)s(x).$$

This function changes sign when $s(x)$ does, where

$$s(x) = -1 + 2\Phi(x) - 4x\psi(x).$$

Again, note $s(0) = 0$, $s(x) \to 0$ as $x \to \infty$ and $s'(0) = -2\psi(0) < 0$. Now $s'(x) = 2\psi(x)(2x^2 - 1)$ so that $s'(x) < 0$ if $x < 1/2^{1/2}$ and $\geqslant 0$ if $x \geqslant 1/2^{1/2}$. Hence $s(x)$ has exactly one change of sign. This implies that $r''(x)$ has one sign change and so $L^{-1}(\Phi(x))$ is convex, and finally $\Phi < L$. The others are much easier to check and are left for the reader to do in Exercise 2.10.31.

Example 2.9.4. Recall, from Definition 2.9.2, $f(x)$ is strongly unimodal if $-\log f(x)$ is convex. In this example we will show that the double exponential distribution is the heaviest tailed strongly unimodal distribution.

Let $D(x)$ be the double exponential cdf. Then for $x \geq 0$, $D(x) = \int_{-\infty}^{x}(1/2)\exp\{-|t|\}\,dt = 1 - (1/2)e^{-x}$. Hence, for $x \geq 0$, $D^{-1}(y) = -\log[2(1-y)]$.

Suppose $F \in \Omega_s$ and f is strongly unimodal, then we wish to show that $F \prec D$. We will show $D^{-1}(F(x)) = -\log[2(1-F(x))]$ is convex for $x \geq 0$ by showing that the derivative $f(x)/(1-F(x))$ is nondecreasing for $x \geq 0$.

The first step is to show that if $-\log f(x)$ is convex, then $f(x-y')/f(x-y) \leq f(x'-y')/f(x'-y)$ for all $x < x'$ and $y < y'$. Let $t = (x'-x)/(x'-x+y'-y)$ then $x-y = t(x-y')+(1-t)(x'-y)$ and $x'-y' = (1-t)(x-y')+t(x'-y')$. Since $-\log f(x)$ is convex we have

$$-\log f(x-y) \leq -t\log f(x-y') - (1-t)\log f(x'-y)$$

and

$$-\log f(x'-y') \leq -(1-t)\log f(x-y') - t\log f(x'-y').$$

Add the two inequalities and exponentiate to establish the result. In fact, the condition is necessary, and this shows that $-\log f(x)$ convex is equivalent to $f(x)$ having monotone likelihood ratio. See Lehmann (1959, p. 330).

We now show $f(x)/(1-F(x))$ is nondecreasing for $x \geq 0$. Let $t_1 < t_2$, then the following statements are equivalent:

$$\frac{f(t_1)}{1-F(t_1)} \leq \frac{f(t_2)}{1-F(t_2)}$$

$$f(t_1)(1-F(t_2)) \leq f(t_2)(1-F(t_1))$$

$$f(t_1)\int_0^\infty f(v+t_2)\,dv \leq f(t_2)\int_0^\infty f(v+t_1)\,dv$$

$$0 \leq \int_0^\infty \left[f(t_2)f(v+t_1) - f(t_1)f(v+t_2)\right]dv.$$

Now identify $t_1 = x - y' < t_2 = x' - y'$ so $x < x'$, and $v = y' - y > 0$ so $y < y'$. Then $t_1 + v = x - y$ and $t_2 + v = x' - y$ and from the first step we have $f(t_1)/f(t_1+v) \leq f(t_2)/f(t_2+v)$. This implies the integral is nonnegative and so $f(x)/(1 - F(x))$ is nondecreasing. Hence $D^{-1}(F(x))$ is convex for $x \geq 0$ and $F < D$.

Theorem 2.9.3. Suppose $F, G \in \Omega_s$. Then $F < G$ if and only if $f(F^{-1}(y))/g(G^{-1}(y))$ is nondecreasing for $y \geq 1/2$.

Proof. Let $q(x) = G^{-1}(F(x))$. Then $q'(x) = f(x)/g(G^{-1}(F(x)))$ is a nondecreasing function for $x \geq 0$. Further, $x = F^{-1}(y)$ is nondecreasing, and $x \geq 0$ when $y \geq 1/2$. Hence $q'(F^{-1}(y))$ is nondecreasing for $y \geq 1/2$.

We now discuss the effect of heavy tails on the efficiency of Wilcoxon versus normal scores procedures. Additional discussion is given in Hodges and Lehmann (1960). First, note from the definition of efficiency that by (2.6.13) and (2.9.9),

$$e_F(T, NS) = 12 \left\{ \frac{\int_{-\infty}^{\infty} f^2(x)\,dx}{\int_{-\infty}^{\infty} \frac{f^2(x)}{\psi(\Phi^{-1}(F(x)))}\,dx} \right\}^2.$$

Since F and Φ are in Ω_s the integrals are 2 times the integrals from 0 to ∞. Using this fact along with a change of variable $u = F^{-1}(x)$, we have

$$e_F(T, NS) = 12 \left\{ \frac{\int_{1/2}^{1} f(F^{-1}(u))\,du}{\int_{1/2}^{1} \frac{f(F^{-1}(u))}{\psi(\Phi^{-1}(u))}\,du} \right\}^2. \qquad (2.9.12)$$

By letting $f(x) = \sigma^{-1} f_1(x\sigma^{-1})$ and computing $e_F(T, NS)$, we see that $e(T, NS)$ is independent of σ.

Theorem 2.9.4. For any $F \in \Omega_s$ for which $e_F(T, NS)$ exists, $0 \leq e_F(T, NS) \leq 1.91$.

Proof. For $\frac{1}{2} < x < y$, $\Phi^{-1}(x) \leq \Phi^{-1}(y)$, and so $(1/\psi(\Phi^{-1}(x))) \leq (1/\psi(\Phi^{-1}(y)))$. From (2.9.12) we have

$$\int_{1/2}^{1} \frac{f(F^{-1}(u))}{\psi(\Phi^{-1}(u))}\,du \geq \frac{1}{\psi(\Phi^{-1}(\frac{1}{2}))} \int_{1/2}^{1} f(F^{-1}(u))\,du$$

and

$$e(T,NS) \leq 12\left\{\psi\left(\Phi^{-1}\left(\tfrac{1}{2}\right)\right)\right\}^2$$

$$= 12\psi^2(0) = 6/\pi \doteq 1.91.$$

Furthermore, let $f(x) = 1/2$ for $|x| < 1$ and 0 otherwise, then the denominator of (2.9.11) is ∞ and $e(T, NS) = 0$. Exercise 2.10.33 shows the upper bound is sharp.

Hence, unlike the t procedures, the normal scores procedures may be much more efficient than the Wilcoxon procedures, at least for light-tailed distributions approaching the uniform. The next theorem describes the relationship between $e_F(T, NS)$ and the tailweight of F.

Theorem 2.9.5. Suppose $F, G \in \Omega_s$ and $F \prec G$. Then $e_F(T, NS) \leq e_G(T, NS)$.

Proof. Since $e_F(T, NS)$ is scale invariant we can choose the scale. Let $\sigma^{-1} = \int g^2(x)\,dx/\int f_1^2(x)\,dx$ where $f(x) = \sigma^{-1}f_1(x\sigma^{-1})$. Then $\int f^2(x)\,dx = \sigma^{-1}\int f_1^2(x)\,dx = \int g^2(x)\,dx$. Hence, without loss of generality, we take $\int f^2(x)\,dx = \int g^2(x)\,dx$.

From (2.9.11) or (2.9.12) it follows that it is sufficient to show

$$\int_{1/2}^{1} \frac{f(F^{-1}(u))}{\psi(\Phi^{-1}(u))}\,du \geq \int_{1/2}^{1} \frac{g(G^{-1}(u))}{\psi(\Phi^{-1}(u))}\,du. \qquad (2.9.13)$$

Since $2\int_{1/2}^{1} f(F^{-1}(u))\,du = \int_{-\infty}^{\infty} f^2(x)\,dx = \int_{-\infty}^{\infty} g^2(x)\,dx = 2\int_{1/2}^{1} g(G^{-1}(u))\,du$, we have

$$0 = \int_{1/2}^{1} g(G^{-1}(u))\,du - \int_{1/2}^{1} f(F^{-1}(u))\,du$$

$$= \int_{1/2}^{1} g(G^{-1}(u))\left[1 - \frac{f(F^{-1}(u))}{g(G^{-1}(u))}\right]du. \qquad (2.9.14)$$

Since $g(G^{-1}(u)) \geq 0$, and since by Theorem 2.9.3, $f(F^{-1}(u))/g(G^{-1}(u))$ is nondecreasing for $u \geq 1/2$, the zero integral implies the existence of a point $c + \tfrac{1}{2}$, $c > 0$, such that $1 - f(F^{-1}(u))/g(G^{-1}(u))$ changes from positive to negative as u crosses $c + 1/2$. Hence $f(F^{-1}(u)) - g(G^{-1}(u)) \leq 0$ or ≥ 0 as $u \leq c + 1/2$ or $\geq u + 1/2$.

Recall from the proof of Theorem 2.9.4 that $1/\psi(\Phi^{-1}(u))$ is nondecreasing for $u \geq 1/2$. Hence, from (2.9.13), and noting where the sign change occurs for the integral, we have

$$\int_{1/2}^{1} \frac{1}{\psi(\Phi^{-1}(u))} \left\{ f(F^{-1}(u)) - g(G^{-1}(u)) \right\} du$$

$$\geq \int_{1/2}^{1/2+c} \frac{1}{\psi(\Phi^{-1}(c+1/2))} \left\{ f(F^{-1}(u)) - g(G^{-1}(u)) \right\} du$$

$$+ \int_{1/2+c}^{1} \frac{1}{\psi(\Phi^{-1}(c+1/2))} \left\{ f(F^{-1}(u)) - g(G^{-1}(u)) \right\} du$$

$$= \frac{1}{\psi(\Phi^{-1}(c+1/2))} \int_{1/2}^{1} \left\{ f(F^{-1}(u)) - g(G^{-1}(u)) \right\} du.$$

But the integral of the difference is 0 as noted in (2.9.14) and the proof is complete.

Hence, as the tails become heavier, the efficiency of T relative to NS increases. At the normal distribution $e_\Phi(T, NS) = 0.955$. Hence $0.955 \leq e_F(T, NS) \leq 1.91$ for any F such that $\Phi \prec F$. Exercise 2.10.34 provides a similar result for $e_F(S, T)$. A more general result for scores functions is given by Gastwirth (1970a).

We now present a result that shows that some Winsorization of the score function is desirable. We consider the symmetric contamination model introduced by Huber (1964) and discussed by Huber (1981, Chapter 4).

Let G in Ω_s be a specified distribution with a strongly unimodal density, Definition 2.9.2. Let $0 < \epsilon < 1$ be fixed and define

$$\Omega(\epsilon) = \left\{ F = (1-\epsilon)G + \epsilon H : H \in \Omega_s, I(h) < \infty \right\}$$

where $I(h)$ is Fisher's information, Definition 2.9.1. We think of G as the assumed model and some F in $\Omega(\epsilon)$ as the true model. Note G may be the true model since it is also in $\Omega(\epsilon)$.

In Theorem 2.9.7 we will show that a Winsorized version of the AMPRT corresponding to G will have maximin asymptotic power over $\Omega(\epsilon)$. Hence this rank test will maximize the minimum of asymptotic local power over the contaminated distributions. The maximin test statistic will be generated by a scores-generating function ϕ_0, (2.9.5), that corresponds to a least favorable distribution F_0 in $\Omega(\epsilon)$.

If $\hat{\theta}_0$ is the Hodges–Lehmann estimate corresponding to $\phi_0(\cdot)$, then we will show that it has minimax asymptotic variance over $\Omega(\epsilon)$. It will thus be seen that the robust procedures are generated by a Winsorized scores-generating function that is similar to the Huber ψ-function generating robust M-estimates.

In general we let $c(\phi, F)$ denote the efficacy, at the distribution F, of a test based on $\bar{V}$ generated by $\phi(\cdot)$. We seek a generating function $\phi_0(\cdot)$ such that

$$\inf_{\Omega(\epsilon)} c(\phi, F) \leqslant \inf_{\Omega(\epsilon)} c(\phi_0, F). \qquad (2.9.15)$$

In this case we say that $\bar{V}_0$, corresponding to $\phi_0(\cdot)$ has maximin efficacy over $\Omega(\epsilon)$. Thus, when the true model is in $\Omega(\epsilon)$, $\bar{V}_0$ is the least adversely affected by contamination in the sense defined by (2.9.15).

In the next theorem we construct a least favorable distribution F_0 in $\Omega(\epsilon)$ in the sense that the asymptotic efficacy of the best test (AMPRT) is a minimum at F_0. Example 2.9.4 suggests that the least favorable distribution will look like g in the middle and have exponential tails. This will produce the heaviest tailed strongly unimodal distribution that is hardest to distinguish from the assumed model G.

Theorem 2.9.6. Let $x_0 > 0$ and $k > 0$ be constants such that $-g'(x_0)/g(x_0) = k$ and $(1 - \epsilon)^{-1} = 2G(x_0) - 1 + 2g(x_0)/k$. Let

$$f_0(x) = \begin{cases} (1 - \epsilon)g(-x_0)\exp\{k(x + x_0)\} & \text{if } x \leqslant -x_0 \\ (1 - \epsilon)g(x) & \text{if } -x_0 \leqslant x \leqslant x_0 \quad (2.9.16) \\ (1 - \epsilon)g(x_0)\exp\{-k(x - x_0)\} & \text{if } x_0 \leqslant x. \end{cases}$$

Then

a. $f_0(x)$ is a density function, its cdf F_0 is in $\Omega(\epsilon)$ and
b. $c(\phi_0, F) \geqslant c(\phi_0, F_0)$, where $\phi_0(\cdot)$ is the optimal scores function defined in (2.9.5) by F_0.

Proof. a. Consider first the integral of $f_0(x)$:

$$\int f_0(x)\,dx = (1 - \epsilon)g(-x_0)/k + (1 - \epsilon)\left[G(x_0) - G(-x_0)\right]$$

$$+ (1 - \epsilon)g(x_0)/k$$

$$= (1 - \epsilon)\left[2G(x_0) - 1\right] + (1 - \epsilon)2g(x_0)/k = 1.$$

Now define $h_0(x)$ by $f_0(x) = (1 - \epsilon)g(x) + \epsilon h_0(x)$ so that $h_0(x) = [f_0(x) - (1 - \epsilon)g(x)]/\epsilon$. Hence $\int h_0(x)\,dx = 1$. We next show $h_0(x) \geq 0$, for all x.

First note that

$$\epsilon h_0(x) = f_0(x) - (1 - \epsilon)g(x)$$

$$= \begin{cases} (1 - \epsilon)\left[g(-x_0)\exp\{k(x + x_0)\} - g(x)\right] & \text{if } x \leq -x_0 \\ 0 & \text{if } -x_0 \leq x \leq x_0 \\ (1 - \epsilon)\left[g(x_0)\exp\{-k(x - x_0)\} - g(x)\right] & \text{if } x_0 \leq x. \end{cases}$$

For $x \leq -x_0$, we first show $g(-x_0)\exp\{k(x + x_0)\} - g(x) \geq 0$. Since $-\log g(x)$ is convex, it lies above its tangent at $-x_0$; hence $-\log g(x) \geq -\log g(-x_0) + (-g'(-x_0)/g(-x_0))(x + x_0)$, for $x \leq -x_0$. Further, by symmetry, $-g'(-x_0)/g(-x_0) = -k$. Hence

$$-\log g(x) \geq -\log g(-x_0) - k(x + x_0)$$

and

$$g(x) \leq g(-x_0)\exp\{k(x + x_0)\}.$$

This establishes the first line in $\epsilon h_0(x)$; the second follows in a similar way. Hence $h_0(x) \geq 0$, and F_0 is in $\Omega(\epsilon)$.

b. Note from (2.9.3) the denominator of $c(\phi_0, F)$ is free of F. Hence we will consider the numerator:

$$J(F) = \int_{-\infty}^{\infty} \phi_0'(2F(x) - 1)f^2(x)\,dx. \tag{2.9.17}$$

From the definition of $\phi_0(u)$ in (2.9.5), using $f_0(x)$ in (2.9.16), we have

$$\phi_0(2F(x) - 1) = -f_0'(x)/f_0(x)$$

$$= \begin{cases} -k & x < -x_0 \\ -g'(x)/g(x) & -x_0 \leq x \leq x_0 \\ k & x < x_0. \end{cases} \tag{2.9.18}$$

Hence $\phi_0'(2F(x) - 1) = 0$ for $|x| > x_0$, and so

$$J(F) = 2\int_0^{x_0} \phi_0'(2F(x) - 1)f^2(x)\,dx$$

$$= 2\int_{1/2}^{F(x_0)} \phi_0'(2u - 1)f\big(F^{-1}(u)\big)\,du. \tag{2.9.19}$$

We now show $f(F^{-1}(u)) \geq f_0(F_0^{-1}(u))$ for $1/2 \leq u \leq F_0(x_0)$. For $0 \leq x \leq x_0$,

$$f_0(x) = (1 - \epsilon) g(x)$$

$$\leq (1 - \epsilon) g(x) + \epsilon h(x) = f(x). \qquad (2.9.20)$$

Hence $F_0(x) \leq F(x)$ and $F_0(F^{-1}(u)) \leq F(F^{-1}(u)) = u$ for $1/2 \leq u \leq F_0(x_0)$. Taking F_0^{-1} of both sides yields

$$0 \leq F^{-1}(u) \leq F_0^{-1}(u) \qquad (2.9.21)$$

for $1/2 \leq u \leq F_0(x_0)$. Now, since $-\log g(x)$ is convex, so is $-\log f_0(x)$. Further, $-f_0'(x)/f_0(x) \geq 0$ implies that $-\log f_0(x)$ is nondecreasing, $\log f_0(x)$ nonincreasing, and finally $f_0(x)$ is nonincreasing. Hence, by (2.9.20) and (2.9.21), $f(F^{-1}(u)) \geq f_0(F^{-1}(u)) \geq f_0(F_0^{-1}(u))$.

The inequalities $f(F^{-1}(u)) \geq f_0(F_0^{-1}(u))$ and $F_0(x_0) \leq F(x_0)$, along with (2.9.19), imply that $J(F) \geq J(F_0)$. This implies $c(\phi_0, F) \geq c(\phi_0, F_0)$, and the proof is complete.

We are now ready to show that ϕ_0, constructed from $f_0(x)$ in (2.9.16), produces the maximin test with property (2.9.15).

Theorem 2.9.7. Suppose $G \in \Omega_s$ has a strongly unimodal density and $F \in \Omega(\epsilon)$. Let ϕ_0 be the score function constructed from $f_0(x)$ in (2.9.16), and let $\overline{V}_0$ be the test generated by ϕ_0. Let $c(\phi, F)$ denote the efficacy of the test generated by ϕ. Then

$$\inf_{\Omega(\epsilon)} \left\{ 1 - \Phi(Z_\alpha - \theta c(\phi_0, F)) \right\} \geq \inf_{\Omega(\epsilon)} \left\{ 1 - \Phi(Z_\alpha - \theta c(\phi, F)) \right\}$$

for any other ϕ. The expression $1 - \Phi(Z_\alpha - \theta c(\phi, F))$ is the asymptotic local power along the sequence of alternatives $\theta_n = \theta/n^{1/2}$ of an asymptotically size α test. See Theorem 2.6.1.

Proof. From the last theorem $c(\phi_0, F) \geq c(\phi_0, F_0)$, and hence

$$\inf_{\Omega(\epsilon)} c(\phi_0, F) \geq c(\phi_0, F_0).$$

From Theorem 2.9.1 we have $c(\phi_f, F) \geq c(\phi, F)$. This holds even if ϕ_f is not a score-generating function since $c^2(\phi_f, F) = I(f)$. Hence

$$c(\phi_0, F_0) \geq \inf_{\Omega(\epsilon)} c(\phi_f, F) \geq \inf_{\Omega(\epsilon)} c(\phi, F).$$

Combining these inequalities we have

$$\inf_{\Omega(\epsilon)} c(\phi_0, F) \geqslant \inf_{\Omega(\epsilon)} c(\phi, F). \qquad (2.9.22)$$

Hence

$$1 - \Phi\left(Z_\alpha - \theta \inf_{\Omega(\epsilon)} c(\phi, F)\right) \leqslant 1 - \Phi\left(Z_\alpha - \theta \inf_{\Omega(\epsilon)} c(\phi_0, F)\right).$$

The result now follows from the fact that $\Phi(\cdot)$ is an increasing function.

Let $\hat{\theta}_\phi$ be the Hodges–Lehmann estimate of θ based on $\overline{V}$ generated by ϕ. Let $\sigma^2(\phi, F)$ denote the asymptotic variance where $n^{1/2}(\hat{\theta}_\phi - \theta)$ is asymptotically $n(0, \sigma^2(\phi, F))$. Recall that $\sigma^2(\phi, F) = 1/c^2(\phi, F)$. Let $\hat{\theta}_0$ correspond to ϕ_0 specified in Theorem 2.9.6. Then (2.9.22) implies

$$\sup_{\Omega(\epsilon)} \sigma^2(\phi_0, F) \leqslant \sup_{\Omega(\epsilon)} \sigma^2(\phi, F). \qquad (2.9.23)$$

Thus, the asymptotic variance of $\hat{\theta}_0$ minimizes the maximum asymptotic variance over $\Omega(\epsilon)$. Jaeckel (1971) developed the result, (2.9.23), when he showed that there exist R and L estimates that solve Huber's minimax problem for $\Omega(\epsilon)$. Huber (1981, p. 97) points out that the minimax result, (2.9.23), cannot always be extended, for rank estimates, to more general types of neighborhoods than $\Omega(\epsilon)$. Sacks and Ylvisaker (1982) provide an example. See Collins (1983) also.

We next discuss the form of the generating function $\phi_0(u)$ corresponding to an assumed strongly unimodal G. Let $\phi_g(u)$ be the generating function constructed from g, in (2.9.5). Note, from Theorem 2.9.6, $-g'(x_0)/g(x_0) = k$. Define u_0 such that $\phi_g(u_0) = -g'(G^{-1}[(u_0 + 1)/2])/g(G^{-1}[(u_0 + 1)/2]) = k$. Then from (2.9.18) we have, for $0 < u < 1$,

$$\phi_0(u) = \min\{\phi_g(u), k\}. \qquad (2.9.24)$$

Thus we see that the optimal score function ϕ_0 is a Winsorized version of $\phi_g(u)$. The amount of Winsorization depends on the amount of contamination.

If the assumed model is normal, then a Winsorized normal scores function is best. Winsorization will result in a bounded influence function (Theorem 2.8.4) and avoid the peculiarity of Example 2.8.6 in which the normal scores estimate has an unbounded influence function. See Huber (1981, p. 99).

Example 2.9.5. We illustrate the calculations involved in the minimax asymptotic variance for the contaminated logistic model. The Winsorized Wilcoxon statistic is best in this case. We have treated various aspects of the Winsorized Wilcoxon in Examples 2.8.2, 2.8.4, 2.8.5, and 2.9.1. We let $g(x)$ denote the logistic pdf, $g(x) = e^{-x}/(1 + e^{-x})^2$, $-\infty < x < \infty$. The cdf is given by $G(x) = 1/(1 + e^{-x})$, $-\infty < x < \infty$, and the relationship $g(x) = G(x)(1 - G(x))$ simplifies many calculations. Note that $-g'(x)/g(x) = 2G(x) - 1$; hence the conditions in Theorem 2.9.6 become

$$\frac{-g'(x_0)}{g(x_0)} = 2G(x_0) - 1 = k$$

and

$$(1 - \epsilon)^{-1} = 2G(x_0) - 1 + \frac{2g(x_0)}{k}$$

$$= k + \frac{(1 - k^2)}{2k}. \tag{2.9.25}$$

From (2.9.24) we have $\phi_0(u) = \min\{u, k\}$ and $k = 1 - \gamma$ in Example 2.8.2. Since $0 < \gamma < 1$, $k < 1$, solving (2.9.25) for k, we take the root

$$k = (1 - \epsilon)^{-1}\left(1 - \sqrt{\epsilon(2 - \epsilon)}\right) = 1 - \gamma. \tag{2.9.26}$$

The efficacy is given in Example 2.9.1, so we can compute $\sigma^2(\phi_0, F_0)$ from $1/c^2$. First note that

$$\int_0^{x(1 - \gamma/2)} g^2(x)\, dx = \int_0^{x(-\gamma/2)} G(x)(1 - G(x))\, dG(x)$$

$$= \frac{(2 - 3\gamma^2 + \gamma^3)}{24}$$

where $x(1 - \gamma/2) = G^{-1}(1 - \gamma/2)$. Hence, using the efficacy equation in Example 2.9.1, after some algebra,

$$\sigma^2(\phi_0, F_0) = \sup_{\Omega(\epsilon)} \sigma^2(\phi_0, F)$$

$$= \frac{12(1 + 2\gamma)}{\left[3 - (1 - \gamma)^2\right]^2}.$$

Table 2.6. $1 - \gamma$ and $\sigma^2(\phi_0, F_0)$ as Functions of ϵ

Function	ϵ					
	.05	.07	.10	.12	.15	.20
$1 - \gamma$	.72	.69	.63	.60	.56	.50
$\sigma^2(\phi_0, F_0)$	3.00	3.00	3.00	3.00	3.01	3.02

In Table 2.6 we show $1 - \gamma$ and $\sigma^2(\phi_0, F_0)$ as a function of ϵ. This suggests that a mixture of roughly 2/3 Wilcoxon scores and 1/3 sign scores should provide adequate protection against mild to moderate amounts of contamination. Table 2.5 compares the Wilcoxon (best at the assumed model) to the Winsorized Wilcoxon [minimax over $\Omega(\epsilon)$].

2.10. EXERCISES

2.10.1. Under the null hypothesis $H_0 : \theta = 0$, $F \in \Omega_s$, show the mean and variance of T, the Wilcoxon signed rank statistic, are $ET = n(n + 1)/4$ and $\text{Var } T = n(n + 1)(2n + 1)/24$, where n is the sample size.

2.10.2. Suppose $X_1, \ldots, X_n$ are i.i.d. $F \in \Omega_s$.
   a. Show that $g(X_1, \ldots, X_n)$ and $g(-X_1, \ldots, -X_n)$ have the same distribution. Hint: Show that $P(g(X_1, \ldots, X_n) \leqslant t) = P(g(-X_1, \ldots, -X_n) \leqslant t)$.
   b. Show that if $g(X_1, \ldots, X_n) + g(-X_1, \ldots, -X_n) = \mu_0$ then $g(X_1, \ldots, X_n)$ is symmetrically distributed about $\mu_0/2$. Hint: Show $P(g(X_1, \ldots, X_n) \leqslant \mu_0/2 - t) = P(g(X_1, \ldots, X_n) \geqslant \mu_0/2 + t)$.
   c. Apply (b) to the Wilcoxon signed rank statistic to show that T has a symmetric distribution under the null hypothesis. What is the point of symmetry?

2.10.3. In this exercise we consider conditions on the marginal distributions of a pair of possibly dependent random variables (T, C) so that the difference $T - C$ will be symmetrically distributed. Prove:
   a. If the marginal distributions are identical (but not necessarily symmetric) then $T - C$ is symmetrically distributed.
   b. If the marginal distributions are symmetric (but may have different scale parameters) the distribution of $T - C$ is symmetric.

2.10.4. Let $V_1 = \sum a_i W_i$ and $V_2 = \sum b_i W_i$, where $W_1, \ldots, W_n$ are i.i.d. $B(1, 1/2)$. Show that

$$EV_1 = \frac{1}{2} \sum a_i$$

$$\operatorname{Var} V_1 = \frac{1}{4} \sum a_i^2$$

$$\operatorname{Cov}(V_1, V_2) = \frac{1}{4} \sum a_i b_i.$$

Use Theorem A9 and prove that, for the Wilcoxon signed rank statistic T, (2.7.2), under H_0,

$$\frac{T - n(n+1)/4}{\sqrt{n(n+1)(2n+1)/24}} \xrightarrow{D} Z \sim n(0,1)$$

Further, find $\lim \rho_n(S, T)$, where $\rho_n(S, T)$ is the correlation between the sign statistic S and T. Finally, using Theorem A13, show that S and T, when properly standardized, have an asymptotic bivariate normal distribution. Hint: Define $T^* = (1/(n+1)) n^{1/2}) \sum j(W_j - 1/2)$ and $S^* = (1/n^{1/2}) \sum (W_j - 1/2)$.

2.10.5. Show that the Wilcoxon signed rank test has tolerances given by

$$\tau_n(\text{accept}) = \frac{2n - 1 - \sqrt{1 + 8C_\alpha}}{2n}$$

$$\tau_n(\text{reject}) = \frac{2n - 1 - \sqrt{(2n+1)^2 - 8C_\alpha}}{2n}.$$

Use the approximation (2.2.9) for C_α, the critical value of the test, to show that both tolerances converge to $\tau = 1 - 1/2^{1/2}$, the estimation tolerance of the median of the Walsh averages in Example 2.4.1.

2.10.6. Prove Theorem 2.4.2 for $\# B = 2q + 1$, odd.

2.10.7. Define $B = \{(i, j) : i \leqslant j, i + j = n + 1\}$. Then $\hat{\theta} = \operatorname{med}(X_i + X_j)/2, (i, j)$ in B is called Galton's estimate by Hodges (1967). Find the tolerance of $\hat{\theta}$.

2.10.8. In (2.5.1), show that $p_3 = (p_1^2 + p_2)/2$.

2.10.9. Compute p_1, p_2, and p_4 in (2.5.1) for a uniform distribution on $(-1, 2)$. Approximate the power of the sign test and Wilcoxon signed rank test in Example 2.5.5, for $n = 40$ and $\alpha = .05$. Use the Central Limit Theorem to make the corresponding approximation for the t test.

2.10.10. Show the asymptotic power, for a sequence of alternatives $\theta_n = a/n^{1/2}$, of the sign test is $1 - \Phi(Z_\alpha - 2f(0)a)$. Compare this to (2.5.27), for the corresponding asymptotic power of the Wilcoxon signed rank test, for $f(x)$ the $n(0, 1)$ density function.

2.10.11. Show that the sample size equation, (2.6.3), for the sign test holds in Example 2.6.1.

2.10.12. In this exercise we outline a proof that the Wilcoxon signed rank statistic, when properly standardized, is asymptotically normal, uniformly near $\theta = 0$, provided the variance is bounded away from 0. Suppose $X_1, \ldots, X_n$ is a sample from $F(x - \theta)$, $F \in \Omega_s$, and suppose $\sigma^2(\theta) = p_4(\theta) - p_2^2(\theta)$ where $p_4(\theta)$ and $p_2(\theta)$ are defined in (2.5.1). Further, suppose there exists a constant K such that $0 < K \leqslant \sigma^2(\theta)$ for all θ in a neighborhood of 0. (This will be true if $\sigma^2(0) > 0$ and $\sigma^2(\theta)$ is continuous.) Define

$$\overline{T} = \frac{1}{n(n + 1)} \sum_{i \leqslant j} \sum T_{ij}$$

$$\overline{T}^* = \frac{1}{n(n + 1)} \sum_{i < j} \sum T_{ij}$$

with T_{ij} given by (2.5.3).

a. Show that in a neighborhood of $\theta = 0$,

$$\frac{\sqrt{n}\left(\overline{T} - p_2(\theta)\right)}{\sigma(\theta)} - \frac{\sqrt{n}\left(\overline{T}^* - p_2(\theta)\right)}{\sigma(\theta)} \leqslant \frac{2\sqrt{n}}{K(n + 1)}.$$

Hence $\overline{T}$ and $\overline{T}^*$ have the same asymptotic behavior; however, $\overline{T}^*$ is easier to deal with.

b. Define

$$V^* = \frac{\sqrt{n}}{\sigma(\theta)n(n - 1)} \sum_{i < j} \sum (T_{ij} - p_2(\theta))$$

and show $EV^* = 0$ and

$$\text{Var } V^* = \frac{p_2(\theta)(1 - p_2(\theta))}{2\sigma^2(\theta)(n-1)} + \frac{n-2}{n-1}.$$

c.   Next let $Y_i = 1 - H(-X_i)$, $H(x) = F(x - \theta)$, and show that the projection of V^* is

$$V_p^* = \frac{1}{\sigma(\theta)\sqrt{n}} \sum (Y_i - p_2(\theta)).$$

Further, show $EV_p^* = 0$ and $\text{Var } V_p^* = 1$.

d.   Let $R = V^* - V_p^*$. Then for $\epsilon > 0$ show

$$P(V_p^* \leqslant t - \epsilon) - P(|R| > \epsilon)$$

$$\leqslant P(V^* \leqslant t) \leqslant P(V_p^* \leqslant t + \epsilon) + P(|R| > \epsilon)$$

e.   Now show, for some constant C and specified $\delta > 0$,

$$P(|R| > \epsilon) \leqslant \frac{1}{2(n-1)K} < \delta$$

$$|P(V_p \leqslant t \pm \epsilon) - \Phi(t \pm \epsilon)| \leqslant \frac{C}{K^3\sqrt{n}} < \delta$$

for sufficiently large n.

f.   Hence conclude that for any $\epsilon > 0$,

$$\Phi(t - \epsilon) - 2\delta - \Phi(t) \leqslant P(V^* \leqslant t) - \Phi(t)$$

$$\leqslant \Phi(t + \epsilon) + 2\delta - \Phi(t).$$

Now, since $\Phi(\cdot)$ is continuous, as $\epsilon \to 0$ we have

$$|P(V^* \leqslant t) - \Phi(t)| < 2\delta$$

for sufficiently large n, uniformly in θ near 0.

2.10.13.   Show that the efficiency equations (2.6.16)–(2.6.18) are scale invariant. Hint: Let $g(x) = \tau^{-1}f(\tau^{-1}x)$ and show the equations do not depend on τ.

2.10.14. a. Compute $e(T, t)$, (2.6.18), when $f(x) = e^{-x}/(1 + e^{-x})^2$, $-\infty < x < \infty$, the logistic distribution; when $f(x) = e^{-|x|}/2$, $-\infty < x < \infty$, the Laplace or double exponential distribution; and $f(x) = 2^{-1}$, $-1 < x < 1$, the uniform distribution.

   b. What happens to $e(T, t)$ if $f(x) = 1/\pi(1 + x^2)$, $-\infty < x < \infty$, the Cauchy distribution?

2.10.15. Let $f(x; \tau) = \tau e^{-|x|^\tau}/2\Gamma(\tau^{-1})$, $-\infty < x < \infty$. The parameter τ determines a family of distributions. For example a normal distribution corresponds to $\tau = 2$ and a double exponential distribution corresponds to $\tau = 1$.

   a. Find $e_\tau(S, T)$, (2.6.16), as a function of τ.

   b. Graph $e_\tau(S, T)$ as a function of τ and find the point where $e_\tau(S, T) = 1$. This indicates how "far" from normality we need to go before S is more efficient than T.

2.10.16. Find $e(S, t)$, (2.6.17), for the normal and double exponential distributions.

2.10.17. Suppose $[\hat{\theta}_L, \hat{\theta}_U]$ is a $(1 - \alpha)$ 100% confidence interval derived from the Wilcoxon signed rank statistic T. If θ_0 is the true parameter value, show that $n^{1/2}(\hat{\theta}_L - \theta_0)$ is asymptotically $n(-Z_{\alpha/2}/c_T, 1/c_T^2)$ where $Z_{\alpha/2}$ is the upper $\alpha/2$ percentile of the standard normal distribution and $c_T^2 = 12(\int f^2(x)\,dx)^2$. Hint: Use the argument of Theorem 2.6.5.

2.10.18. Argue that equations for S and t similar to (2.7.8) can be derived. In other words, $n^{1/2}(\hat{\theta}_U - \hat{\theta}_L)2Z_{\alpha/2}$ is approximately (in probability) equal to $1/2f(0)$ and σ_f for S and t, respectively. In particular, prove that if $F \in \Omega_0$, $f(0) < \infty$, then

$$\lim P_0\left\{ \sup_{|b| \leqslant B} \left| \sqrt{n}\left[\bar{S}(b/\sqrt{n}) - \bar{S}(0) \right] + bf(0) \right| > \epsilon \right\} = 0,$$

where $\bar{S} = n^{-1}\sum s(X_i)$.

2.10.19. Apply the result in Exercise 2.10.2 to show that V, defined in (2.8.1) has a symmetric distribution under the null hypothesis when we are sampling from a population symmetric about 0. What is the point of symmetry for the distribution of V?

2.10.20. Under the null hypothesis, show that $\bar{V}_1$ and $\bar{V}_2$, defined by (2.8.1), have an asymptotic bivariate normal distribution when properly standardized. What is the asymptotic variance-covariance matrix? Hint: Use Theorem A13 in the Appendix.

2.10.21. Show that the normal scores-generating function in Example 2.8.1 satisfies the conditions on $\phi(u)$ in Definition 2.8.2 by showing that $\int_0^1 \Phi_+^{-1}(u)\,du = (2/\pi)^{1/2}$ and $\int_0^1 [\Phi_+^{-1}(u)]^2\,du = 1$.

2.10.22. Show, for the Winsorized signed rank statistic in Example 2.8.2, that

$$E\overline{V} \to (1 - \gamma^2)/4$$

$$n\,\text{Var}\,\overline{V} \to (1 - \gamma)^2(1 + 2\gamma)/12.$$

2.10.23. Construct a table of $E\overline{V}$ and $n\,\text{Var}\,\overline{V}$ for various values of n that illustrates the rate of convergence to the asymptotic parameters in Exercise 2.10.22.

2.10.24. Compute the exact and asymptotic mean and variance, under the null hypothesis, of the modified sign statistic in Example 2.8.3 and discuss the asymptotic distribution. Find the asymptotic testing tolerances.

2.10.25. a. Develop the Hodges–Lehmann estimate from the modified sign statistic of Example 2.8.3. Show that it is possible to compute the estimate without constructing the Walsh averages.

b. Write the formulas for the endpoints of the confidence interval derived from the modified sign statistic.

c. Find the asymptotic estimation tolerance as a function of γ.

2.10.26. Suppose the score function $\phi(\cdot)$ is bounded and $\phi(0) = 0$. Suppose also that $f'(x)$ exists. Show that c in (2.9.3) can be written as

$$c = \frac{-\int_0^\infty \phi(2F(x) - 1)f'(x)\,dx}{\sqrt{\dfrac{1}{4}\int_0^1 \phi^2(u)\,du}}.$$

2.10.27. a. Use Exercise 2.10.26 to find the efficacy of the modified sign statistic.

b. When f is the $n(0, 1)$ pdf, find the value of γ that maximizes the efficacy. Compare the efficiency of this test relative to the t test with that of the regular sign test relative to the t test.

2.10.28. Recent studies show that hypertension drugs such as propranolol may alleviate the symptoms of stage fright (*Time Magazine*, July 5, 1982, p. 58). To test this hypothesis, 29 professionals and students gave two solo recitals before an audience of critics and faculty members. Ninety minutes before the recitals they were given either propranolol or a placebo. Heatbeat rate was measured by remote electrocardiogram monitoring during the performance. Normal resting heartbeat rate is 70 beats per minute. Artificial data on 8 performers is as follows:

	Performance							
Treatment	1	2	3	4	5	6	7	8
Drug	85	107	69	122	106	121	137	87
Placebo	126	140	95	148	142	172	133	143

Let θ denote the median of the distribution of differences, placebo $-$ drug. Use the Wilcoxon signed rank procedures to test the hypothesis $H_0 : \theta = 0$ versus $H_A : \theta > 0$ at $\alpha \doteq .05$, construct a point estimate of θ, and construct an approximate 95% confidence interval for θ.

2.10.29. Let Ω_s be as given by (2.1.1) and let Ω_u in Ω_s be the subclass of unimodal symmetric distributions. Find the following infimums:

$$\inf_{\Omega_s} e(S, T) \quad \text{and} \quad \inf_{\Omega_u} e(S, T).$$

2.10.30. Show the asymptotic testing tolerance for $\overline{V}_f$ generated by $\phi_f(u)$ in (2.9.5) is given by ϵ such that $f(F^{-1}(1 - \epsilon/2)) = f(0)/2$.

2.10.31. Show the influence curve for $\hat{\theta}_f$, the estimate corresponding to $\overline{V}_f$ generated by $\phi_f(u)$ in (2.9.5) is

$$\Omega(y) = -\frac{f'(y)}{I(f)f(y)}$$

where $I(f)$ is Fisher's information, Definition 2.9.1.

2.10.32. Find $\phi_f(u)$, (2.9.5), for logistic and double exponential distributions.

2.10.33. Let $F_{a,\epsilon}(x) = \Phi(x)$ if $|x| \leqslant \epsilon$ and $\Phi(\epsilon + a(x - \epsilon))$ if $|x| > \epsilon$. Use this distribution to show that the upper bound in Theorem 2.9.4 is sharp.

2.10.34. Suppose $F, G \in \Omega_s$ and $F \prec G$. Prove that $e_F(S, T) \leqslant e_G(S, T)$.

2.10.35. Suppose $X_1, X_2, \ldots, X_n$ are independent with $X_i \sim F(x - \theta_i)$, $F \in \Omega_s$. For testing $H_0 : \theta_1 = \cdots = \theta_n$ versus $H_A : \theta_1 \leqslant \cdots \leqslant \theta_n$ with at least one strict inequality, we consider Mann's (1945) test for trend: $S^* = \sum\sum_{i<j} s(X_j - X_i)$. The statistic S^* compares X_i to all observations that come later, and hence makes more comparisons than the Cox–Stuart test in Exercise 1.8.13.

  a. Under H_0, it is easy to see that $ES^* = n(n - 1)/4$. A rather intricate counting argument, similar to the development of the variance of the Wilcoxon signed rank statistic in Theorem 2.5.1, can be used to show that under H_0, $\mathrm{Var}\, S^* = n(n - 1)(2n + 5)/72$. Use the Projection Theorem 2.5.2 to show that S^*, when properly standardized, is asymptotically normally distributed under H_0; then construct the approximate critical value for a size α test.

  b. Suppose $(X_1, Y_1), \ldots, (X_n, Y_n)$ is a bivariate random sample from $F(x, y)$, an absolutely continuous distribution. Two observations (X_i, Y_i) and (X_j, Y_j) are concordant if both members of one pair exceed the corresponding members of the other pair. Concordance can be expressed as $(X_i - X_j)(Y_i - Y_j) > 0$. If $(X_i - X_j)(Y_i - Y_j) < 0$ the observations are called discordant. Let P and Q be the number of concordant and discordant pairs and define

  $$\tau = \frac{2(P - Q)}{n(n - 1)},$$

  called Kendall's τ. Show that $-1 \leqslant \tau \leqslant 1$. For testing H_0: (X, Y) independent, reject H_0 if $|\tau| \geqslant k$. Find the mean and variance of τ under H_0. Prove that $(\tau - E\tau)/(\mathrm{Var}\,\tau)^{1/2}$ is asymptotically $n(0, 1)$ under H_0 and construct the approximate size α critical value. (See Theorem 4.4.2 for an alternative approach.)

2.10.36. Suppose $X_1, \ldots, X_n$ i.i.d. $F(x - \theta)$, $F \in \Omega_0$, with $f(0) < \infty$ and $\sigma^2 = \int x^2 f(x)\, dx < \infty$. To test $H_0 : F \in \Omega_s$ versus $H_A : F \in \Omega_0 - \Omega_s$, Gastwirth (1971) proposed the sign test for symmetry: Reject H_0 if $|S(\overline{X}) - n/2| \geqslant k$ where $S(\overline{X}) = \sum s(X_i - \overline{X})$.

a. Let $\bar{S}(\bar{X}) = n^{-1}S(\bar{X})$. Under H_0 prove that

$$\sqrt{n}\left(\bar{S}(\bar{X}) - 1/2\right) \xrightarrow{D} Z$$

$$\sim n\left(0, 1/4 + \sigma^2 f^2(0) - f(0)\int_{-\infty}^{\infty} |x| f(x)\, dx\right).$$

Outline: (1) Without loss of generality let $\theta = 0$ and use Exercise 2.10.18 to write $n^{1/2}(\bar{S} - 1/2) = n^{1/2}(\bar{S}(0) - 1/2) - n^{1/2}f(0)\bar{X} + o_p(1)$, (2) Use Theorem A11 in the Appendix to show that

$$\frac{1}{\sqrt{n}}\left(\frac{\sum[s(X_i) - 1/2]}{f(0)\sum X_i}\right) \xrightarrow{D} Z$$

$$\sim MVN\left[\begin{pmatrix} 0 \\ 0 \end{pmatrix}, \begin{pmatrix} \sigma_{11} & \sigma_{12} \\ \sigma_{12} & \sigma_{22} \end{pmatrix}\right]$$

with $\sigma_{11} = 1/4$, $\sigma_{12} = f(0)\int_0^\infty xf(x)\,dx$, $\sigma_{22} = f^2(0)\sigma^2$, (3) Use Theorem A2(b) for vectors.

b. Part (a) shows the test based on $\bar{S}(\bar{X})$ is not distribution free under H_0. Suppose the naive user constructs a nominal 5% test by rejecting H_0 if $|S(\bar{X}) - n/2| > 1.96n^{1/2}/2$. Hence the critical region is constructed as if $S(\bar{X})$ is roughly $B(n, 1/2)$. Use part (a) to approximate the true level of the test if the underlying distribution is the normal, the double exponential and the symmetric Pareto distribution with $f(x) = 3/2(1 + |x|)^4$, $-\infty < x < \infty$.

2.10.37. Let $X_1, \ldots, X_n$ be i.i.d. $F(x - \theta)$, $F \in \Omega_s$. Split the sample of n observations into p groups each of size α, α fixed. Let $T_\alpha = \sum_1^p T_i$ and reject $H_0: \theta = 0$ for $H_A: \theta > 0$ if $T_\alpha \geqslant k$. Feustal and Davisson (1967) call T_α a mixed Wilcoxon test.

a. Prove that $(T_\alpha - ET_\alpha)/(\mathrm{Var}\, T_\alpha)^{1/2}$ is asymptotically $n(0, 1)$; hence k can be approximated from the normal table.

b. It takes on the order of n^2 operations to compute T whereas it only takes on the order of $p\alpha^2$ for T_α. For large n, the savings may be significant. In order to assess the efficiency loss due to grouping, find the efficiency e_α of T_α relative to T, and plot e_α as a function of α.

c. Find and interpret the $\lim_{\alpha \to 1} e_\alpha$.

CHAPTER 3

The Two-Sample Location Model

3.1. INTRODUCTION

In this chapter we consider the problem of comparing two populations. Methods, based on ranks, are developed for testing and estimation of differences in location when the two populations have the same, but arbitrary, shape. The main emphasis is on finite samples rather than asymptotic results. We discuss the asymptotic distributions and describe the extensions of efficiency from one- to two-sample test comparisons. We do not discuss robustness aspects of the two-sample procedures since they are similar to the one-sample results. Discussion can be found in the work of Rieder (1982) and his references, and in the work of Lambert (1982) and her references.

The sampling model is defined by random samples $X_1, \ldots, X_m$ from $F(x - \theta_x)$ and $Y_1, \ldots, Y_n$ from $F(x - \theta_y)$, $F \in \Omega_0$. Hence, F is not assumed to be symmetric, but it does provide the same shape for the two populations. Next let $\Delta = \theta_y - \theta_x$, the difference in the population medians. We wish to test $H_0: \Delta = 0$ versus $H_A: \Delta > 0$ and to construct point and interval estimates of Δ. Without loss of generality we may take $\theta_x = 0$; hence the sampled populations can be written $F(x)$ and $F(y - \Delta)$, respectively. The difference in locations Δ, not the locations themselves, are considered.

Example 3.1.1. The randomization model provides an example. We begin with N subjects who are randomly assigned to a treatment group or a control group. Suppose X represents the control measurement and Y the treatment measurement. If the treatment acts to add a nonnegative constant Δ to the control effect, then we wish to test $H_0: \Delta = 0$ versus $H_A: \Delta > 0$. Because the N subjects originally came from a single population, the foregoing model specifying a common shape under H_0 is appropriate. The treatment may act to alter the distribution in some way other than a location change. Whether the location test is sensitive to the other changes is an issue discussed under consistency. See Example 3.5.1 for a discussion of the Mann–Whitney–Wilcoxon test which is introduced in the next section.

Finally we note that, in a comparison of attributes where randomization is impossible, the assumption of a common distribution shape is not automatically satisfied under H_0. For example in comparing the grade point averages of two schools we cannot assign subjects at random to the schools. The next example also illustrates this point. A Behrens–Fisher rank test is introduced in Section 3.5.

Example 3.1.2. Salk (1973) describes a study that analyzes the soothing effect of the mother's heartbeat on her newborn infant. The infants were placed in a nursery immediately after birth and they remained there for four days except for normal feedings by their mothers. The treatment group ($n = 102$) was continuously exposed to the sound of an adult's heartbeat (72 beats per minute at 85 decibels). The control group ($m = 112$) consisted of another group of infants in the same nursery. Hence it appears that the babies were not assigned at random to treatment or control.

One of the measurements consisted of the weight change in grams from the day following birth to the fourth day. It was hypothesized that the heartbeat group would gain more weight than the control group since the heartbeat group was expected to cry less. Let $X(Y)$ denote the control (treatment) measurement. Suppose $X_1, \ldots, X_m$ and $Y_1, \ldots, Y_n$ are samples from $F(x)$ and $F(y - \Delta)$, $F \in \Omega_0$. Then we wish to test $H_0 : \Delta = 0$ versus $H_A : \Delta > 0$. Note the similar shape assumption is an integral part of the model and is not automatically satisfied under H_0 as it would be for a randomized experiment. By monitoring the nursery, Salk estimated that the heartbeat group cried 38% of the time and the control group 60% of the time. Moreover, 70% of the heartbeat group, as opposed to 33% of the control group, gained weight. The heartbeat group showed a median gain of 40 grams; the control group showed a median loss of 20 grams. In Example 3.2.2 we provide some data and illustrate a formal test of $H_0 : \Delta = 0$ versus $H_A : \Delta > 0$.

3.2. THE MANN–WHITNEY–WILCOXON RANK STATISTIC

To test $H_0 : \Delta = 0$ versus $H_A : \Delta > 0$, Wilcoxon (1945) proposed the following simple procedure: First rank the combined data from smallest to largest and compute U, the sum of ranks of the Y observations, then reject H_0 if U is "too large." Large values of U indicate a shift of the Y sample to the right of the X sample. It remains then to find the distribution (exact or approximate) of U under $H_0: \Delta = 0$ in order to determine the critical value of the test. In Section 6.3 this test is extended to the two-sample multivariate location model.

We first discuss the joint distribution of the ranks of the Y observations in the combined data and then apply the results to U. Let $N = m + n$ and let R_i denote the rank of Y_i in the combined data; hence there are R_i observations less than or equal to Y_i, and $N - R_i$ greater than Y_i.

Theorem 3.2.1. Under $H_0: \Delta = 0$, if $R_1, \ldots, R_n$ denote the ranks of $Y_1, \ldots, Y_n$ in the combined data then

$$P(R_i = s) = \frac{1}{N}, \qquad s = 1, \ldots, N,$$

$$P(R_i = s, R_j = t) = \begin{cases} \dfrac{1}{N(N-1)} & \text{if } s \neq t \\ 0 & \text{if } s = t \end{cases}$$

where $N = m + n$ and $i \neq j$.

Proof. Under $H_0: \Delta = 0$, the combined data constitute a random sample of size $N = m + n$ from $F(x)$, $F \in \Omega_0$. Hence every permutation of the combined sample is equally likely. The event $R_i = s$ is determined by the $(N - 1)!$ sequences of mX's and nY's in which Y_i is always in the sth position. Thus,

$$P(R_i = s) = \frac{(N-1)!}{N!} = \frac{1}{N}.$$

The $P(R_i = s, R_j = t)$ is determined in exactly the same way. Since $F \in \Omega_0$, the probability of a tie, that is, $P(R_i = s, R_j = s)$, is zero.

In Exercise 3.7.1 you are asked to show that under $H_0: \Delta = 0$, $ER_i = (N+1)/2$, $\operatorname{Var} R_i = (N^2 - 1)/12$ and $\operatorname{Cov}(R_i, R_j) = -(N+1)/12$ for $i \neq j$. Hence we have for $U = \sum_1^n R_i$,

$$EU = n(N+1)/2$$

$$\operatorname{Var} U = mn(N+1)/12. \tag{3.2.1}$$

A counting form of the statistic can also be given. Recall that R_i is the number of combined sample items less than or equal to Y_i; hence

$$R_i = \#(X_j < Y_i) + \#(Y_k \leqslant Y_i), \qquad j = 1, \ldots, m, \quad k = 1, \ldots, n,$$

and

$$\sum_1^n R_i = \#(X_j < Y_i) + n(n+1)/2, \qquad j = 1, \ldots, m, \quad i = 1, \ldots, n.$$

The $n(n+1)/2$ represents the number of times Y observations are less than or equal to other Y observations and can be easily determined by putting the Y's in order. Thus,

$$U = \sum_1^n R_i = \#(Y_i - X_j > 0) + n(n+1)/2, \quad i = 1, \ldots, n, j = 1, \ldots, m,$$

and we let $W = \#(Y_i - X_j > 0)$ so that

$$U = W + n(n+1)/2. \tag{3.2.2}$$

The two statistics U and W have the same statistical properties; U is the rank sum form proposed by Wilcoxon, and W is the counting form proposed by Mann and Whitney (1947). This form of the statistic has been traced by Kruskal (1957) to the work of Gustav Deuchler published in German in 1914. Note also that

$$EW = mn/2$$
$$\tag{3.2.3}$$
$$\operatorname{Var} W = mn(N+1)/12.$$

Now that we have the mean and variance for both U and W we need the distribution under $H_0 : \Delta = 0$ in order to construct tests and confidence intervals. We consider the exact distribution first and then use the Projection Theorem 2.5.2 to develop the asymptotic distribution. The following example is similar to Example 2.2.1 for the one-sample case.

Example 3.2.1. Take $m = n = 2$ and suppose the null hypothesis is true so that the $2x$'s and $2y$'s are a sample of 4 observations. There are $\binom{4}{2} = 6$

equally likely arrangements which can be listed as follows:

Ranks:	1	2	3	4	W	U
Arrangements:	y	y	x	x	0	3
	y	x	y	x	1	4
	y	x	x	y	2	5
	x	y	y	x	2	5
	x	y	x	y	3	6
	x	x	y	y	4	7

The distribution of W (or U) under $H_0 : \Delta = 0$ can be listed as

w	0	1	2	3	4
$P(W = w)$	$\frac{1}{6}$	$\frac{1}{6}$	$\frac{2}{6}$	$\frac{1}{6}$	$\frac{1}{6}$.

Note the distribution-free nature of W is illustrated since no assumption on F was necessary to determine the probabilities of W. As in the one-sample case, the statistic is no longer distribution free under alternatives. This example shows that the distribution problem is reduced to counting the number of sequences such that $W = w$. The symmetry of the null distribution of W is also illustrated.

Define $\bar{P}_{m,n}(k)$ to be the number of sequences of mx's and ny's such that $W = k$. In the foregoing example $\bar{P}_{2,2}(2) = 2$. A recursive equation can now be developed.

Theorem 3.2.2. Given mx's and ny's,

$$\bar{P}_{m,n}(k) = \bar{P}_{m,n-1}(k - m) + \bar{P}_{m-1,n}(k)$$

where $\bar{P}_{i,j}(k) = 0$ if $k < 0$, $\bar{P}_{i,0}(k)$ and $\bar{P}_{0,i}(k)$ are 1 or 0 as $k = 0$ or $k \neq 0$.

Proof. Divide the sequences such that $W = k$ into two groups: those ending in x and those ending in y. In the former case, W computed on all but the last x must equal k, since attaching an x does not change the count $W = \#(Y_i > X_j)$. Hence, for these sequences of $m - 1x$'s and ny's, $W = k$. In the latter case, W computed on all but the last y must be $k - m$ because the last y, when attached, will exceed mx's. Hence, for these sequences of mx's and $n - 1y$'s, $W = k - m$. Putting the two cases together yields the equation for $\bar{P}_{m,n}$.

Theorem 3.2.3. Under $H_0 : \Delta = 0$ let $P_{m,n}(k) = P_{H_0}(W = k)$. Then

$$P_{m,n}(k) = \frac{n}{m+n} P_{m,n-1}(k-m) + \frac{m}{m+n} P_{m-1,n}(k)$$

with the boundary conditions given in the previous theorem.

Proof. Since the $\binom{m+n}{m}$ sequences of mx's and ny's are equally likely under $H_0 : \Delta = 0$, we have, from Theorem 3.2.2,

$$P_{m,n}(k) = \frac{\bar{P}_{m,n}(k)}{\binom{m+n}{m}}$$

$$= \frac{m! \, n!}{(m+n)!} \left[\bar{P}_{m,n-1}(k-m) + \bar{P}_{m-1,n}(k) \right]$$

$$= \frac{n}{m+n} \frac{\bar{P}_{m,n-1}(k-m)}{\binom{m+n-1}{m}} + \frac{m}{m+n} \frac{\bar{P}_{m-1,n}(k)}{\binom{m+n-1}{n}}$$

$$= \frac{n}{m+n} P_{m,n-1}(k-m) + \frac{m}{m+n} P_{m-1,n}(k).$$

This equation can be used for computer generation of tables of probabilities of W (or U). A similar equation for the Wilcoxon signed rank statistic is given in Exercise 3.7.3.

In the next theorem we use the Projection Theorem 2.5.2 to develop the limiting distribution of W under H_0 so that critical values can be approximated.

Theorem 3.2.4. Suppose $m, n \to \infty$ in such a way that $m/N \to \lambda$, $0 < \lambda < 1$, where $N = m + n$. Then $(W - EW)/(\mathrm{Var}\, W)^{1/2}$ and $(U - EU)/(\mathrm{Var}\, U)^{1/2}$ have limiting $n(0, 1)$ distributions under $H_0 : \Delta = 0$.

Proof. Let $T_{ij} = 1$ if $Y_j > X_i$ and 0 otherwise, then $W = \sum\sum T_{ij}$. Under H_0, $ET_{ij} = 1/2$, so we will let $W^* = \sum\sum(T_{ij} - 1/2)$ and note that $EW^* = 0$.

Next note that $E[(T_{ij} - 1/2)|X_k = x] = P(Y_j > X_i | X_k = x) - 1/2$ and

$$E\left[(T_{ij} - \tfrac{1}{2}) | X_k = x \right] = \begin{cases} 0 & \text{if } k \neq i \\ P(Y > x) - \tfrac{1}{2} & \text{if } k = i \end{cases}$$

and

$$E\left[(T_{ij} - \tfrac{1}{2}) \mid Y_k = y\right] = \begin{cases} 0 & \text{if } k \neq j \\ P(y > X) - \tfrac{1}{2} & \text{if } k = j. \end{cases}$$

Since X and Y have the same distribution F, under H_0, we have:

$$E\left[\sum\sum(T_{ij} - 1/2) \mid X_k = x\right] = n\{[1 - F(x)] - 1/2\}$$

$$E\left[\sum\sum(T_{ij} - 1/2) \mid Y_k = y\right] = m\{F(y) - 1/2\}.$$

The projection of W^* is

$$V_p = n\sum_{i=1}^{m}\left[1/2 - F(X_i)\right] + m\sum_{j=1}^{n}\left[F(Y_j) - 1/2\right].$$

Consider $(N^{1/2}/mn)V_p = (N^{1/2}/m)\sum_1^m V_i + (N^{1/2}/n)\sum_1^n V_i^*$ where V_i (and V_i^*) is uniformly distributed on $(-1/2, 1/2)$ with mean 0 and variance $1/12$. We can apply the Central Limit Theorem to each of the two terms on the right side. For example $(N^{1/2}/m)\sum_1^m V_i \overset{D}{\to} (1/\lambda)^{1/2}Z_1$, $Z_1 \sim n(0, 1/12)$. Now note that if $V_n \overset{D}{\to} Z_1$, $V_n^* \overset{D}{\to} Z_2$ and V_n, V_n^* are independent, then the characteristic function of $V_n + V_n^*$ converges to the characteristic function of $Z_1 + Z_2$. Hence

$$\frac{\sqrt{N}}{mn}V_p \overset{D}{\to} \sqrt{\frac{1}{\lambda}}\,Z_1 + \sqrt{\frac{1}{(1-\lambda)}}\,Z_2$$

where Z_1, Z_2 are independent $n(0, 1/12)$, and

$$\frac{\sqrt{N}}{mn}V_p \overset{D}{\to} Z \sim n\left(0, 1/[12\lambda(1-\lambda)]\right).$$

Further, $\text{Var}(N^{1/2}V_p/mn) \to 1/[12\lambda(1-\lambda)]$. From (3.2.3),

$$\text{Var}(N^{1/2}W^*/mn) = N(N+1)/(12mn) \to 1/[12\lambda(1-\lambda)].$$

From the Projection Theorem 2.5.2, (2.5.7),

$$E\left(\sqrt{N}\,W^*/(mn) - \sqrt{N}\,V_p/(mn)\right)^2 \to 0,$$

so by Theorem 2.5.3, $N^{1/2}W^*/(mn)$ has the same normal limiting distribution as $N^{1/2}V_p/(mn)$.

The proof is completed by writing out $[W - EW]/(\operatorname{Var} W)^{1/2}$ and applying Slutsky's Theorem A3 (in the Appendix) along with the limiting normality of $N^{1/2}W^*/(mn)$.

Hence $P(W \leqslant w) \doteq \Phi(t)$, where $t = (w + 0.5 - mn/2)/[mn(m + n + 1)/12]^{1/2}$. The standardized value t contains a continuity correction. The approximation can be improved by using an Edgeworth expansion similar to the one presented for the Wilcoxon signed rank statistic in (2.2.11). Fix and Hodges (1955) show that

$$P(W \leqslant w) \doteq \Phi(t) + \left[\frac{m^2 + n^2 + mn + m + n}{20mn(m + n + 1)} \right](t^3 - 3t)\psi(t),$$

where $\psi(\cdot)$ is the $n(0, 1)$ pdf. Fix and Hodges present a table comparing the normal approximation to the Edgeworth approximation. The simple normal approximation, with continuity correction, is adequate for most purposes. Bickel (1974) discusses the error of the Edgeworth approximation.

For testing, it is more convenient to use $U = \sum_1^n R_i$ since it only requires ranking (ordering) $m + n$ observations. The counting form $W = \#(Y_i - X_j > 0)$ requires the computation of mn differences, which can become unmanageable for moderate values of m and n. The counting form is important for the estimation of Δ.

Note that $W = \#(Y_i - X_j > 0)$ $i = 1, \ldots, n$, $j = 1, \ldots, m$, is the sign statistic computed on the mn differences. Hence, following the discussion of Section 1.5, the Hodges–Lehmann estimate of Δ is

$$\hat{\Delta} = \operatorname*{med}_{i,j}(Y_i - X_j). \qquad (3.2.4)$$

Furthermore, if $P_{H_0}(W \leqslant k) = \alpha/2$, then since the distribution of W is symmetric under $H_0: \Delta = 0$ from Exercise 3.7.5, we have

$$[D_{(k+1)}, D_{(mn-k)}) \qquad (3.2.5)$$

is a $(1 - \alpha)$ 100% confidence interval for Δ where $D_{(1)} \leqslant \cdots \leqslant D_{(mn)}$ are the ordered differences $Y_i - X_j$, $i = 1, \ldots, n$, $j = 1, \ldots, m$. Using the normal approximation, k in (3.2.5) can be approximated (using a continuity correction) by

$$k = \frac{mn}{2} - 0.5 - Z_{\alpha/2}\sqrt{\frac{mn(m + n + 1)}{12}} \qquad (3.2.6)$$

where $Z_{\alpha/2}$ is the upper $\alpha/2$ percentile of the standard normal distribution.

The Minitab statistical computing system contains commands that compute the Mann–Whitney–Wilcoxon test, point estimate, and confidence interval.

Example 3.2.2.   We return to the discussion of Example 3.1.2. The babies were separated into three groups according to birth weight. We consider the group of larger babies with birth weight of at least 3510 grams. There were $n = 20$ babies in the treatment group (Y) and $m = 36$ in the control group (X). The data is given in Table 3.1. The data was reconstructed from a dot graph in the article by Salk (1973).

For testing $H_0 : \Delta = 0$ versus $H_A : \Delta > 0$, $\alpha = .05$ was specified and a normal approximation was used to determine the critical value $c = n(m + n + 1)/2 + 0.5 + 1.645[mn(m + n + 1)/12]^{1/2} = 666.7$, to use with $U = \sum_1^{20} R_i$, where $R_1, \ldots, R_{20}$ are the ranks of the treatment observations in the combined sample. For the data in Table 3.1 we have $U = 762.5 > c$; hence we reject $H_0 : \Delta = 0$ at $\alpha = .05$ and declare the treatment effect to be significant. The value of U was determined by assigning the average rank to tied observations.

To assess the magnitude of the treatment effect we constructed a point and interval estimate of Δ. There are $mn = 720$ differences, $Y_j - X_i$, $j = 1, \ldots, 20$, $i = 1, \ldots, 36$, and $\hat{\Delta} = \text{med}(Y_j - X_i) = 60.0$. From (3.2.6), using $Z_{\alpha/2} = 1.96$, $k = 244.8$. If we use $k = 244$, then $[D_{(245)}, D_{(476)}) = [29.9, 100.0)$ is a slightly more than 95% confidence interval for Δ. Note how far the confidence interval misses 0, reflecting a strong rejection of $H_0 : \Delta = 0$. The computations were carried out using the Mann command in Minitab. See Ryan et al. (1981).

It should be noted that ties are generally broken by assigning the average rank to all observations in the group of tied observations. This corresponds to assigning $\frac{1}{2}$ to $Y - X$ differences that are zero. This is called the midrank method and is discussed at some length by Lehmann (1975, Chapter 1, Section 4). See also Putter (1955).

We can also study the behavior of the true significance level of W on serially correlated data. Similar to the serial correlation model introduced at the end of Section 1.7 and discussed again at the end of Section 2.7, let (X_i, X_{i+1}), $i = 1, 2, \ldots$, have a bivariate normal distribution with means 0, variances 1, and correlation ρ_x. Likewise, define a sequence (Y_i, Y_{i+1}), $i = 1, 2, \ldots$, independent from the X sequence, for which the bivariate normal distribution has means 0, variances 1, and correlation ρ_y. The results of Serfling (1968b) show that the projection V_p, given in the proof of Theorem 3.2.4, still determines the asymptotic distribution of W. In particular, using Theorem A16, $(N^{1/2}/mn)V_p$ converges in distribution to $(1/\lambda)^{1/2}Z_1 + [1/(1 - \lambda)]^{1/2}Z_2$ where Z_1 and Z_2 are independent normal random variables with distributions $n(0, (1/12) + (1/\pi)\sin^{-1}(\rho_x/2))$ and

Table 3.1. Weight Gains for the Large Babies

Treatment (n = 20)	Control (m = 36)
190.	140.
80.	100.
80.	100.
75.	70.
50.	25.
40.	20.
30.	10.
20.	0.
20.	−10.
10.	−10.
10.	−25.
10.	−25.
0.	−25.
0.	−30.
−10.	−30.
−25.	−30.
−30.	−45.
−45.	−45.
−60.	−45.
−85.	−50.
	−50.
	−50.
	−60.
	−75.
	−75.
	−85.
	−85.
	−100.
	−110.
	−130.
	−130.
	−155.
	−155.
	−180.
	−240.
	−290.

$n(0, (1/12) + (1/\pi)\sin^{-1}(\rho_y/2))$, respectively. See (2.7.11). Then, as in Theorem 3.2.4, $(N^{1/2}/mn)(W - EW)$ is asymptotically $n(0, [12\lambda(1 - \lambda)]^{-1} + (\lambda\pi)^{-1}\sin^{-1}(\rho_x/2) + [(1 - \lambda)\pi]^{-1}\sin^{-1}(\rho_y/2))$. Hence, as in the case of the Wilcoxon signed rank statistic T, the Mann–Whitney–Wilcoxon statistic W is no longer distribution free, even asymptotically, in the presence of serial correlation. The true level of W is affected in the same way as that of T.

3.3. THE DISTRIBUTION OF RANKS UNDER ALTERNATIVES

In this section we consider the problem of computing the joint probabilities of the ranks in general. Although the results can be used for computing exact power of rank tests, our main goal is the construction of locally most powerful rank tests for a given underlying model. These locally optimal tests can then be compared to two-sample versions of the asymptotically most power rank tests discussed after Theorem 2.9.1 for the one-sample rank tests.

Let $X_1, \ldots, X_m$ and $Y_1, \ldots, Y_n$ be random samples from arbitrary, absolutely continuous distributions denoted by $G(x)$ and $H(y)$, respectively, with densities $g(x)$ and $h(y)$. Let $R_{(1)} < \cdots < R_{(n)}$ denote the ranks of $Y_{(1)} < \cdots < Y_{(n)}$ in the combined data. The following result is due to Hoeffding (1951).

Theorem 3.3.1. Suppose $h(x) > 0$ implies $g(x) > 0$. Then

$$P(R_{(1)} = r_1, \ldots, R_{(n)} = r_n) = \frac{1}{\binom{m+n}{m}} E\left[\prod_{i=1}^{n} \frac{h(V_{(r_i)})}{g(V_{(r_i)})} \right]$$

where $V_{(1)} < \cdots < V_{(m+n)}$ are the order statistics of a sample of size $m + n$ from G.

Proof. We first consider the event $R_{(1)} = r_1, \ldots, R_{(n)} = r_n$ given $y_{(1)} < \cdots < y_{(n)}$. Then there are $r_1 - 1$ x's less than $y_{(1)}$ and $r_2 - r_1 - 1$ x's between $y_{(1)}$ and $y_{(2)}$. Let $r_0 = 0$ and $r_{n+1} = m + n + 1$. The conditional distribution can be written as a multinomial probability by thinking of tossing the m x's into the $n + 1$ cells defined by the fixed, ordered y's. Hence

$$P(R_{(1)} = r_1, \ldots, R_{(n)} = r_n | y_{(1)} < \cdots < y_{(n)})$$

$$= \frac{m!}{\prod_{j=0}^{n} (r_{j+1} - r_j - 1)!} G(y_{(1)})^{r_1 - 1} [G(y_{(2)}) - G(y_{(1)})]^{r_2 - r_1 - 1} \cdots$$

$$\times [1 - G(y_{(n)})]^{m+n-r_n}$$

where, for example, the probability that an x falls into the second cell is $P(y_{(1)} < X < y_{(2)}) = G(y_{(2)}) - G(y_{(1)})$. Now multiply by $n! \prod h(y_{(i)})$, the joint density of $Y_{(1)} < \cdots < Y_{(n)}$, and integrate with respect to $y_{(1)}$, $\ldots, y_{(n)}$ to get the marginal distribution of $R_{(1)} < \cdots < R_{(n)}$. In the following we also multiply and divide by $(m + n)! \prod g(y_{(i)})$.

$$P(R_{(1)} = r_1, \ldots, R_{(n)} = r_n)$$

$$= \int \cdots \int \frac{m! \, n!}{(m + n)!} \frac{\prod h(y_{(i)})}{\prod g(y_{(i)})}$$

$$\times \left\{ \frac{(m + n)!}{\prod (r_{j+1} - r_j - 1)!} G(y_{(1)})^{r_1 - 1} \cdots \right.$$

$$\left. \times [1 - G(y_{(n)})]^{m + n - r_n} \prod g(y_{(i)}) \right\} dy_{(1)} \cdots dy_{(n)}$$

Recognize the function in $\{ \ \}$ as the joint marginal density of $V_{(r_1)} < \cdots < V_{(r_n)}$ from $V_{(1)} < \cdots < V_{(m+n)}$; see Wilks (1962, p. 237). Hence the integral can be rewritten as the expectation stated in the theorem.

This result has been used for the calculation of probabilities in various special cases. For most distributions of interest, the expression is intractable. See Lehmann (1953) or Hayman and Govindarajulu (1966). We use the equation to develop locally optimal rank tests in the location model. The corresponding result for the one sample model is given in Exercise 3.7.18. In Exercise 3.7.20 locally optimal rank tests in the scale model are discussed.

Restrict attention to the location model with $G(x) = F(x)$ and $H(y) = F(y - \Delta)$, $F \in \Omega_0$. Under $H_0 : \Delta = 0$, the distribution of $R_{(1)} < \cdots < R_{(n)}$ is uniform over the $\binom{m+n}{m}$ equally likely sequences. This can be seen immediately from Theorem 3.3.1 with $g = h$. Hence, if k is an integer such that $k / \binom{m+n}{m} = \alpha$, then any set C of k rank vectors $(r_1, \ldots, r_n)$ is a size α critical region for $H_0 : \Delta = 0$. Our problem is to determine the best critical region under some stated criterion.

The power of a size α critical region C is given by

$$\beta(\Delta) = \sum_{(r_1, \ldots, r_n) \in C} \frac{1}{\binom{m+n}{m}} E\left[\prod_1^n \frac{f(V_{(r_i)} - \Delta)}{f(V_{(r_i)})} \right] \tag{3.3.1}$$

where $V_{(1)} < \cdots < V_{(m+n)}$ are order statistics of a sample of size $m + n$ from $F(x)$.

Definition 3.3.1. The locally most powerful size α rank test (LMPRT) is given by the size α critical region C^* such that $\beta'(0)$ is a maximum. Here $\beta'(0)$ is the derivative of $\beta(\Delta)$ evaluated at $\Delta = 0$.

This definition differs somewhat from the usual definition of a LMPRT; see Lehmann (1959). The more stringent definition requires the rank test to be uniformly most powerful among all rank tests in a sufficiently small neighborhood of $\Delta = 0$. In our definition we only require the slope of the power function be maximized at $\Delta = 0$; hence our LMPRT has the most rapidly increasing power function at the null hypothesis. Unless the situation is pathological, by continuity of the power function, the LMPRT of Definition 3.3.1 will be optimal in a neighborhood. In the next theorem we suppose that F has a differentiable density, and derivatives with respect to Δ may be passed through expectations.

Theorem 3.3.2. Given $F \in \Omega_0$, suppose differentiation under the expectation is valid, then the LMPRT rejects $H_0 : \Delta = 0$ in favor of $H_A : \Delta > 0$ if

$$V = -\sum_{j=1}^{n} E\left[\frac{f'(V_{(r_i)})}{f(V_{(r_i)})} \right] \geqslant c$$

where c is determined by $P_{H_0}(V \geqslant c) = \alpha$.

Proof. From (3.3.1), differentiating under the expectation yields:

$$\frac{d}{d\Delta} \beta(\Delta) = \frac{1}{\binom{m+n}{m}} \sum_C E\left\{ \frac{\sum_i \left[-f'(V_{(r_i)} - \Delta) \prod_{j \neq i} f(V_{(r_j)} - \Delta) \right]}{\prod f(V_{(r_i)})} \right\}$$

and

$$\beta'(0) = \frac{1}{\binom{m+n}{m}} \sum_C \left(-\sum_i E\left\{ \frac{f'(V_{(r_i)})}{f(V_{(r_i)})} \right\} \right) \tag{3.3.2}$$

where C is any set of k rank vectors $(r_1, \ldots, r_n)$ such that $k / \binom{m+n}{m} = \alpha$.

To maximize $\beta'(0)$ we build up C by including those rank vectors $(r_1, \ldots, r_n)$ which yield the k largest values of $V = -\sum E\{f'(V_{(r_i)}) / f(V_{(r_i)})\}$. Hence we can find a constant c such that $V \geqslant c$ yields the size α critical region that maximizes $\beta'(0)$, and this completes the proof.

Suppose we generate a score $a(i)$ by defining

$$a(i) = -E\left\{\frac{f'(V_{(i)})}{f(V_{(i)})}\right\} \tag{3.3.3}$$

where $V_{(1)} < \cdots < V_{(m+n)}$ are the order statistics from $F(x)$. Then the LMPRT is provided by the statistic $V = \sum_{i=1}^{n} a(R_i)$ where $R_1, \ldots, R_n$ are the ranks of the Y observations in the combined data. We now consider three examples and then turn to a discussion of general score statistics for the two sample location model.

Example 3.3.1. Suppose $f(x)$ is the standard normal density; hence $-f'(x)/f(x) = x$. The LMPRT is determined by $V = \sum_{1}^{n} a(R_i)$, called the normal scores statistic, where $a(i) = E(V_{(i)})$ and $V_{(1)} < \cdots < V_{(m+n)}$ are the $m + n$ order statistics from the $n(0, 1)$ distribution. For selected sample sizes there are tables of expected values of normal order statistics available for use in the computation of V on specific data; see Fisher and Yates (1938), who initially proposed the test. A natural approximation to $E(V_{(i)})$ is $\Phi^{-1}(i/(m + n + 1))$ where $\Phi(\cdot)$ is the standard normal distribution function. Hence an approximate normal scores statistic is given by $V^* = \sum_{1}^{n}\Phi^{-1}[R_i/(N + 1)]$, first proposed by van der Waerden (1952). Compare V^* to the one-sample statistic in Example 2.8.1. Because of the smoothness of $\Phi(\cdot)$, V and V^* are quite close. Computer packages may return V^* as the normal scores statistic. A simple equation, accurate to four decimal places, for computing the normal scores (see Example 2.8.1) is given by

$$EV_{(i)} = 4.91\left[p^{0.14} - (1 - p)^{0.14}\right], \qquad p = \frac{i - 3/8}{n + 1/4}. \tag{3.3.4}$$

For additional discussion see the papers by Terry (1952) and Klotz (1964).

Example 3.3.2. Suppose $f(x)$ is the logistic density given by $f(x) = e^{-x}/(1 + e^{-x})^2$, $-\infty < x < \infty$. The distribution function is $F(x) = 1/(1 + e^{-x})$, $-\infty < x < \infty$, and it is easy to check that $f(x) = F(x)[1 - F(x)]$. Further, $-f'(x)/f(x) = 2F(x) - 1$. Applying (3.3.3) we have $a(i) = E[2F(V_{(i)}) - 1]$. Exercise 3.7.7 implies that $EF(V_{(i)}) = i/(m + n + 1)$, and hence

$$V = \sum_{1}^{n}\left[\frac{2R_i}{m + n + 1} - 1\right]$$

$$= \frac{2}{m + n + 1}\sum_{1}^{n}\left[R_i - \frac{m + n + 1}{2}\right].$$

We thus see that the LMPRT is determined by the Mann–Whitney–Wilcoxon rank score test when the underlying distribution is logistic.

Example 3.3.3. Suppose $f(x)$ is the double exponential density and $f(x) = e^{-|x|}/2$, $-\infty < x < \infty$. Except at $x = 0$ where the derivative is not defined, $-f'(x)/f(x) = \text{sgn}(x)$. Hence the LMPRT is determined by $a(i) = E \, \text{sgn} \, V_{(i)}$, with $V_{(1)} < \cdots < V_{(m+n)}$ the order statistics from $f(x)$. As in Example 3.3.1, we need tables of values of the expected signs of the double exponential order statistics and these are less numerous than normal scores in the literature. Some tables can be found in Govindarajulu and Eisenstat (1965). We can also consider an approximation to $a(i)$ by passing the expectation through the function. Note that $x > 0$ if and only if $2F(x) - 1 > 0$, and hence

$$V = \sum E \, \text{sgn} \, V_{(R_i)}$$

$$= \sum E \, \text{sgn} \big[2F(V_{(R_i)}) - 1 \big].$$

Now define

$$V^* = \sum \text{sgn} \big[2EF(V_{(R_i)}) - 1 \big]$$

$$= \sum \text{sgn} \left[2 \frac{R_i}{m + n + 1} - 1 \right]$$

$$= \sum \text{sgn} \left[R_i - \frac{m + n + 1}{2} \right].$$

Now note that $\text{sgn}[R_i - (m + n + 1)/2] = 1$ if and only if $R_i > (m + n + 1)/2$ if and only if Y_i is greater than the median of the combined sample. Let V_+^* (V_-^*) be the number of Y observations greater (less) than the median of the combined sample, then $V^* = V_+^* - V_-^*$. Finally, when $m + n$ is even, $V_+^* + V_-^* = n$, $V^* = 2V_+^* - n$. A bit more care is needed for $m + n$ odd but it does not matter much for large samples. The statistic V_+^* is called Mood's median statistic and was discussed by Mood (1950). The statistic V_+^* provides an approximation to V, the LMPRT. In this case, however, the expectation was passed through the nonsmooth sign function. This can result in some loss in power for small sample sizes. Conover et al. (1978) show the LMPRT is a bit better than V_+^* for small samples even at the double exponential distribution. Asymptotically they can be shown to be equivalent and in practice most people will use V_+^* since tables for the locally most powerful scores are not readily available. Later in this chapter we will see that Mood's test is a two-sample analog of the sign test since it has the same two-sample test efficiency as the sign test in the one sample model.

In general, if $a(i) = -E[f'(V_{(i)})/f(V_{(i)})]$, the score for the LMPRT, an approximate score can be defined by

$$a^*(i) = -\frac{f'(F^{-1}[i/(m+n+1)])}{f(F^{-1}[i/(m+n+1)])} \tag{3.3.5}$$

where we first approximate $V_{(i)}$ by $F^{-1}(U_{(i)})$ where $U_{(1)} < \cdots < U_{(m+n)}$ are the order statistics from the uniform distribution on $(0,1)$ and then we pass the expectation through the function and note $EU_{(i)} = i/(m+n+1)$ from Exercise 3.7.7. Hence, in addition to the Mann–Whitney–Wilcoxon statistic of Section 3.2, by selecting different density functions f, we can generate a multitude of rank tests. Generally the approximation $V^* = \sum a^*(R_i)$ is more practical than the actual LMPRT, $V = \sum a(R_i)$. In the next section we study the distribution of general scores statistics under the null hypothesis. The limiting normality is established so that critical points for the tests can be approximated.

3.4. GENERAL SCORES

In this section we outline the major results for the asymptotic distribution theory under the null hypothesis. The situation is more complicated than that of the one-sample location model where independence implied the Central Limit Theorem was sufficient to establish asymptotic normality of the general scores statistic.

In the two-sample model we no longer have the required independence, and it is necessary to resort to projection techniques. The basic results are proved by Hajek and Sidak (1967, Chapter V). They were able to prove a two-sample analog to Theorem 2.8.1 and it is described later. Before stating these results we develop the moments of the general scores statistics.

As usual, we have m X observations and n Y observations with R_1, $\ldots, R_n$ the ranks of $Y_1, \ldots, Y_n$ in the combined data. Similar to Definition 2.8.1, we have:

Definition 3.4.1. Let $0 = a(0) \leqslant a(1) \leqslant \cdots \leqslant a(N)$, $N = m + n$, be a nonconstant sequence, then $V = \sum_1^n a(R_i)$ is called a general scores statistic. We may also write $a_i = a(i)$.

Theorem 3.4.1. Under the null hypothesis $H_0 : \Delta = 0$, in the location model,

$$EV = \frac{n}{N} \sum_1^N a_i = n\bar{a}$$

$$\text{Var } V = \frac{mn}{N(N-1)} \sum_1^N (a_i - \bar{a})^2$$

Proof. First note $EV = \sum_1^n Ea(R_i) = nEa(R_1)$. Now, from Theorem 3.2.1, $Ea(R_1) = \sum_1^N a(i)P(R_1 = i) = \sum_1^N a_i/N$. Further, $Ea^2(R_1) = \sum_1^N a_i^2/N$ and $\text{Var } a(R_1) = Ea^2(R_1) - [Ea(R_1)]^2 = \sum_1^N a_i^2/N - \bar{a}^2 = \sum_1^N (a_i - \bar{a})^2/N$. From Theorem 3.2.1

$$\text{Cov}(a(R_1), a(R_2)) = E\left[(a(R_1) - \bar{a})(a(R_2) - \bar{a}) \right]$$

$$= \sum\sum_{i \neq j} \frac{1}{N(N-1)} (a_i - \bar{a})(a_j - \bar{a}).$$

Since $\sum_{i \neq j}(a_i - \bar{a}) = \sum_i (a_i - \bar{a}) - (a_j - \bar{a}) = -(a_j - \bar{a})$, we have

$$\text{Cov}(a(R_1), a(R_2)) = -\frac{1}{N(N-1)} \sum_1^N (a_j - \bar{a})^2$$

and

$$\text{Var } V = \text{Var} \sum_1^n a(R_i)$$

$$= n \text{Var}\left[a(R_1) \right] + \frac{2n(n-1)}{2} \text{Cov}(a(R_1), a(R_2))$$

$$= \frac{n}{N} \sum_1^N (a_i - \bar{a})^2 - \frac{n(n-1)}{N(N-1)} \sum_1^N (a_i - \bar{a})^2,$$

which combines to yield the desired result.

In addition, in Exercise 3.7.4 it is pointed out that V has a symmetric distribution under the null hypothesis, provided $a_i + a_{N-i+1}$ has the same value for all $i = 1, \ldots, N$. The point of symmetry is $n\bar{a}$.

We now introduce the score-generating function for the two-sample model.

Definition 3.4.2. Suppose $\phi(u)$, $0 < u < 1$, is nondecreasing. Suppose further that $0 < \int_0^1 (\phi(u) - \bar{\phi})^2 du < \infty$ where $\bar{\phi} = \int_0^1 \phi(u) du$. Define $a(i) = \phi(i/(N+1))$, $N = m + n$, then $\bar{V} = \sum_1^n a(R_i)/N$ is the general scores statistic generated by the scores-generating function.

Hajek and Sidak (1967, p. 163) prove that if V satisfies Definition 3.4.2 and if $\min(m,n) \to \infty$, then $[\bar{V} - E\bar{V}]/(\text{Var } \bar{V})^{1/2}$ has a standard normal limiting distribution. Now we suppose that $N \to \infty$ and $m/N \to \lambda$, $0 < \lambda < 1$, so that neither sample size dominates asymptotically. Then we have

$$E\bar{V} = \frac{n}{N} \bar{a} = \frac{n}{N} \cdot \frac{1}{N} \sum_1^N \phi\left(\frac{i}{N+1}\right) \to (1 - \lambda)\bar{\phi}, \qquad (3.4.1)$$

and

$$N \text{ Var } \bar{V} = \frac{mn}{N^2(N-1)} \sum_1^N (a_i - \bar{a})^2 \to \lambda(1 - \lambda) \int_0^1 \left(\phi(u) - \bar{\phi}\right)^2 du. \quad (3.4.2)$$

Hence the asymptotic normality can be expressed in terms of the asymptotic parameters as:

$$\frac{\sqrt{N}\left(\bar{V} - (1 - \lambda)\bar{\phi}\right)}{\sqrt{\lambda(1-\lambda)\int\left(\phi(u) - \bar{\phi}\right)^2 du}} \qquad (3.4.3)$$

has a standard normal limiting distribution. For additional discussion and development of this result see Chapter 8 of Randles and Wolfe (1979).

If we let $\phi(u) = -f'(F^{-1}(u))/f(F^{-1}(u))$, then, provided f has finite Fisher information, Definition 2.9.1, Hajek and Sidak (1967) show that Definition 3.4.2 is satisfied. Hence, from (3.3.5), $V^* = \sum a^*(R_i)$, the approximation to the LMPRT, is asymptotically normally distributed. The scores (3.3.3) which define the LMPRT V do not have a score-generating function in the sense of Definition 3.4.2. However, Hajek and Sidak (1967, p. 165) show that V and V^*, when properly standardized, have the same limiting distribution. In fact, they have a stronger result that corresponds directly to (3.4.1), (3.4.2), and (3.4.3). Namely, if ϕ satisfies Definition 3.4.2 and if we define $a_i = E\phi(U_{(i)})$ where $U_{(1)} < \cdots < U_{(N)}$ are the order statistics from a uniform distribution on $(0,1)$ then (3.4.1)–(3.4.3) hold for $\bar{V} =$

$\sum_1^n a(R_i)/N$. The point is that we can use either $E\phi(U_{(i)})$ or $\phi(EU_{(i)})$ to define the statistic $\overline{V}$ and it does not matter asymptotically.

Example 3.4.1. Let $\phi(u) = \Phi^{-1}(u)$, where $\Phi(\cdot)$ is the standard normal distribution function. Then

$$\overline{V} = \frac{1}{N} \sum_1^n \Phi^{-1}\left(\frac{R_i}{N+1}\right)$$

is the approximation to the LMPRT for underlying normal distributions discussed in Example 3.3.1. We now check the square integrability condition in Definition 3.4.2. Note

$$\overline{\phi} = \int_0^1 \Phi^{-1}(u)\,du$$

$$= \int_{-\infty}^\infty t\,d\Phi(t) = 0,$$

hence

$$\int_0^1 \left(\phi(u) - \overline{\phi}\right)^2 du = \int_0^1 \left[\Phi^{-1}(u)\right]^2 du$$

$$= \int_{-\infty}^\infty t^2\,d\Phi(t) = 1.$$

Thus, if we replace λ by m/N and $1 - \lambda$ by n/N, we have $(N^3/mn)^{1/2}\overline{V}$ is approximately $n(0, 1)$. This approximation, used in conjunction with (3.3.4) for computing $\overline{V}$, makes the normal scores test practical.

Example 3.4.2. We now consider Mood's statistic, first mentioned in Example 3.3.3. The statistic V_+^*, the number of Y observations that exceed the median of the combined sample, can be written $V_+^* = \sum_1^n \phi(R_i/(N+1))$ where

$$\phi(u) = \begin{cases} 0 & 0 < u \leqslant 1/2 \\ 1 & 1/2 < u < 1. \end{cases}$$

We consider the case $N = m + n = 2r$, $n \leqslant m$ in detail. Since

$$a_i = \phi(i/(N+1)) = 0$$

if $i = 1, 2, \ldots, r$ and 1 if $i = r + 1, \ldots, N$ it is easy to check

$$EV_+^* = n\bar{a} = \frac{nr}{N} = \frac{n}{2}$$

$$\text{Var } V_+^* = \frac{mn}{N(N-1)} \sum_1^N (a_i - \bar{a})^2 = \frac{mn}{4(N-1)} \, .$$

With $n \leqslant m$, the exact distribution under the null hypothesis is hypergeometric with

$$P(V_+^* = k) = \frac{\binom{r}{k}\binom{r}{n-k}}{\binom{m+n}{m}}$$

for $k = 0, 1, \ldots, n$. Further, since $\phi(\cdot)$ is bounded, Definition 3.4.2 applies and $[V_+^* - n/2]/[mn/4(m+n-1)]^{1/2}$ is limiting $n(0, 1)$.

In Section 2.8 we introduced several examples of one-sample scores statistics; recall Examples 2.8.2 and 2.8.3. We now discuss the construction of two-sample scores statistics from one-sample scores statistics.

Example 3.4.3. Suppose we have a one-sample score-generating function $\phi^+(u)$, $0 < u < 1$, satisfying Definition 2.8.2. We will let $\bar{V}^+$ denote the one-sample statistic defined in (2.8.4). Now extend $\phi^+(u)$ to $(-1, 1)$ by $\phi^+(-u) = -\phi^+(u)$ and define $\phi(u) = \phi^+(2u - 1)$, $0 < u < 1$. Alternately, define $\phi(u) = \phi^+(2u - 1)$ if $1/2 < u < 1$ and $-\phi^+(1 - 2u)$ if $0 < u \leqslant 1/2$. Then $\bar{V}$ denotes the corresponding two-sample statistic. Further,

$$\bar{\phi} = \int_0^1 \phi(u)\,du$$

$$= \int_0^1 \phi^+(2u - 1)\,du$$

$$= \frac{1}{2} \int_{-1}^1 \phi^+(v)\,dv$$

$$= \frac{1}{2} \left\{ -\int_{-1}^0 \phi^+(-v)\,dv + \int_0^1 \phi^+(v)\,dv \right\}$$

$$= \frac{1}{2} \left\{ -\int_0^1 \phi^+(v)\,dv + \int_0^1 \phi^+(v)\,dv \right\} = 0$$

and

$$\int_0^1 \left(\phi(u) - \bar{\phi}\right)^2 du = \int_0^1 \left[\phi^+ (2u - 1)\right]^2 du$$

$$= \frac{1}{2} \int_{-1}^1 \left[\phi^+ (v)\right]^2 dv$$

$$= \frac{1}{2} \left\{ \int_{-1}^0 \left[\phi^+ (-v)\right]^2 dv + \int_0^1 \left[\phi^+ (v)\right]^2 dv \right\}$$

$$= \int_0^1 \left[\phi^+ (v)\right]^2 dv.$$

As an immediate application, let $\phi^+ (u) = u$, $0 < u < 1$, so that $\bar{V}^+$ is the Wilcoxon signed rank statistic. Then $\phi(u) = 2u - 1$, $0 < u < 1$, and $\bar{V}$ is the centered Mann–Whitney–Wilcoxon statistic. If we take $\phi^+ (u) = 1$, $0 < u < 1$, corresponding to $\bar{V}^+$, the sign statistic, then $\phi(u) = 1$ if $1/2 < u < 1$ and -1 if $0 < u \leqslant 1/2$, and hence $\bar{V}$ is the centered Mood's statistic. If $\phi^+ (u) = \Phi^{-1}[(u + 1)/2]$ the generating function for the one-sample normal scores statistic then $\phi(u) = \Phi^{-1}(u)$, generating the two-sample normal scores statistic. In Exercise 3.7.8, the two-sample statistics corresponding to the Winsorized signed rank statistics and the modified sign statistics of Examples 2.8.2 and 2.8.3 are discussed. Later we will see that $\bar{V}^+$ and $\bar{V}$ have the same efficiency properties. In particular, if $\bar{V}^+$ is the AMPRT (see the discussion following Theorem 2.9.1), then $\bar{V}$ will be asymptotically most powerful in the two-sample model (see Section 3.5).

The asymptotic theory of Hajek and Sidak (1967) discussed previously makes it possible to approximate the critical point for a broad range of two-sample tests. Provided the expectation in (3.3.3) can be found or is tabled, the critical point for the LMPRT can be approximated. The approximation to the locally most powerful score, (3.3.5), is one of several applications of the use of score-generating functions. The asymptotic normality is critical in the case of general scores since, unlike the Mann–Whitney–Wilcoxon statistic, the exact distribution, under the null hypothesis, is seldom tabled or available. In most cases the null distribution of the test statistic is symmetric, and the normal approximation provides a good approximation even for small sample sizes.

We next discuss a counting form for the two-sample general scores statistic and use this to construct a Hodges–Lehmann estimate of Δ. Recall from Section 3.2 that $U = \sum_1^n R_i = W + n(n + 1)/2$ where $W = \#(Y_i - X_j$

> 0), $i = 1, \ldots, n$, $j = 1, \ldots, m$. For the general scores statistic, define $V(\Delta) = \sum_1^n a(R_i(\Delta))$ where $R_i(\Delta)$ is the rank of $Y_i - \Delta$ among $X_1, \ldots, X_m$, $Y_1 - \Delta, \ldots, Y_n - \Delta$. In Exercise 3.7.9 you are asked to show that $V(\Delta)$ decreases by $a_{i+j} - a_{i+j-1}$ as Δ crosses $Y_{(j)} - X_{(i)}$. Hence the mn pairwise differences play the same role in the two-sample problem as the Walsh averages in the one-sample case. Further, if $T_{ij}(\Delta) = 1$ when $Y_{(j)} - X_{(i)} > 0$ and 0 otherwise, then

$$V(\Delta) = Q(\Delta) + \sum_1^n a_i \qquad (3.4.4)$$

where

$$Q(\Delta) = \sum_{i=1}^m \sum_{j=1}^n (a_{i+j} - a_{i+j-1}) T_{ij}(\Delta). \qquad (3.4.5)$$

Here $Q = W$ of Section 3.2 when $a_i = i$, $i = 1, \ldots, m + n$. Compare these definitions and results to those of Section 2.8. In the same spirit as Section 2.8, if we restrict attention to scores such that $a_{i+j} - a_{i+j-1} = 1/(N + 1)$ or 0, then it is simple to describe the estimate $\hat{\Delta}$. Let

$$B = \{(i, j) : a_{i+j} - a_{i+j-1} = 1/(N + 1)\}, \qquad (3.4.6)$$

then, when the distribution of V is symmetric (see Exercise 3.7.5),

$$\hat{\Delta} = \operatorname*{med}_{(i,j) \in B} \left[Y_{(j)} - X_{(i)} \right]. \qquad (3.4.7)$$

We complete this section with an application of these ideas to Mood's median statistic; additional examples can be found in the exercises.

Example 3.4.4. This example is a continuation of Example 3.4.2 on Mood's median statistic. We consider the case $N = m + n = 2r$, $n \leqslant m$. Let $a_i = 0$ for $i = 1, \ldots, r$ and 1 for $i = r + 1, \ldots, N$. Mood's statistic is $V_+^* = \sum_1^n a(R_i)$ and $a_{i+j} - a_{i+j-1} = 1$ only for $j + i = r + 1$. Hence, since $\sum_1^n a_i = 0$, $V_+^* = \#(Y_{(j)} - X_{(i)} > 0)$ for $j + i = r + 1$ and the estimate of Δ is

$$\hat{\Delta} = \operatorname*{med}_{j+i=r+1} \left[Y_{(j)} - X_{(i)} \right]$$

$$= \operatorname*{med}_{j=1, \ldots, n} \left[Y_{(j)} - X_{(r+1-j)} \right].$$

In this case, we can determine the ordering of the relevant differences by only knowing the order statistics for the individual samples. We simply note that:

$$Y_{(1)} - X_{(r)} < Y_{(2)} - X_{(r-1)} < \cdots < Y_{(n)} - X_{(r-n+1)} .$$

For the case $n = 2_{n^*} - 1$ and $m = 2_{m^*} - 1$, we see at once that $\hat{\Delta} = Y_{(n^*)} - X_{(r-n^*+1)} = Y_{(n^*)} - X_{(m^*)} = \text{med } Y_j - \text{med } X_i$. Recall that $\sum_1^n a_i = 0$, $V = Q$, and the counting form does not differ from the rank form. Hence, if $P_{H_0}(V_+^* \leqslant k) = \alpha/2$, determined from the hypergeometric distribution or the normal approximation, then the lower and upper ends of the $(1 - \alpha)$ 100% confidence interval for Δ are

$$\hat{\Delta}_L = Y_{(k+1)} - X_{(r-k)} \tag{3.4.8}$$

and

$$\hat{\Delta}_U = Y_{(n-k)} - X_{(r-n+k+1)} .$$

Recall the equivalence between hypothesis testing and confidence intervals. To test $H_0 : \Delta = 0$ versus $H_A : \Delta \neq 0$, we reject H_0 at level α if $V_+^* \leqslant k$ or $V_+^* \geqslant n - k$ where $P_{H_0}(V_+^* \leqslant k) = \alpha/2$. This is equivalent to rejecting H_0 at level α if 0 is not contained in $[\hat{\Delta}_L, \hat{\Delta}_U]$, the $(1 - \alpha)$ 100% confidence interval.

This formulation shows that when the observations are taken sequentially—that is, we observe the order statistics—we may be able to terminate Mood's test before observing all $m + n$ observations. This has applications in life testing in which time until death is observed. From the last paragraph, if $P_{H_0}(V_+^* \leqslant k) = \alpha/2$, then Mood's test rejects $H_0 : \Delta = 0$ if

$$0 < \hat{\Delta}_L = Y_{(k+1)} - X_{(r-k)} \quad \text{or} \quad X_{(r-k)} < Y_{(k+1)}$$

or

$$0 > \hat{\Delta}_U = Y_{(n-k)} - X_{(r-n+k+1)} \quad \text{or} \quad Y_{(n-k)} > X_{(r-n+k+1)} .$$

Hence, for example, if we observe $X_{(r-k)}$ before $Y_{(k+1)}$ we can terminate the experiment and reject $H_0 : \Delta = 0$. For more discussion see Gastwirth (1968). Comparison with another median test is given in Exercise 3.7.11. A discussion of early termination with the Mann–Whitney–Wilcoxon test is given by Alling (1963).

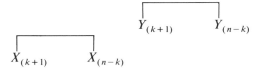

Figure 3.1. Sign-type confidence intervals for θ_x and θ_y.

We now relate this inference on Δ, based on Mood's procedures, to the sign procedures discussed in Chapter 1. We consider the case $m = n$ for simplicity; hence $m + n = 2n = 2r$ and $r = n$. We then have $\hat{\Delta}_L = Y_{(k+1)} - X_{(n-k)}$ and $\hat{\Delta}_U = Y_{(n-k)} - X_{(k+1)}$. This interval can be broken apart as follows: consider the separate intervals $[X_{(k+1)}, X_{(n-k)}]$ and $[Y_{(k+1)}, Y_{(n-k)}]$. Then $\hat{\Delta}_L$ is the difference in lower and upper ends of the Y and X intervals, respectively, likewise for $\hat{\Delta}_U$; see Fig. 3.1. A little thought and Fig. 3.1 shows that 0 is not contained in $[\hat{\Delta}_L, \hat{\Delta}_U]$ if and only if the two intervals $[X_{(k+1)}, X_{(n-k)}]$ and $[Y_{(k+1)}, Y_{(n-k)}]$ are disjoint. That is, Mood's test rejects $H_0 : \Delta = 0$ if and only if the intervals derived from the sign statistic are disjoint. Finally, we need the confidence coefficients for the sign intervals. This is determined by $P_{H_0}(S \leq k)$ where k is originally determined from $P_{H_0}(V_+^* \leq k) = \alpha/2$. For example, using normal approximations, if we take $1 - \alpha = 0.95$ so $k = n/2 - 2[n^2/4(2n - 1)]^{1/2}$ from Example 3.4.2 with equal sample sizes, then $P_{H_0}(S \leq k) \doteq \Phi(-2^{1/2}) = 0.08$. Hence, the sign intervals have approximately 84% confidence coefficients. In summary, for equal sample sizes, if we reject $H_0 : \Delta = 0$ when the two 84% confidence intervals are disjoint, then this is equivalent to a two-sided 5% Mood's median test. The Hodges–Lehmann estimate of $\Delta = \theta_y - \theta_x$ is the difference in medians, and a 95% confidence interval for Δ is found by taking the appropriate differences in the ends of the 84% sign confidence intervals. Exercise 3.7.14 outlines the case of unequal sample sizes.

Example 3.4.5. In this example we construct the Kolmogorov–Smirnov two-sample statistic.

First, define a score

$$a_j(i) = 0 \quad \text{if} \quad 1 \leq i \leq j$$
$$1 \quad \text{if} \quad j < i \leq N.$$

This score can be generated by $\phi_t(u) = 0$ if $0 < u \leq t$ and 1 if $t < u < 1$, where $j = [t(N + 1)]$ and $[\cdot]$ denotes the greatest integer function. Now let $V_j = \sum_{i=1}^{n} a_j(R_i)$. Note that if $N = 2r$ then $V_r = V_+^*$, Mood's statistic. Hence, V_j is the two-sample analog of the one-sample modified sign statistic described in Example 2.8.3. Exercise 3.7.8 asks for the centered version of V_j.

The statistic V_j counts the number of observations in the second (y) sample that have ranks greater than j. Under the null hypothesis that the two samples come from the same distribution, $EV_j = (N - j)n/N$. Let $Z_{(1)} \leqslant \cdots \leqslant Z_{(N)}$ denote the order statistics of the combined sample and let $I_i = 1$ if $Z_{(i)}$ is from the second (y) sample and 0 otherwise. Then we have

$$V_j = \sum_{i=j+1}^{N} I_i = n - \sum_{i=1}^{j} I_i .$$

Hence

$$V_j - EV_j = n - \sum_{i=1}^{j} I_i - n(N - j)/N = nj/N - \sum_{i=1}^{j} I_i .$$

We now introduce the empirical cdfs of the two samples, $G_m(x)$ and $H_n(y)$. If we evaluate their difference at $Z_{(j)}$, we have

$$G_m(Z_{(j)}) - H_n(Z_{(j)}) = \frac{1}{m} \sum_{i=1}^{j} (1 - I_i) - \frac{1}{n} \sum_{i=1}^{j} I_i = \frac{j}{m} - \left(\frac{1}{m} + \frac{1}{n} \right) \sum_{i=1}^{j} I_i$$

$$= \frac{m+n}{mn} \left[\frac{jn}{m+n} - \sum_{i=1}^{j} I_i \right] = \frac{N}{mn} \left[V_j - EV_j \right].$$

The Kolmogorov–Smirnov statistic is

$$D = \max_x |G_m(z) - H_n(z)| = \max_{1 \leqslant j \leqslant N} |G_m(Z_{(j)}) - H_n(Z_{(j)})|$$

$$= \frac{N}{mn} \max_{1 \leqslant j \leqslant N} |V_j - EV_j|.$$

The second equality follows from the fact that G_m and H_n are step functions with jumps at most at $Z_{(1)}, \ldots, Z_{(N)}$.

This shows that the Kolmogorov–Smirnov statistic is not a linear rank statistic. However, it can be expressed as the maximum of a finite number of linear rank statistics: the modified Mood's statistics. For that reason D is distribution free under the null hypothesis; the null distribution has been tabled. The null distribution of D is not symmetric and the limiting distribution is not normal. Hajek and Sidak (1967, Chapter V) show that

$$\lim_{N \to \infty} P\left(\sqrt{\frac{mn}{N}}\, D \leqslant t\right) = 1 - 2 \sum_{k=1}^{\infty} (-1)^{k+1} e^{-2k^2 t^2}.$$

The statistic D measures the discrepancy between the empirical cdfs. Hence the test is designed to reject $H_0 : G = H$ in favor of general alternatives $H_A : G \neq H$. This is in contrast to the rank tests studied previously which were designed for location alternatives. The test based on D is useful for detecting nonlocation alternatives and is also useful in situations in which Mood's test has high power. There are also one-sided versions of this test (see Exercise 3.7.19) and one-sample versions.

In the one-sample versions, the empirical cdf is compared to a theoretical cdf. If the theoretical cdf is parameterized, then the parameters can be estimated by minimizing the $\max_x |F_n(x) - F(x, \theta)|$. Parr (1981) published a bibliography on these methods.

In the two-sample case, Doksum (1977) reviewed several graphical methods for comparing two samples based on the Kolmogorov–Smirnov statistic. If $G(t) = F(t)$ and $H(t) = F(t - \Delta(t))$ then $\Delta(t)$ represents a general shift function and can be used to measure the difference between two populations in the absence of location and scale parameters. Doksum describes how to construct confidence bands for $\Delta(t)$, based on D.

3.5. ASYMPTOTIC DISTRIBUTION THEORY UNDER ALTERNATIVES

We first discuss the Mann–Whitney–Wilcoxon statistic W. The asymptotic distribution under the null hypothesis was derived in Theorem 3.2.4 using the Projection Theorem. The derivation of the asymptotic normality for general alternatives using the Projection Theorem is outlined in Exercise 3.7.15. The moments of W are derived later. Using the asymptotic normality under the null hypothesis and moments in general, we establish the consistency of the Mann–Whitney–Wilcoxon test and describe the consistency class as discussed in Section 1.3. For the case of unequal distribution shapes, a Behrens–Fisher type test is developed using W. Finally, we derive the Pitman efficacy of W and complete the section with a discussion of efficiency for the general scores statistics.

Let $X_1, \ldots, X_m$ and $Y_1, \ldots, Y_n$ be random samples from arbitrary, continuous distributions $G(x)$ and $H(y)$, respectively. Recall that $W = \sum\sum T_{ij}$, where $T_{ij} = 1$ if $Y_j - X_i > 0$ and 0 otherwise, is the counting form of the Mann–Whitney–Wilcoxon statistic. Next, define the following parameters:

$$p_1 = P(Y > X) = \int_{-\infty}^{\infty} G(y)h(y)\,dy = \int_{-\infty}^{\infty} \left[1 - H(x)\right] g(x)\,dx$$

$$p_2 = P(Y_1 > X_1, Y_2 > X_1) = \int_{-\infty}^{\infty} \left[1 - H(x)\right]^2 g(x)\,dx \qquad (3.5.1)$$

$$p_3 = P(Y_1 > X_1, Y_1 > X_2) = \int_{-\infty}^{\infty} G^2(y)h(y)\,dy.$$

Theorem 3.5.1. The mean and variance of W are given by

$$EW = mn p_1$$

$$\text{Var } W = mn\left(p_1 - p_1^2\right) + mn(n - 1)\left(p_2 - p_1^2\right) + mn(m - 1)\left(p_3 - p_1^2\right).$$

Proof. First note that $EW = mnET_{11} = mnP(Y > X)$, and $P(Y > X) = \int_{-\infty}^{\infty}\int_x^{\infty} g(x)h(y)\,dy\,dx = \int_{-\infty}^{\infty}[1 - H(x)]g(x)\,dx$. Now,

$$EW^2 = E\left[\sum_i \sum_j T_{ij}\right]^2 = E\left[\sum_i \sum_j \sum_k \sum_t T_{ij}T_{kt}\right]$$

$$= E\left\{ \sum_i \sum_j T_{ij}^2 + \sum_i \sum_{j \neq t} T_{ij}T_{it} \right.$$

$$\left. + \sum_j \sum_{i \neq k} T_{ij}T_{kj} + \sum_{i \neq j} \sum_{k \neq t} T_{ij}T_{kt} \right\}$$

$$= mnET_{11}^2 + mn(n - 1)ET_{11}T_{12} + mn(m - 1)ET_{12}T_{22}$$

$$+ mn(m - 1)(n - 1)ET_{11}T_{22}.$$

For example,

$$ET_{11}T_{12} = P(Y_1 > X_1, Y_2 > X_1)$$

$$= \int_{-\infty}^{\infty} \int_x^{\infty} \int_x^{\infty} g(x)h(y_1)h(y_2)\,dy_1\,dy_2\,dx$$

$$= \int_{-\infty}^{\infty} \left[1 - H(x)\right]^2 g(x)\,dx$$

$$= p_2.$$

The others are computed in a like manner. Then $\operatorname{Var} W = EW^2 - [EW]^2$ which results in the given equation, when simplfied.

Now, when $m, n \to \infty$ in such a way that $m/N \to \lambda$, $0 < \lambda < 1$, Exercise 3.7.15 shows that, provided $0 < p_1 < 1$, $[W - EW]/(\operatorname{Var} W)^{1/2}$ is asymptotically $n(0, 1)$. More precisely, if $\overline{W} = W/mn$, then we have from the exercise that $N^{1/2}(\overline{W} - p_1)$ is asymptotically $n(0, (p_2 - p_1^2)/\lambda + (p_3 - p_1^2)/(1 - \lambda))$.

Definition 3.5.1. We will say that $H(x)$ is stochastically larger than $G(x)$ if $G(x) \geqslant H(x)$ for all x, with strict inequality for at least one x. See Exercise 3.7.19 for a test for stochastic ordering.

Example 3.5.1. We now consider the consistency of the Mann–Whitney–Wilcoxon test. Define $\mu(G, H) = E\overline{W} = \int [1 - H(x)] g(x) \, dx$. Further, let Ω_{SO} denote the subclass of stochastically ordered distributions. Then

$$\mu(G, H) = \tfrac{1}{2} \quad \text{if} \quad G(x) = H(x)$$
$$> \tfrac{1}{2} \quad \text{if} \quad G, H \in \Omega_{SO} \qquad (3.5.2)$$

where the inequality occurs because G, H are continuous distribution functions with $G(x) > H(x)$ for some x and hence for an interval of x values. Hence $\overline{W}$ separates the null hypothesis of $G(x) = H(x)$ for all x from the subclass of stochastically ordered distributions. It is easy to check that the $\operatorname{Var} \overline{W}$ tends to 0 so that $\overline{W}$ converges in probability to $\mu(G, H)$. Further, we have the required asymptotic normality under the null hypothesis; so, Theorem 1.3.1 implies that $\overline{W}$ provides a consistent test for stochastically ordered alternatives. Note that if $G(x) = F(x)$ and $H(y) = F(y - \Delta)$, $\Delta > 0$, then $G, H \in \Omega_{SO}$. Hence, the test is consistent for a change in location. It should be noted that stochastic ordering is more general than a location change. Compare this to stochastically positive distributions in the one-sample model discussed in Example 2.4.1.

The sampling model that we have considered thus far supposes that the two populations have the same shape. We now consider the model $X_1, \ldots, X_m$ i.i.d. $G(x)$, $G \in \Omega_0$ and $Y_1, \ldots, Y_n$ i.i.d. $H(y - \Delta)$, $H \in \Omega_0$. Testing $H_0 : \Delta = 0$ provides a nonparametric analog of the Behrens–Fisher problem. When G and H are two normal distributions with different variances, Welch's (1937) modfication of the usual pooled two-sample t test performs quite well. We are now in a position to consider the problem for general $G, H \in \Omega_0$. The idea is to use the limiting normality of W for arbitrary G and H and introduce a consistent estimate of the variance of W.

From Exercise 3.7.15 we know that $(W - EW)/(\mathrm{Var}\, W)^{1/2}$ is limiting $n(0,1)$. When $G = H$ and $\Delta = 0$, EW and $\mathrm{Var}\, W$ are given by (3.2.3) and are independent of the common cdf $G = H$. We seek a consistent estimate of $\mathrm{Var}\, W$. From (3.5.1), with a little algebra, we have

$$p_1 - p_1^2 = \left\{ \int G(t)\, dH(t) \right\} \times \left\{ \int H(t)\, dG(t) \right\}$$

$$p_2 - p_1^2 = \int H^2(t)\, dG(t) - \left\{ \int H(t)\, dG(t) \right\}^2 \qquad (3.5.3)$$

$$p_3 - p_1^2 = \int G^2(t)\, dH(t) - \left\{ \int G(t)\, dH(t) \right\}^2$$

These equations suggest forming the natural estimates by replacing G and H by G_m and H_n, the empirical cdfs. These estimates are most simply expressed in terms of the placements, defined next. See Orban and Wolfe (1982) for a discussion of the use of placements in the construction of test statistics.

Definition 3.5.2. Given $X_1, \ldots, X_m$ and $Y_1, \ldots, Y_n$, the placement of X_i among $Y_1, \ldots, Y_n$ is the count $\rho_i(x) = \#\, Y_j < X_i, j = 1, \ldots, n$. Likewise, the placement of Y_k is $\rho_k(y) = \#\, X_i < Y_k, i = 1, \ldots, m$.

Now,

$$\mathrm{Est}\left\{ \int H(t)\, dG(t) \right\} = \int H_n(t)\, dG_m(t)$$

$$= \frac{1}{m} \sum_{i=1}^{m} H_n(X_i)$$

$$= \frac{1}{mn} \sum_{i=1}^{m} \rho_i(x)$$

$$= \frac{1}{n} \bar{\rho}(x), \qquad (3.5.4)$$

and

$$\mathrm{Est}\left\{ p_1 - p_1^2 \right\} = \frac{1}{mn} \bar{\rho}(x)\bar{\rho}(y). \qquad (3.5.5)$$

Further,

$$\text{Est}\left\{ \int H^2(t)\,dG(t) \right\} = \int H_n^2(t)\,dG_m(t)$$

$$= \frac{1}{mn^2} \sum_{i=1}^m \rho_i^2(x) \tag{3.5.6}$$

and

$$\text{Est}\{ p_2 - p_1^2 \} = \frac{1}{mn^2} \sum_{i=1}^m \rho_i^2(x) - \left[\frac{1}{n}\bar{\rho}(x) \right]^2$$

$$= \frac{1}{mn^2} \left\{ \sum_{i=1}^m \rho_i^2(x) - m\bar{\rho}^2(x) \right\}$$

$$= \frac{1}{mn^2} \sum_{i=1}^m (\rho_i(x) - \bar{\rho}(x))^2$$

$$= \frac{1}{mn^2} S^2(x), \tag{3.5.7}$$

where $S^2(x) = \sum_{i=1}^m (\rho_i(x) - \bar{\rho}(x))^2$, the centered sum of squares of the placements.

In Theorem 3.5.1, if we replace $m - 1$ and $n - 1$ by m and n, then a computationally simple, consistent estimate of $\text{Var } W$ is given by

$$\widehat{\text{Var }} W = \bar{\rho}(x)\bar{\rho}(y) + S^2(x) + S^2(y). \tag{3.5.8}$$

Note that since the placements are functions of the ranks, that the statistic

$$\hat{W} = \frac{W - mn/2}{\sqrt{\widehat{\text{Var }} W}} \tag{3.5.9}$$

is a rank statistic. When $\Delta = 0$ and $G = H$, $\hat{W}$ is distribution free and its permutation distribution has been tabled, for selected sample sizes, by Fligner and Policello (1981). However, to insure that $\hat{W}$ is even asymptotically distribution free when $\Delta = 0$ and $G \neq H$ we must assume $G, H \in \Omega_s$. The following theorem due to Fligner and Policello (1981) is the basis for the test.

Theorem 3.5.2.   Suppose $\Delta = 0$ and $G, H \in \Omega_s$. Then $\hat{W}$ is asymptotically $n(0, 1)$.

Proof.   When $G, H \in \Omega_s$, from Theorem 3.5.1,

$$EW = mn \int G(y)h(y)\,dy$$

$$= mn \int [1 - G(-y)]h(y)\,dy$$

$$= mn \int [1 - G(-y)]h(-y)\,dy$$

$$= mn \left\{ 1 - \int G(y)h(y)\,dy \right\}.$$

Comparing the first to last line shows $\int G(y)h(y)\,dy = 1/2$, and hence $EW = mn/2$ when $G, H \in \Omega_s$. The theorem now follows from Slutsky's Theorem A3 since we have $(W - EW)/(\text{Var } W)^{1/2}$ is limiting $n(0, 1)$ by Exercise 3.7.15, and $\widehat{\text{Var }W}$ is consistent. ·

The test rejects $H_0 : \Delta = 0$ for $H_A : \Delta > 0$ when $\hat{W} \geqslant Z_\alpha$, where $1 - \Phi(Z_\alpha) = \alpha$.

In Table 3.2, from Fligner and Policello (1981), the empirical levels, based on 10,000 simulations, are given for W, $\hat{W}$, t, and t_W (Welch's test).

Table 3.2.   Empirical Levels Times 1000 for Nominal $\alpha = .05$, $m = 11$, $n = 10$, σ is the scale of Y

Distribution	σ	W	$\hat{W}$	t	t_W
Normal	0.1	81	48	48	48
	0.25	69	54	50	52
	1	50	48	48	47
	4	71	54	60	47
	10	82	62	69	52
Contaminated	0.1	76	51	33	34
Normal	0.25	65	52	33	33
	1	48	46	35	33
	4	68	52	43	32
	10	83	63	50	35

Note: t_W is Welch's t. The contaminated normal is from Example 2.6.4 with $\epsilon = 0.1$. Each level is based on 10,000 simulations. Reprinted with permission of the American Statistical Association.

Both the normal and contaminated normal (Example 2.6.4) distributions were simulated for $H(t) = G(\sigma^{-1}t)$, σ varying from 0.1 to 10. The test based on $\hat{W}$ is the most stable, outperforming t_W for the contaminated normal. Fligner and Policello (1981) simulated other underlying distributions and included a study of the power of the tests. The test based on $\hat{W}$ emerged as superior in maintaining its level, and achieved high power at the same time. Exercise 3.7.16 investigates the behavior of the level of W, the regular Mann–Whitney–Wilcoxon test, when $G \neq H$. See Hettmansperger and Malin (1975) and Fligner and Rust (1982) for modifications of Mood's test in case G and H are not assumed to be in Ω_s.

We will not discuss the regularity conditions needed to rigorously develop the efficacy of W. These conditions, such as the uniform convergence to normality, are achieved in a way similar to that in the one-sample model; see the discussion following the Pitman regularity conditions in Section 2.6. The formal calculation of the efficacy will now be given. Recall that we need the asymptotic variance under the null hypothesis and the asymptotic mean under the alternative. Now $N^{1/2}(\overline{W} - p_1)$ is asymptotically $n(0, \sigma^2)$ where

$$\sigma^2 = (p_2 - p_1^2)/\lambda + (p_3 - p_1^2)/(1 - \lambda).$$

Under the null hypothesis $\sigma^2(0) = 1/12\lambda(1 - \lambda)$. Furthermore,

$$\mu(\Delta) = p_1 = \int_{-\infty}^{\infty} \left[1 - F(x - \Delta)\right] f(x) \, dx$$

and

$$\mu'(0) = \int_{-\infty}^{\infty} f^2(x) \, dx.$$

Hence the efficacy of W is

$$c = \mu'(0)/\sigma(0) = \sqrt{12\lambda(1 - \lambda)} \int f^2(x) \, dx. \tag{3.5.10}$$

Note that c is simply $[\lambda(1 - \lambda)]^{1/2}$ times the efficacy of the corresponding one-sample test. In Table 2.4 we illustrated the finite sample efficiency of the Wilcoxon signed rank test relative to the t test. A variation on this approach is discussed by Witting (1960) for the two-sample Mann–Whitney–Wilcoxon relative to the two-sample t test. Using numerical approximations to the small-sample power functions, Witting approximates the efficiency for small sample sizes. In the cases he studied, the efficiencies, for samples as small as $m = n = 10$, never fall below 0.94 for an underlying normal model. Hence, in agreement with Klotz (1963), it would appear that the excellent asymptotic efficiency properties of the rank test relative to the t test at a normal model hold for small samples as well. Further, note that c is maximized by $\lambda = 1/2$; hence it is best to take equal sample sizes. We may also conclude from Theorem 2.5.5 that if $\hat{\Delta}$ = med($Y_i - X_j$), the Hodges–Lehmann estimate of Δ, then $N^{1/2}(\hat{\Delta} - \Delta)$ is asymptotically $n(0, 1/c^2)$ with c given by (3.5.10).

When deriving properties of the Mann–Whitney–Wilcoxon statistic, it is usually easier to use the counting form W. For example, we apply the Projection Theorem 2.5.2 to W rather than U, the rank sum form. However, U has exactly the same properties as W because they are linearly related. When discussing the general scores statistics, we will consider the rank form given in Definition 3.4.2, namely,

$$\overline{V} = \frac{1}{N} \sum_{i=1}^{n} \phi\left(\frac{R_i}{N+1} \right). \tag{3.5.11}$$

From the discussion preceding (3.4.1) we have $N \operatorname{Var} \overline{V} \to \lambda(1 - \lambda)$ $\int_0^1 (\phi(u) - \bar{\phi})^2 du$. The asymptotic mean of $\overline{V}$, under the alternative, is developed like (2.9.2) in the one-sample case. The rank of Y_i is R_i, the number of observations in the combined sample less than or equal to Y_i, and hence $R_i = mG_m(Y_i) + nH_n(Y_i)$ where $G_m(\cdot)$ and $H_n(\cdot)$ are the empirical distribution functions for the X and Y samples, respectively. The statistic can now be written as

$$\begin{aligned}
\overline{V} &= \frac{1}{N} \sum_{1}^{n} \phi\left(\frac{R_i}{N+1} \right) \\
&= \frac{1}{N} \sum_{1}^{n} \phi\left(\frac{m}{N+1} G_m(Y_i) + \frac{n}{N+1} H_n(Y_i) \right) \\
&= \frac{n}{N} \int_{-\infty}^{\infty} \phi\left(\frac{m}{N+1} G_m(y) + \frac{n}{N+1} H_n(y) \right) dH_n(y).
\end{aligned}$$

Since $G_m(x)$ and $H_n(y)$ converge in probability to $G(x)$ and $H(y)$, respectively, we expect, under some regularity conditions, that

$$\overline{V} \xrightarrow{P} (1 - \lambda) \int_{-\infty}^{\infty} \phi(\lambda G(y) + (1 - \lambda)H(y)) dH(y). \tag{3.5.12}$$

For the location model with $G(x) = F(x)$ and $H(y) = F(y - \Delta)$, we have

$$\mu(\Delta) = (1 - \lambda) \int_{-\infty}^{\infty} \phi(\lambda F(y) + (1 - \lambda)F(y - \Delta))f(y - \Delta) dy.$$

Now make the change of variable $t = y - \Delta$ and then differentiate with respect to Δ to get

$$\mu'(0) = \frac{d}{d\Delta}\,\mu(\Delta)\Big|_{\Delta=0} = \lambda(1-\lambda)\int_{-\infty}^{\infty}\phi'(F(t))f^2(t)\,dt.$$

Hence the Pitman efficacy is

$$c = \frac{\sqrt{\lambda(1-\lambda)}\int_{-\infty}^{\infty}\phi'(F(t))f^2(t)\,dt}{\sqrt{\int_0^1(\phi(u)-\bar\phi)^2\,du}}. \tag{3.5.13}$$

The regularity conditions for establishing this result rigorously are similar to the one-sample case discussed after (2.9.3).

The same type of change of variable and integration by parts used to derive (2.9.4) from (2.9.3) can be used on (3.5.13). The result is

$$c = \frac{\sqrt{\lambda(1-\lambda)}\int_0^1\phi(u)\phi_f(u)\,du}{\sqrt{\int_0^1(\phi(u)-\bar\phi)^2\,du}} \tag{3.5.14}$$

with

$$\phi_f(u) = -\frac{f'(F^{-1}(u))}{f(F^{-1}(u))}.$$

Since $\int_0^1\phi_f(u)\,du = 0$ and $\int_0^1\phi_f^2(u)\,du = I(f)$, Fisher's information given in Definition 2.9.1, Theorem 2.9.1 can be extended immediately to the two-sample case with $[\lambda(1-\lambda)I(f)]^{1/2}$ in place of $[I(f)]^{1/2}$. Hence if $\bar V_f$ is the two-sample statistic generated by $\phi_f(u)$, then it is the asymptotically most powerful rank test statistic when $F \in \Omega_0$ is the underlying distribution. Note, also, that $\bar V_f$ is the approximation to the locally most powerful rank test statistic. See (3.3.3) and (3.3.5).

In the next example we show that (3.5.10), which relates the one- and two-sample Wilcoxon efficacies, holds for general scores statistics. In Example 3.4.3 we constructed two-sample score-generating functions from one-sample score-generating functions. We now reverse the process and discuss the construction of a one-sample score-generating function from a given two-sample score-generating function. It then follows that the efficiency properties are common to the one- and two-sample tests generated by these score functions.

Example 3.5.2. In this example suppose that $F \in \Omega_s$ and define

$$\phi^+(u) = \phi\left(\frac{u+1}{2}\right), \qquad 0 < u < 1,$$

for some given two-sample score-generating function $\phi(u)$ which satisfies $\phi(u) = -\phi(1-u)$. Hence ϕ is an odd function centered at $u = \frac{1}{2}$. This implies that $\bar{\phi} = 0$ and

$$\int_0^1 \left[\phi^+(u)\right]^2 du = 2 \int_{1/2}^1 \phi^2(u)\, du = \int_0^1 \phi^2(u)\, du.$$

Furthermore, using

$$\phi^{+\prime}(u) = \frac{1}{2}\phi'\left(\frac{u+1}{2}\right),$$

we have

$$2 \int_0^\infty \phi^{+\prime}(2F(x) - 1)f^2(x)\, dx$$

$$= 2 \int_0^\infty \frac{1}{2}\phi'(F(x))f^2(x)\, dx$$

$$= \frac{1}{2} \int_{-\infty}^\infty \phi'(F(x))f^2(x)\, dx$$

Now the efficacy of the one-sample test generated from $\phi^+(\cdot)$ is given by (2.9.2) as

$$c = \frac{2 \int_0^\infty \phi^{+\prime}(2F(x) - 1)f^2(x)\, dx}{\sqrt{\dfrac{1}{4} \int_0^1 (\phi^+(u))^2\, du}}$$

$$= \frac{\dfrac{1}{2} \int_{-\infty}^\infty \phi'(F(x))f^2(x)\, dx}{\sqrt{\dfrac{1}{4} \int_0^1 \phi^2(u)\, du}}.$$

Cancelling the $1/2$, and since $\bar{\phi} = 0$, we have the two-sample efficacy is $[\lambda(1-\lambda)]^{1/2}$ times the one-sample efficacy. Since the efficiency is the ratio of the squares of the efficacies and since the factor $\lambda(1 - \lambda)$ cancels, we see that the efficiency properties of the one-sample and two-sample tests are identical.

The Mann–Whitney–Wilcoxon test corresponds to the Wilcoxon signed rank test, and Mood's median test corresponds to the sign test; other examples are given in Example 3.4.3. The efficiency calculations for the one-sample tests in Section 2.5 can be interpreted now in the two-sample case.

We complete this section by using Example 3.5.2 to anticipate the form of the locally most powerful rank test in the one-sample location model. We begin with the LMPRT score, (3.3.3),

$$a(i) = - E\left\{ \frac{f'(V_{(i)})}{f(V_{(i)})} \right\} = - E\left\{ \frac{f'[F^{-1}(U_{(i)})]}{f[F^{-1}(U_{(i)})]} \right\}$$

where $U_{(1)} < \cdots < U_{(N)}$ are the order statistics from a uniform distribution on $(0, 1)$. Example 3.5.2 suggests that the one-sample score should be

$$a^+(i) = - E\left\{ \frac{f'[F^{-1}(\{U_{(i)} + 1\}/2)]}{f[F^{-1}(\{U_{(i)} + 1\}/2)]} \right\}. \tag{3.5.15}$$

Suppose $F \in \Omega_s$ and define

$$F_+(x) = P(|X| \leqslant x) = \begin{cases} 0 & x \leqslant 0 \\ 2F(x) - 1 & x > 0. \end{cases}$$

Then $f_+(x) = 2f(x)$ if $x > 0$ and 0 otherwise. Let $y = F_+(x)$, then $x = F_+^{-1}(y)$ and $x = F^{-1}[(y + 1)/2]$. Hence $F_+^{-1}(y) = F^{-1}[(y + 1)/2]$, and (3.5.15) becomes

$$a^+(i) = - E\left\{ \frac{f_+'[F_+^{-1}(U_{(i)})]}{f_+[F_+^{-1}(U_{(i)})]} \right\} = - E\left\{ \frac{f_+'(V_{(i)}^+)}{f_+(V_{(i)}^+)} \right\} \tag{3.5.16}$$

where $V_{(1)}^+ < \cdots < V_{(N)}^+$ are the order statistics from the distribution $F_+(x)$.

The one-sample test based on $V^+ = \sum a^+(R_i^+)s(X_i)$ where R_i^+ is the rank of $|X_i|$ among $|X_1|, \ldots, |X_N|$ is, in fact, the locally most powerful rank test; see Hajek and Sidak (1967, Chapter II) or Randles and Wolfe (1979, Chapter 10). Either (3.5.15) or (3.5.16) can be used to construct V^+.

The AMPRT, discussed after Theorem 2.9.1, is the approximation to the LMPRT.

In both the one- and two-sample location models the AMPRT and LMPRT have the same efficiency properties. From the point of view of Pitman efficiency the tests are equivalent; however, as mentioned after (3.3.5), the AMPRT is more practical. A direct development of locally most powerful rank tests in the one-sample model, similar to the development in Section 3.3, can be based on Exercise 3.7.18.

3.6. COMPARISON OF DESIGNS

We complete this chapter with a comparison, proposed by Hodges and Lehmann (1973), of two types of randomization in testing a treatment (T) against a control (C). The presence of a known type effect such as right and left handedness creates the setting for the construction of both one- and two-sample tests. The proposed designs are compared by comparing the two tests; hence ideas from Chapters 1 and 2 are brought together in this example.

In the Completely Randomized Design (CRD) we have N pairs of subjects. Treatment and control are randomly assigned within each pair, perhaps by the toss of a coin. The usual analysis proceeds by applying a one-sample test to the $T - C$ differences, that is, to the T excesses in Fig. 3.2.

Now suppose that each pair consists of a type A and type B member, for example right and left hand, and suppose there is a type effect in the data. In a Restrictedly Randomized Design (RRD) we randomly choose a pairs from the N and apply the treatment to the type A member. In the remaining $b = N - a$ pairs, type B gets the treatment. A two-sample test is then applied to the two sets of A excesses in the figure. The two designs are illustrated in Fig. 3.2.

We next develop a model for the RRD. Let τ represent the type effect and let $X_1, \ldots, X_N$ denote the A excesses. Then before the treatment is applied we have a random sample of size N from $F(x - \tau)$, $F \in \Omega_s$. Let δ denote the treatment effect. Then $X_1, \ldots, X_a$ is a sample from $F(x - \tau - \delta)$ and $X_{a+1}, \ldots, X_N$, denoted $X_1', \ldots, X_b'$, is a sample from $F(x - \tau + \delta)$. Hence 2δ represents the difference in locations due to a treatment effect, and we wish to test $H_0 : \delta = 0$ versus $H_A : \delta > 0$. We suppose that as $N \to \infty$, $a/N \to \alpha$, $0 < \alpha < 1$ and $b/N \to \beta$, $0 < \beta < 1$. The efficacy of the Mann–Whitney–Wilcoxon test is $c_{MW} = (\alpha\beta)^{1/2}2(12)^{1/2}\int f^2(x)\,dx$. The extra factor of 2 appears because the difference in locations is 2δ. The efficacy of the two-sample t is $c_{2t} = (\alpha\beta)^{1/2}2/\sigma_F$ where σ_F^2 is the variance of

CRD	Pairs	Member Types A	Member Types B	Treatment Excesses
	1	T	C	$T - C$
	2	C	T	$\vdots$
	$\vdots$	$\vdots$	$\vdots$	$\vdots$
	N	C	T	$T - C$

RRD	Pairs	Member Types A	Member Types B	A Excesses
	1	T	C	$T - C$
	2	T	C	$\vdots$
	$\vdots$	$\vdots$	$\vdots$	$\vdots$
	a	T	C	$T - C$
	1	C	T	$C - T$
	$\vdots$	$\vdots$	$\vdots$	$\vdots$
	b	C	T	$C - T$

Figure 3.2. Two kinds of randomized designs.

$F \in \Omega_s$. The efficacy is maximized by taking $\alpha = \beta = 1/2$; hence from now on we will only consider the case in which N is split equally and $a = b$.

In the CRD, with probability $1/2$, a type A subject is treated. Let $Z_1, \ldots, Z_N$ denote the treatment excesses. Then Z_i is an A excess with probability $1/2$ and a B excess with probability $1/2$. Given that Z_i is an A excess, $P(Z_i \leqslant z) = F(z - \tau - \delta)$ and given that Z_i is a B excess, $P(Z_i \leqslant z) = F(z + \tau - \delta)$ when $F \in \Omega_s$. This follows from the preceding discussion on the model for the RRD. Hence if $G(\cdot)$ is the distribution function for the T excesses,

$$G(z) = P(Z_i \leqslant z) = \tfrac{1}{2}F(z - \tau - \delta) + \tfrac{1}{2}F(z + \tau - \delta)$$

with density

$$g(z) = \tfrac{1}{2}f(z - \tau - \delta) + \tfrac{1}{2}f(z + \tau - \delta).$$

The mean and variance of $G(\cdot)$ are δ and $\sigma_F^2 + \tau^2$, respectively. Hence the efficacy of the one-sample t test is $c_{1t} = 1/(\sigma_F^2 + \tau^2)^{1/2}$. In order to find the efficacy of the Wilcoxon signed rank test, we compute

$$\int_{-\infty}^{\infty} g^2(z)\,dz = \int_{-\infty}^{\infty} \frac{1}{2} f^2(z)\,dz + \int_{-\infty}^{\infty} \frac{1}{2} f(z)f(z+2\tau)\,dz,$$

and $c_W = 12^{1/2}\int g^2(z)\,dz$.

We now compare the CRD to the RRD through the Pitman efficiency. In the case of the t tests, the efficiency of the one-sample t (CRD) relative to the two-sample t (RRD) is

$$e(1t, 2t) = \frac{c_{1t}^2}{c_{2t}^2} = \frac{\sigma_F^2}{\sigma_F^2 + \tau^2} \leqslant 1 \qquad \text{for all } \tau. \qquad (3.6.1)$$

In the rank case we compute the efficiency of the Wilcoxon signed rank test (CRD) relative to the Mann–Whitney–Wilcoxon test (RRD) and find

$$e(W, MW) = \frac{1}{4}\left\{1 + \frac{\int f(z)f(z+2\tau)\,dz}{\int f^2(z)\,dz}\right\}^2. \qquad (3.6.2)$$

From the Cauchy–Schwarz inequality, $\{\int f(z)f(z+2\tau)\,dz\}^2 \leqslant \int f^2(z)\,dz \int f^2(z+2\tau)\,dz = \{\int f^2(z)\,dz\}^2$. Hence we have

$$\tfrac{1}{4} \leqslant e(W, MW) \leqslant 1, \qquad \text{for all } \tau.$$

In either case the RRD is superior to the CRD. If there is no type effect present, $\tau = 0$, the efficiency is 1 so nothing is lost by using the RRD. The actual savings are computed, in the case of an underlying normal population, in Exercise 3.7.17.

3.7. EXERCISES

3.7.1. Let $R_1, \ldots, R_n$ denote the ranks of $Y_1, \ldots, Y_n$ in the combined sample of size $N = m + n$. Under $H_0 : \Delta = 0$ show that $ER_i = (N+1)/2$, $\operatorname{Var} R_i = (N^2 - 1)/12$ and $\operatorname{Cov}(R_i, R_j) = -(N+1)/12$ for $i \neq j$. Then show that if $U = \sum_1^n R_i$, $EU = n(N+1)/2$ and $\operatorname{Var} U = mn(N+1)/12$.

3.7.2. Suppose $X_1, \ldots, X_n$ are i.i.d. F, an arbitrary continuous distribution with mean μ and variance σ^2. Let $R_1, \ldots, R_n$ be the ranks of

$X_1, \ldots, X_n$. Using the fact that the conditional distribution of X_1, given $R_1 = j$ is the distribution of the jth order statistic, show that

$$EX_1 R_1 = \frac{1}{n} \sum_{j=1}^{n} jE(X_1 \mid R_1 = j)$$

$$= \int_{-\infty}^{\infty} yf(y) \, dy + (n-1) \int_{-\infty}^{\infty} yf(y)F(y) \, dy.$$

Hence show that the correlation between X_1 and R_1 is

$$\rho_n = \frac{\mu + (n-1)\int_{-\infty}^{\infty} yf(y)F(y) \, dy - \mu(n+1)/2}{\sigma(n^2 - 1)/12}.$$

Find the correlation ρ_n and $\lim_{n \to \infty} \rho_n$ when f is $n(0,1)$.

3.7.3. Recall T is the Wilcoxon signed rank statistic discussed in Section 2.2. Derive the following recurrence equation for the distribution of T under H_0:

$$P(T = k) = \frac{\bar{P}_n(k)}{2^n}$$

for $k = 0, 1, \ldots, n(n+1)/2$, where

$$\bar{P}_n(k) = \bar{P}_{n-1}(k) + \bar{P}_{n-1}(k - n)$$

with $\bar{P}_0(0) = 1$, $\bar{P}_0(k) = 0$, and $\bar{P}_n(k) = 0$ for $k < 0$. Hint: Let $\bar{P}_n(k)$ be the number of subsets $\{r_1, \ldots, r_j\}$ in $\{1, \ldots, n\}$ for which $\sum_{1}^{j} r_i = k$.

3.7.4. Suppose $X_{(1)} < \cdots < X_{(n)}$ and $Y_{(1)} < \cdots < Y_{(n)}$ are the order statistics for two samples each of size n. Galton's statistic, which can be used to compare two teams with ranked members, is defined as $V = \#(Y_{(i)} > X_{(i)})$, $i = 1, \ldots, n$. Let $\bar{P}_{mn}(k)$ be the number of sequences of m x's and n y's such that $V = k$, and derive a recurrence formula for $\bar{P}_{mn}(k)$. By computing the distribution of V for various values of n, see if you can guess the form of the distribution of V.

3.7.5. Let $V = \sum_{1}^{n} a(R_i)$, given in Definition 3.4.1. Prove that the distribution of V, under H_0, is symmetric about $n\bar{a}$ provided that $a(i) + a(N - i + 1) = K$ for all $i = 1, 2, \ldots, N$, where K is a constant.

3.7.6. In Example 1.1.1 it was hypothesized that homing pigeons used the sun to navigate. In Example 1.5.1 and Exercise 1.8.11 the data on birds released on sunny and cloudy days suggested that homing pigeons home better on sunny days. The hypothesis can be formally tested using the Mann–Whitney–Wilcoxon test. Following is a portion of the data, randomly selected.

$$\text{Sunny } \frac{17 \quad 32 \quad 42 \quad 42 \quad 55 \quad 72 \quad 97}{10 \quad 38 \quad 105 \quad 126 \quad 141}$$

The measure is the error angle made with the homing line when the bird disappears over the horizon. Let Δ be the difference of the sunny and cloudy population medians. Test $H_0 : \Delta = 0$ versus $H_A : \Delta < 0$ at approximate level .05, construct a point estimate of Δ, and construct an approximate 95% confidence interval for Δ. Carry out a similar analysis based on Mood's procedure described in Examples 3.4.2 and 3.4.4.

3.7.7. Suppose $X_1, \ldots, X_n$ are i.i.d. $F(x)$, $F \in \Omega_0$. Show that $Y_1 = F(X_1), \ldots, Y_n = F(X_n)$ are i.i.d. $U(0, 1)$. Show that the pdf of $Y_{(i)}$, the ith order statistic, is

$$g(y) = \frac{\Gamma(n + 1)}{\Gamma(i)\Gamma(m + n - i + 1)} y^{i-1}(1 - y)^{m + n - i}$$

for $0 < y < 1$. Further, show that $EY_i = i/(n + 1)$.

3.7.8. Construct the two-sample score functions that correspond to the Winsorized Wilcoxon signed rank statistic and the modified sign statistic in Examples 2.8.2 and 2.8.3, respectively. Find the asymptotic moments (3.4.1) and (3.4.2) for use in the limiting distribution.

3.7.9. Suppose $V = \sum_1^n a(R_i)$ is a general score statistic given in Definition 3.4.1. Let $V(\Delta) = \sum_1^n a(R_i(\Delta))$, (3.4.4). Prove that $V(\Delta)$ decreases by $a_{i+j} - a_{i+j-1}$ as Δ crosses $Y_{(j)} - X_{(i)}$. Hence the properties of the general score statistic are determined by its behavior at the mn pairwise differences. Verify (3.4.4) and (3.4.5).

3.7.10. Using (3.4.6), discuss how to construct estimates of Δ that correspond to the statistics in Exercise 3.7.8.

3.7.11. Let $X_1, \ldots, X_m$ and $Y_1, \ldots, Y_n$ be random samples from $F(x)$ and $F(y - \Delta)$, $F \in \Omega_0$. We will consider an alternative to Mood's test, due to Mathisen (1943), for testing $H_0 : \Delta = 0$ versus H_A:

$\Delta > 0$. Suppose for simplicity that $m = 2q - 1$ and define $M = \#(Y_i > X_{(q)})$, $i = 1, \ldots, n$. Hence M counts the number of Y values that exceed the median of the X's. Under $H_0: \Delta = 0$, show that

$$P(M = t) = \frac{\binom{q - 1 + t}{q - 1}\binom{n - t + q - 1}{q - 1}}{\binom{m + n}{m}}$$

for $t = 0, 1, \ldots, n$. Further, $EM = n/2$ and $\operatorname{Var} M = n(m + n + 1)/[4(m + 2)]$.

Using the notation of the sign statistic, note that $M = S(X_{(q)}) = \sum s(Y_i - X_{(q)})$. Under H_0, use Exercise 2.10.18 to write

$$\frac{1}{\sqrt{n}}\left[S(X_{(q)}) - \frac{n}{2} \right]$$

$$= \frac{1}{\sqrt{n}}\left[S(0) - \frac{n}{2} \right] + \sqrt{n}\, f(0) X_{(q)} + o_p(1).$$

Now show that, if $m, n \to \infty$ so that $m/(m + n) \to \lambda$, $0 < \lambda < 1$, then

$$\frac{1}{\sqrt{n}}\left[M - \frac{n}{2} \right] \xrightarrow{D} Z \sim n(0, 1/[4\lambda]).$$

3.7.12. This exercise continues Exercise 3.7.11. We reject $H_0: \Delta = 0$ for $H_A: \Delta > 0$ at approximate size α if $M \geqslant c$ where $c \doteq n/2 + Z_\alpha(n/[4\lambda])^{1/2}$. Show that $M \geqslant c$ if and only if $Y_{(n-c+1)} > X_{(q)}$. Hence we can terminate Mathisen's test as soon as we observe $Y_{(n-c+1)}$ or $X_{(q)}$. Show that Mood's test, discussed in Example 3.4.4, terminates the one-sided test as soon as we observe $X_{(r-n+d)}$ or $Y_{(n-d+1)}$, where $P(V_+^* \geqslant d) = \alpha$ and $d \doteq n/2 + Z_\alpha(n\lambda/4)^{1/2}$. Show that for large n, Mathisen's test will always terminate before Mood's test. Hence Mathisen's test achieves a greater savings.

3.7.13. Carry out a 5% Mood's test and find a 95% confidence interval and point estimate for Δ, for the data in Example 3.2.2.

3.7.14. Recall $V_+^* = \#(Y_i > \text{median of combined sample})$ is Mood's statistic in Example 3.4.4. Suppose $n \leqslant m$ and $P(V_+^* \leqslant k) \doteq \alpha/2$, where $k = n/2 - Z_{\alpha/2}\{mn/[4(N-1)]\}^{1/2}$ and $N = m + n$, from Example 3.4.2. From (3.4.8) the two sign-confidence intervals that determine Mood's test are $[X_{(d_x+1)}, X_{(m-d_x)}]$ and $[Y_{(d_y+1)}, Y_{(n-d_y)}]$ where $d_x = r - n + k + 1$ and $d_y = k$. Note that $d_x = d_y + (m - n)/2$. Show that the confidence coefficients for the two individual intervals can be approximated as $\gamma_x \doteq \Phi(-Z_{\alpha/2}[n/(N-1)]^{1/2})$ and $\gamma_y \doteq \Phi(-Z_{\alpha/2}[m/(N-1)]^{1/2})$.

3.7.15. Suppose $X_1, \ldots, X_m$ are i.i.d. $G(x)$ and $Y_1, \ldots, Y_n$ are i.i.d. $H(y)$, where G and H are arbitrary continuous cdfs. Let

$$W' = \frac{\sqrt{N}}{mn} \sum_{i=1}^{m} \sum_{j=1}^{n} (T_{ij} - p_1),$$

where $T_{ij} = 1$ if $Y_j > X_i$ and 0 otherwise. Suppose $m, n \to \infty$ in such a way that $m/N \to \lambda$, $0 < \lambda < 1$, $N = m + n$. Prove that W' is asymptotically $n(0, \sigma^2)$, where

$$\sigma^2 = \frac{1}{\lambda}(p_2 - p_1^2) + \frac{1}{(1-\lambda)}(p_3 - p_1^2).$$

Show, further, that $\operatorname{Var} W' \to \sigma^2$ and when $0 < p_1 < 1$, $(W' - EW')/(\operatorname{Var} W')^{1/2}$ is asymptotically $n(0, 1)$. (This exercise generalizes Theorem 3.2.4.)

3.7.16. Suppose $X_1, \ldots, X_m$ are i.i.d. $G(x)$ and $Y_1, \ldots, Y_n$ are i.i.d. $H(y - \Delta)$, where $G, H \in \Omega_0$. Let $W^* = (W - mn/2)/[mn(m + n + 1)/12]^{1/2}$, the standardized Mann–Whitney–Wilcoxon statistic. When $\Delta = 0$ and $G = H$, W^* is limiting $n(0, 1)$. We wish to investigate the effect on the significance level of W^* when $G \neq H$.

Let $\alpha_T = P(W^* \geqslant Z_\alpha)$, then when $\Delta = 0$ and $G = H$, $\alpha_T \doteq \alpha$. Here α is called the nominal level and α_T, the true level. Suppose $m, n \to \infty$ in such a way that $m/N \to \lambda$, $0 < \lambda < 1$. For arbitrary G, $H \in \Omega_0$, show that

$$\alpha_T \to \begin{cases} 0, & \text{if } p_1 < \frac{1}{2} \\ 1, & \text{if } p_1 > \frac{1}{2} \\ 1 - \Phi\left(\dfrac{Z_\alpha}{\sqrt{12\{(1-\lambda)(p_2 - p_1^2) + \lambda(p_3 - p_1^2)\}}}\right), & \text{if } p_1 = \frac{1}{2} \end{cases}$$

where p_1, p_2, and p_3 are given in (3.5.1). To have any control over α_T, we must have $p_1 = \frac{1}{2}$. A sufficient condition for $p_1 = \frac{1}{2}$ is $G, H \in \Omega_s$.

Consider the case where $H(y) = G(y/\sigma)$, $G \in \Omega_s$, so the distributions differ by the scale factor σ. Show that

$$p_2 - p_1^2 = \int \left[1 - G(x/\sigma)\right]^2 dG(x) - \frac{1}{4} \to \begin{cases} \frac{1}{4} & \text{if } \sigma \to 0 \\ 0 & \text{if } \sigma \to \infty \end{cases}$$

Likewise $p_3 - p_1^2 \to 0$ or $\frac{1}{4}$ as $\sigma \to 0$ or ∞. Now show that

$$\alpha_T \to \begin{cases} 1 - \Phi\left(Z_\alpha/\sqrt{3(1-\lambda)}\right) & \text{if } \sigma \to 0 \\ 1 - \Phi\left(Z_\alpha/\sqrt{3\lambda}\right) & \text{if } \sigma \to \infty. \end{cases}$$

Finally, show for the case of equal sample sizes and $\alpha = .05$, that α_T ranges from .05 to .087 as $\sigma \to 0$ or ∞. This indicates that when the distributions are symmetric and differ perhaps by a scale factor, and when the sample sizes are the same, the true level of W^* is not that much different from the nominal level.

3.7.17. Suppose that F is $n(0, \sigma^2)$. Find $e(W, MW)$, (3.6.2), as a function of σ^2 and τ. Table and graph $e(W, MW)$ as a function of τ/σ.

3.7.18. Suppose $X_1, \ldots, X_N$ are i.i.d. $F \in \Omega_s$. Let $F^+(x) = P(|X| \leqslant x)$, the cdf of $|X|$. From Section 1.4, the distribution F^+ can be specified by the triple (p, G, H) where $p = P(X > 0)$, $G(x) = P(|X| \leqslant x \mid X \leqslant 0)$ and $H(x) = P(|X| \leqslant x \mid X > 0)$. Let N^+ be the number of positive X's. Given $N^+ = n$, let $R_{(1)}^+, \ldots, R_{(n)}^+$ be the ranks of the positive observations among the absolute values. Show that

$$P(R_{(1)}^+ = r_1, \ldots, R_{(n)}^+ = r_n, N^+ = n)$$

$$= \binom{N}{n} p^n (1-p)^{N-n} P(R_{(1)}^+ = r_1, \ldots, R_{(n)}^+ = r_n \mid N^+ = n)$$

where $P(R_{(1)}^+ = r_1, \ldots, R_{(n)}^+ = r_n \mid N^+ = n)$ is given by the result in Theorem 3.3.1.

3.17.19. The one-sided Kolmogorov–Smirnov statistic is $D^+ = \max_x \{G_m(z) - H_n(z)\}$, see Example 3.4.5. This statistic could be used to test for stochastic ordering; Definition 3.5.1. The null hypothesis $H_0: G = H$ would be rejected for $H_A: G \geqslant H$, with strict inequality somewhere when $D^+ \geqslant c$.

a. Suppose $m = n$, equal sample sizes. Use the reflection principle in Feller (1968, p. 68) to show that, under the null hypothesis,

$$P(D^+ \geqslant h/n) = \frac{\binom{2n}{n-h}}{\binom{2n}{n}},$$

$h = 1, 2, \ldots, n.$

b. In general, it can be shown that

$$\lim_{m,n \to \infty} P\left(\sqrt{\frac{mn}{m+n}} \, D^+ \leqslant t\right) = 1 - e^{-2t^2}$$

(See Hajek and Sidak, 1967, Chapter V.) If $m = n$, $n \to \infty$, verify the limit using the result in (a). Hint: Use Sterling's formula $k! \sim (2\pi)^{1/2} k^{k+1/2} e^{-k}$ on the factorials.

. **3.7.20.** The result in Theorem 3.3.1 can be used to construct LMPRT's in the scale model. Suppose that $X_1, \ldots, X_m$ is a random sample from $F(x)$ and $Y_1, \ldots, Y_n$ is a random sample from $G(y) = F(y/\tau)$, $F \in \Omega_0$. Suppose the model is sufficiently regular to interchange differentiation and expectation.

a. Show that the LMPRT for testing $H_0: \tau = 1$ versus $H_A: \tau > 1$ can be based on the statistic

$$V = -\sum_{j=1}^{n} E\left\{V_{(R_j)} \frac{f'(V_{(R_j)})}{f(V_{(R_j)})}\right\}.$$

where $V_{(1)} < \cdots < V_{(m+n)}$ are the order statistics from F.

b. An important scale model, used in life testing and reliability, is the exponential model. The pdf is $f(x) = e^{-x}$ if $x > 0$ and 0 otherwise. Suppose $X_1, \ldots, X_n$ is a sample from this exponential distribution. The joint pdf of $X_{(1)} < \cdots < X_{(n)}$ is $n! \, e^{-\Sigma_1^n x_i}$, $0 \leqslant x_1 \leqslant x_2 \leqslant \cdots \leqslant x_n < \infty$, and 0 otherwise. Let $W_1 = X_{(1)}, W_2 = X_{(2)} - X_{(1)}, \ldots, W_n = X_{(n)} - X_{(n-1)}$. Find the joint pdf of $W_1, \ldots, W_n$. Argue that $W_1, \ldots, W_n$ are independent and for $i = 1, \ldots, n$, W_i has an exponential

distribution with pdf $(n - i + 1)e^{-(n-i+1)w}$, with mean EW_i
$= (n - i + 1)^{-1}$. Now note that $X_{(i)} = W_1 + \cdots + W_i$ and

$$EX_{(i)} = \sum_{j=1}^{i} (n - j + 1)^{-1}.$$

c. Using (b), show that the LMPRT for $H_0 : \tau = 1$ versus $H_A : \tau$
   > 1 when $F(x) = 1 - e^{-x}$, $x > 0$, and 0 otherwise defined by

$$V = \sum_{j=1}^{n} a(R_j)$$

where $a(i) = \sum_{k=1}^{i}(n - k + 1)^{-1}$. Show that $EV = n$ and

$$\text{Var } V = \frac{mn}{N - 1}\left[1 - \frac{1}{N} \sum_{j=1}^{N} j^{-1} \right].$$

The normal approximation can be used to approximate critical
values. The test is due to Savage (1956).

CHAPTER 4

The One- and Two-Way Layouts and Rank Correlation

4.1. INTRODUCTION

In the last chapter, methods were developed to investigate the difference in locations of two populations. In this chapter we consider the extension to more than two populations. We develop the methods only for ranks (not for general rank scores). Asymptotic distribution theory is developed for approximations under the null hypothesis and for the computation of asymptotic efficiency.

The one- and two-way layout designs are treated in detail. In the former we have k samples and wish to test the null hypothesis that the samples all came from the same population. In the latter case we wish to compare k populations but the data is labeled by an additional variable. In this case we have data classified in two ways: the population and the block. The one- and two-way layouts are examples of the more general linear model which is the subject of Chapter 5. We have chosen to treat these special cases in the present chapter because the tests due to Kruskal and Wallis (1952), for the one-way layout, and Friedman (1937), for the two-way layout, are simple to use, do not require a computer, and are widely available in applied texts. The present chapter serves as an introduction for the following chapter which deals with the general linear model.

Along with the tests that are introduced, we discuss multiple comparisons. This is an important follow-up to any significant comparison of k populations since the comparisons help identify the sources of significance. Further, we develop tests for the null hypothesis against an alternative consisting of a prespecified ordering of the k population locations. These tests for ordered alternatives are generalizations of one-sided tests in the simple location models.

The final section of this chapter contains a discussion of rank correlation. We then interpret the Kruskal–Wallis and Friedman tests in terms of rank correlation. This provides further motivation and rationale for the two tests. In addition, we discuss measures of concordance or agreement among and between groups of judges. These measures are constructed from rank correlations and are also connected to the Friedman test statistic.

4.2. THE ONE-WAY LAYOUT: THE KRUSKAL–WALLIS TEST

The sampling model consists of k samples $X_{11}, \ldots, X_{n_1 1}, \ldots, X_{1k}, \ldots,$ $X_{n_k k}$ from $F(x - \theta_1), \ldots, F(x - \theta_k)$, repectively, where $F \in \Omega_0$. This can also be written $X_{ij}, i = 1, \ldots, n_j; j = 1, \ldots, k$ with cdf $F(x - \theta_j), F \in \Omega_0$. We wish to construct a test of $H_0 : \theta_1 = \cdots = \theta_k$ versus $H_A : \theta_1, \ldots, \theta_k$ not all equal. This null hypothesis simply specifies that the locations are all equal without specifiying the common location. Without loss of generality, we could let $\theta_1 = 0$ and define $\Delta_j = \theta_{j+1} - \theta_j, j = 1, \ldots, k - 1$. Then the null hypothesis becomes $H_0 : \Delta_1 = \cdots = \Delta_{k-1} = 0$.

Example 4.2.1. In a study of the humoral (blood) basis of behavior, Terkel and Rosenblatt (1968) induced maternal behavior in virgin female rats by injecting them with blood plasma drawn from females that had just given birth. Virgin rats were exposed to young pups, and the time it took them to begin retrieving the pups was recorded. Retrieving is a recognized maternal behavior that generally appears within 48 hours after giving birth. It is known that virgin females will begin to show maternal behavior when they are continuously exposed to pups for about 5 days. Hence, the issue is whether maternal blood plasma will reduce the time.

The experiment used 32 virgin female rats, each 60 days old. They were randomly assigned to four groups of size 8. The groups consisted in: (1) rats injected with maternal blood plasma, (2) rats in proestrus (prior to heat) that received plasma from rats in proestrus, (3) rats in diestus (heat) that received plasma from rats in diestus, and (4) rats that were injected with a saline solution (placebo). We will consider the data as arising from four populations with respective cdfs $F(x - \theta_i), i = 1, \ldots, 4, F \in \Omega_0$, and we wish to test $H_0 : \theta_1 = \theta_2 = \theta_3 = \theta_4$ versus $H_A : \theta_1, \theta_2, \theta_3, \theta_4$ not all equal. If the test rejects H_0, we then wish to determine which groups are significantly different. Data for this example is analyzed in Example 4.2.2.

The data can be visualized in a two-way array in which each column is a sample. Hence there are k columns and the jth column has n_j observations from $F(x - \theta_j)$, $F \in \Omega_0$. The basic strategy is to rank the combined data set of size $N = \sum_1^k n_j$ and compare the column rank sums or averages. Let R_{ij} denote the rank of X_{ij} in the combined data and let

$$R_{\cdot j} = \sum_{i=1}^{n_j} R_{ij} \quad \text{and} \quad \overline{R}_{\cdot j} = R_{\cdot j}/n_j. \tag{4.2.1}$$

If $k = 2$, $R_{\cdot 1}$ is the sum of ranks in the first sample in a two-sample problem. Since $R_{\cdot 1} + R_{\cdot 2} = N(N + 1)/2$, there is no additional information in the second sum of ranks $R_{\cdot 2}$, and it could be dropped.

The same is true in the general k sample problem since $\sum_{j=1}^k R_{\cdot j} = N(N + 1)/2$. However, we will retain the entire set, $R_{\cdot 1}, \ldots, R_{\cdot k}$, and take the dependence into account later in the equations for test statistics. This means that we do not have to specify which rank sum is to be dropped.

Under the null hypothesis, R_{ij} has the distributional properties specified by Theorem 3.2.1. Hence Exercise 3.7.1 in Chapter 3 provides the means, variances, and covariances for the ranks in the k-sample case. The needed results are given in the following theorem.

Theorem 4.2.1. Suppose the k samples come from a common distribution so the null hypothesis holds, and let $R_{\cdot j}$ and $\overline{R}_{\cdot j}$ be given in (4.2.1). Then

$$ER_{\cdot j} = n_j(N + 1)/2, \qquad E\overline{R}_{\cdot j} = (N + 1)/2$$

$$\text{Var}\,R_{\cdot j} = n_j(N - n_j)(N + 1)/12, \qquad \text{Var}\,\overline{R}_{\cdot j} = (N - n_j)(N + 1)/(12n_j)$$

$$\text{Cov}(R_{\cdot i}, R_{\cdot j}) = -n_i n_j(N + 1)/12, \qquad \text{Cov}(\overline{R}_{\cdot i}, \overline{R}_{\cdot j}) = -(N + 1)/12.$$

Proof. We treat $X_{1j}, \ldots, X_{n_j j}$ as one sample and the rest of the data as a second sample. Then we have immediately from Exercise 3.7.1 that $ER_{\cdot j} = n_j(N + 1)/2$ and $\text{Var}\,R_{\cdot j} = n_j(N - n_j)(N + 1)/12$ where n_j and $N - n_j$ are the two sample sizes. The $\text{Cov}(R_{\cdot i}, R_{\cdot j})$ is determined as follows: Let $R_{\cdot(ij)}$ denote the sum of ranks of the combined ith and jth samples. Then $\text{Var}\,R_{\cdot(ij)} = (n_i + n_j)(N - n_i - n_j)(N + 1)/12$. But $R_{\cdot(ij)} = R_{\cdot i} + R_{\cdot j}$, so

$$\text{Var}\,R_{\cdot(ij)} = \text{Var}(R_{\cdot i} + R_{\cdot j})$$

$$= \text{Var}\,R_{\cdot i} + \text{Var}\,R_{\cdot j} + 2\text{Cov}(R_{\cdot i}, R_{\cdot j}).$$

Hence

$$\text{Cov}(R_{\cdot i}, R_{\cdot j}) = \tfrac{1}{2}\{\text{Var } R_{\cdot(ij)} - \text{Var } R_{\cdot i} - \text{Var } R_{\cdot j}\},$$

and when the foregoing given formulas for the variances on the right side are substituted and the expression simplified, we find $\text{Cov}(R_{\cdot i}, R_{\cdot j}) = -n_i n_j (N+1)/12$. The formulas for $\overline{R}_{\cdot i}$ follow immediately from those for $R_{\cdot i}$ by the properties of expectation.

Now the difference $\overline{R}_{\cdot j} - (N+1)/2$ represents the departure from that expected under the null hypothesis. When the accumulated departures are too large we wish to reject $H_0 : \theta_1 = \cdots = \theta_k$. This suggests a test statistic of the form:

$$H = \sum_{j=1}^{k} c_{jN}^2 \left\{ \frac{\overline{R}_{\cdot j} - (N+1)/2}{\sqrt{\text{Var } \overline{R}_{\cdot j}}} \right\}^2. \tag{4.2.2}$$

The weighting constants $c_{1N}, \ldots, c_{kN}$ are chosen so that H is asymptotically chi-squared with $k-1$ degrees of freedom. The use of c_{jN}^2 rather than c_{jN} will be notationally convenient later. At first thought, a natural choice for c_{jN} would seem to be 1. Then H would be the sum of squares standardized rank averages. However, as pointed out earlier, $\overline{R}_{\cdot 1}, \ldots, \overline{R}_{\cdot k}$ are correlated and this will require some adjustment, which is achieved by proper choice of $c_{1N}, \ldots, c_{kN}$.

Following the same argument as in Example 3.2.1, since the $N!$ $/(n_1! \cdots n_k!)$ sequences of combined sample observations are equally likely under the null hypothesis, the distribution of H, for any fixed choice of $c_{1N}, \ldots, c_{kN}$, can be tabled. However, a different table is required for every k and every configuration of sample sizes. Hence, this approach is not very practical, and we turn to the asymptotic distribution as a workable alternative. The following theorem forms the basis for the asymptotic distribution theory needed in the one-way layout.

Theorem 4.2.2. Suppose the k samples come from a common distribution. Suppose $n_j \to \infty$, $j = 1, \ldots, k$ in such a way that $n_j/N \to \lambda_j$, $0 < \lambda_j < 1$, where $N = \sum_1^k n_j$. Suppose $c_{jN} \to c_j$ for $j = 1, \ldots, k$. Define $\mathbf{T}' = (T_1, \ldots, T_k)$ where

$$T_j = c_{jN} \frac{1}{\sqrt{N}} \left(\overline{R}_{\cdot j} - \frac{N+1}{2} \right).$$

Then, **T** is asymptotically $MVN(\mathbf{0}, \mathbf{B})$ where

$$b_{ij} = \begin{cases} c_i^2(1-\lambda_i)/(12\lambda_i) & \text{if } i = j \\ -c_i c_j/12 & \text{if } i \neq j. \end{cases}$$

Proof. First, we relabel the observations X_{ij} $i = 1, \ldots, n_j, j = 1, \ldots, k$, as $Y_1, \ldots, Y_N$, $N = \sum n_i$, where the first n_1 Y's represent the first sample X_{i1}, $i = 1, \ldots, n_1$, and so forth. Let J represent the indices of the jth sample and $\bar{J}$ the complement; hence $Y_j, j \in J$, denotes the jth sample and Y_i, $i \in \bar{J}$, the remainder of the combined sample.

Now $R_j = \#(Y_v > Y_u) + n_j(n_j + 1)/2$, $u \in \bar{J}$, $v \in J$. Let $T_{uv} = 1$ if $Y_v > Y_u$ and 0 otherwise, then

$$R_j - ER_j = \sum_{v \in J} \sum_{u \in \bar{J}} (T_{uv} - 1/2).$$

Under the null hypothesis,

$$E(T_{uv} - 1/2 \mid Y_i = y) = \begin{cases} 0, & u \neq i \text{ and } v \neq i \\ F(y) - 1/2, & v = i \\ 1/2 - F(y), & u = i. \end{cases}$$

Hence we have

$$E(R_j - ER_j \mid Y_i = y) = \sum_{v \in J} \sum_{u \in \bar{J}} E(T_{uv} - 1/2 \mid Y_i = y)$$

$$= \begin{cases} (N - n_j)[F(y) - 1/2] & i \in J \\ n_j[1/2 - F(y)] & i \in \bar{J}. \end{cases}$$

Since $\bar{R}_j - (N + 1)/2 = (R_j - ER_j)/n_j$, the projection of T_j is given by

$$V_j = c_{jN} \frac{1}{\sqrt{N}} \left\{ \frac{(N - n_j)}{n_j} \sum_{i \in J} [F(Y_i) - 1/2] + \sum_{i \in \bar{J}} [1/2 - F(Y_i)] \right\}$$

$$= c_{jN} \frac{1}{\sqrt{N}} \sum_{i=1}^{N} a_i [F(Y_i) - 1/2] \tag{4.2.3}$$

where $a_i = (N - n_j)/n_j$ if $i \in J$ and $a_i = -1$ if $i \in \bar{J}$. Further, from Exercise

3.7.7,

$$\operatorname{Var} V_j = c_{jN}^2 \frac{1}{N} \cdot \frac{1}{12} \left\{ n_j \frac{(N-n_j)^2}{n_j^2} + (N-n_j) \right\}$$

$$= c_{jN}^2 \frac{N(N-n_j)}{12 n_j N} \to c_j^2 \frac{(1-\lambda_j)}{12\lambda_j} .$$

Likewise, from Exercise 4.5.1,

$$\operatorname{Cov}(V_i, V_j) \to - c_i c_j / 12.$$

Theorem A10 applies to show that V_j, $j = 1, \ldots, k$ are asymptotically normally distributed and Theorem A11 applies to show that $\mathbf{V}' = (V_1, \ldots, V_k)$ is asymptotically multivariate normal with mean vector $\mathbf{0}$ and variance-covariance matrix $\mathbf{B}$, given in the statement of the theorem.

From Theorem 4.2.1

$$\operatorname{Var} T_j = c_{jN}^2 \frac{1}{N} \frac{(N-n_j)(N+1)}{(12 n_j)} \to c_j^2 \frac{(1-\lambda_j)}{12\lambda_j} .$$

By the Projection Theorem 2.5.2, $E(T_j - V_j)^2 = \operatorname{Var} T_j - \operatorname{Var} V_j \to 0$ for $j = 1, \ldots, k$, and hence the difference in vectors $\mathbf{T} - \mathbf{V}$ converges to zero in probability. This implies that $\mathbf{T}$ has the same limiting distribution as $\mathbf{V}$ and the proof is complete.

Before discussing the choice of $c_{1N}, \ldots, c_{kN}$ for which H, given by (4.2.2), has an asymptotic chi-square distribution, we present a result from normal distribution theory. Recall from matrix theory that a matrix $\mathbf{A}$ is idempotent if $\mathbf{A}^2 = \mathbf{A}$ and that the rank of an idempotent matrix is the sum of the diagonal elements (the trace).

Theorem 4.2.3. Suppose $(Z_1, \ldots, Z_k)'$ has a $MVN(\mathbf{0}, \mathbf{A})$ distribution. Suppose that $\mathbf{A}$ is idempotent with rank r. Then $\sum_1^k Z_i^2$ has a chi-square distribution with r degrees of freedom, denoted $\chi^2(r)$.

Proof. There exists an orthogonal matrix $\mathbf{G}$ such that $\mathbf{G}'\mathbf{A}\mathbf{G}$ is a diagonal matrix $\mathbf{D}$ with r ones and $k - r$ zeros. Define $\mathbf{U} = \mathbf{G}'\mathbf{Z}$ where $\mathbf{Z}' = (Z_1, \ldots, Z_k)$, then $\mathbf{U}$ has a $MVN(\mathbf{0}, \mathbf{D})$ distribution (Arnold, 1981, p. 46). Now $\sum_1^k Z_i^2 = \mathbf{Z}'\mathbf{Z} = \mathbf{U}'\mathbf{G}'\mathbf{G}\mathbf{U} = \mathbf{U}'\mathbf{U} = \sum_1^r U_i^2$, which is the sum of squares of r i.i.d. $n(0,1)$ random variables. Thus $\sum_1^k Z_i^2$ has a $\chi^2(r)$ distribution.

The next theorem provides the correction values $c_{1N}, \ldots, c_{kN}$ so that H, (4.2.2), has an asymptotic chi-square distribution.

Theorem 4.2.4. Under the null hypothesis that the k samples come from a common distribution,

$$H^* = \sum_{j=1}^{k} \left(1 - \frac{n_j}{N}\right) \left\{ \frac{\bar{R}_{.j} - (N+1)/2}{\sqrt{(N-n_j)(N+1)/(12n_j)}} \right\}^2$$

$$= \frac{12}{N(N+1)} \sum_{j=1}^{k} n_j \left[\bar{R}_{.j} - (N+1)/2\right]^2$$

$$= \left(\frac{12}{N(N+1)} \sum_{j=1}^{k} \frac{R_{.j}^2}{n_j}\right) - 3(N+1)$$

has an asymptotic $\chi^2(k-1)$ distribution.

Proof. Define, from the statement of Theorem 4.2.2,

$$T_j^* = T_j / \sqrt{(1 - \lambda_j)/(12\lambda_j)} ,$$

then it follows that $\mathbf{T}^{*\prime} = (T^*_1, \ldots, T_k^*)$ has an asymptotic $MVN(\mathbf{0}, \mathbf{B}^*)$ distribution with

$$b_{ij}^* = \begin{cases} c_i^2 , & i = j \\ -c_i c_j \sqrt{\lambda_i \lambda_j / [(1-\lambda_i)(1-\lambda_j)]} , & i \neq j. \end{cases}$$

Since $n_j/N \to \lambda_j$, we have, approximately, from (2.4.2),

$$H = \sum_{j=1}^{k} c_{jN}^2 \left\{ \frac{\bar{R}_{.j} - (N+1)/2}{\sqrt{(N-n_j)(N+1)/(12n_j)}} \right\}^2$$

$$\doteq \sum_{j=1}^{k} c_{jN}^2 \left\{ \frac{\bar{R}_{.j} - (N+1)/2}{\sqrt{N(1-\lambda_j)/(12\lambda_j)}} \right\}^2$$

$$= \sum_{j=1}^{k} (T_j^*)^2.$$

Now, Theorem A2(b) implies that H converges in distribution to $\sum_1^k Z_i^2$ where $(Z_1, \ldots, Z_k)'$ has a $MVN(\mathbf{0}, \mathbf{B}^*)$ distribution. Further, from Theorem 4.2.3, $\sum_1^k Z_i^2$ will have a chi-square distribution provided $\mathbf{B}^*$ is idempotent. Hence, we seek the constants $c_1, \ldots, c_k$ such that $(\mathbf{B}^*)^2 = \mathbf{B}^*$.

By comparing $(\mathbf{B}^*)^2$ to $\mathbf{B}^*$ it is possible to anticipate the solution. We state the solution and verify that $\mathbf{B}^*$ is idempotent. Let

$$c_{jN}^2 = \frac{N - n_j}{N} \to 1 - \lambda_j.$$

Hence $b_{ij}^* = 1 - \lambda_i$ if $i = j$ and $b_{ij}^* = -(\lambda_i \lambda_j)^{1/2}$ when $i \neq j$. We can then write

$$\mathbf{B}^* = \mathbf{I} - \boldsymbol{\delta}\boldsymbol{\delta}'$$

where $\boldsymbol{\delta}' = [(\lambda_1)^{1/2}, \ldots, (\lambda_k)^{1/2}]$ and $\mathbf{I}$ is the $k \times k$ identity. Since $\boldsymbol{\delta}'\boldsymbol{\delta} = \sum_1^k \lambda_i = 1$, it is easy to check that $\mathbf{B}^*$ is idempotent. The rank of $\mathbf{B}^* = $ trace of $\mathbf{B}^* = \sum_1^k b_{ii}^* = \sum_1^k (1 - \lambda_i) = k - \sum_1^k \lambda_i = k - 1$. Hence, we have $k - 1$ degrees of freedom and the proof is complete. In Exercise 4.5.2 you are asked to carry out the algebraic reduction to verify the other equations for H^*. The third equation is the most practical.

The statistic H^* is called the Kruskal–Wallis (1952) statistic and rejects $H_0: \theta_1 = \cdots = \theta_k$ at approximate level α when $H^* \geqslant \chi_\alpha^2(k - 1)$, where $\chi_\alpha^2(k - 1)$ is the $1 - \alpha$ percentile of the chi-square distribution with $k - 1$ degrees of freedom.

When the Kruskal–Wallis test rejects H_0, we can construct pairwise multiple comparisons to locate the source of significance. There are $k(k - 1)/2$ pairwise comparisons, each based on the difference in the column average ranks. Let

$$D_{ij} = \frac{1}{\sqrt{N}} \left(\bar{R}_{.j} - \bar{R}_{.i} \right). \tag{4.2.4}$$

Under the null hypothesis, $ED_{ij} = 0$ and from Theorem 4.2.1,

$$\operatorname{Var} D_{ij} = \frac{1}{N} \left[\operatorname{Var} \bar{R}_{.j} + \operatorname{Var} \bar{R}_{.i} - 2 \operatorname{Cov}\left(\bar{R}_{.j}, \bar{R}_{.i} \right) \right]$$

$$= \frac{N + 1}{12} \left[\frac{1}{n_i} + \frac{1}{n_j} \right]$$

$$\to \frac{1}{12} \left[\frac{1}{\lambda_i} + \frac{1}{\lambda_j} \right]. \tag{4.2.5}$$

In Theorem 4.2.2 take $c_{jN} = 1$, $j = 1, \ldots, k$, then an application of Theorem A2(b) to the differences implies that D_{ij} is asymptotically $n(0, \sigma^2)$ with σ^2 given by (4.2.5).

Let α denote a prescribed overall error rate for the experiment, and let $\alpha' = 2\alpha/[k(k-1)]$ denote the pairwise comparison error rate. We will declare θ_i and θ_j significantly different at overall level α if

$$|D_{ij}| \geqslant Z_{\alpha'/2}\sqrt{\operatorname{Var} D_{ij}}$$

or

$$|\bar{R}_{.j} - \bar{R}_{.i}| \geqslant Z_{\alpha'/2}\sqrt{\frac{N(N+1)}{12}\left(\frac{1}{n_i} + \frac{1}{n_j}\right)}. \qquad (4.2.6)$$

The next theorem provides the interpretation of α as the overall error rate.

Theorem 4.2.5. Under the null hypothesis, the probability of committing at least one error with (4.2.6), for $1 \leqslant i \leqslant j \leqslant k$, is bounded above by α, when N is large.

Proof. Let E_{ij} be the event $|D_{ij}| \geqslant Z_{\alpha'/2}(\operatorname{Var} D_{ij})^{1/2}$. From the asymptotic normality of D_{ij} we have, under the null hypothesis, $P(E_{ij}) \doteq \alpha'$. Hence the probability that D_{ij} commits an error is approximately α'. The probability of at least one error is

$$P\left[\bigcup_{i<j} E_{ij}\right] \leqslant \sum\sum_{i<j} P(E_{ij}) = \frac{k(k-1)}{2}\alpha' = \alpha.$$

The inequality is known as Bonferroni's inequality.

The preceding multiple comparisons were first suggested by Dunn (1964). For other approaches see the discussion in Miller (1981).

Example 4.2.2. Data from the experiment performed on 32 virgin rats (Example 4.2.1) is given in Table 4.1. The measurement is time until retrieving behavior is established. The unit of time is the length of an observation session. Hence, a value of 0.5 means the behavior began half way into the first observation session. In ranking the data, we have assigned the average rank to tied observations.

Table 4.1. Latency Times (Artificial Data)

	Maternal Plasma	Proestrus Plasma	Diestrus Plasma	Saline
	0.5(2)[a]	1.1(6)	0.4(1)	0.9(4)
	0.7(3)	1.6(8)	1.9(10)	2.1(11)
	1.0(5)	3.7(18)	2.4(13.5)	3.0(16)
	1.2(7)	4.3(20)	2.8(15)	4.7(21.5)
	1.7(9)	4.7(21.5)	3.9(19)	6.4(25)
	2.3(12)	5.6(24)	5.4(23)	6.6(26.5)
	2.4(13.5)	6.6(26.5)	11.4(31)	8.5(28)
	3.1(17)	8.8(29)	20.4(32)	10.0(30)
$R_{.j}$	68.5	153	144.5	162
$ER_{.j}$	132	132	132	132

[a] Number in parentheses is the rank.

To test $H_0 : \theta_1 = \cdots = \theta_4$ versus $H_A : \theta_1, \theta_2, \theta_3, \theta_4$ not all equal at $\alpha = .05$, we reject H_0 if $H^* \geqslant \chi^2_{.05}(3)$, where H^* is the Kruskal–Wallis statistic, given in Theorem 4.2.4 and $\chi^2_{.05}(3) = 7.81$. The data yields $H^* = 7.85$, and hence we reject H_0 and conclude there are differences among the populations.

In Table 4.2, we have recorded the absolute differences $|R_{.j} - R_{.i}|$. Since the sample sizes are equal, (4.2.6) can be converted to rank sums and becomes

$$|R_{.j} - R_{.i}| \geqslant Z_{\alpha'/2} \sqrt{\frac{nN(N+1)}{6}}$$

It is clear from Table 4.2 that the significance arises from the differences between the maternal plasma group and all other groups. Further, there is not much difference among the others. If we take an overall error rate of

Table 4.2. Absolute Differences of Rank Sums

	MP	PP	DP
PP	84.5		
DP	76	8.5	
S	93.5	9	17.5

$\alpha = .12$, then $\alpha' = .02$, $Z_{\alpha'/2} = 2.326$ and we would use $|R_{.j} - R_{.i}| \geqslant 87.3$. If $\alpha = .24$, then $|R_{.j} - R_{.i}| \geqslant 77.1$. Hence, to declare significant differences, we must assign a fairly large overall error rate. This is not surprising in this example since H^* is just barely significant at the 5% level, and we have small sample sizes on which to base the multiple comparisons.

The idea of Pitman efficiency can be extended to tests whose statistics have asymptotic chi-square distributions. A multivariate version of Theorem 2.6.1 is needed. We present only a heuristic development; a rigorous treatment can be found in Hajek and Sidak (1967, Chapter VI). Recall, from the discussion following Theorem 2.6.1, that the asymptotic distribution of a test statistic, under a sequence of alternatives converging to the null hypothesis, changes only in the mean. Hence, when we know the asymptotic distribution under H_0, we need to investigate the behavior of the mean of the test statistic under a sequence of alternatives.

In Theorem 4.2.4, we showed that the vector $\mathbf{T}^{*'} = (T_1^*, \ldots, T_k^*)$, where

$$T_j^* = c_{jN} \left\{ \frac{\bar{R}_{.j} - E\bar{R}_{.j}}{\sqrt{\operatorname{Var} \bar{R}_{.j}}} \right\},$$

is asymptotically $MVN(\mathbf{0}, \mathbf{B}^*)$. We have replaced λ_j by n_j/N in the definition of T_j^*. The following theorem describes the behavior of ET_j^* under a sequence of alternatives. The calculations are similar to those in Example 2.5.7.

Theorem 4.2.6. Suppose the Pitman regularity conditions in Section 2.6 hold. Let $\boldsymbol{\theta}_N' = (\theta_1, \ldots, \theta_k) = (\alpha + \beta_1/N^{1/2}, \ldots, \alpha + \beta_k/N^{1/2})$. Then, as $N \to \infty$,

$$E_{\boldsymbol{\theta}_N} T_j^* \to c_j \sqrt{12\lambda_j/(1 - \lambda_j)} \int f^2(x)\, dx \left(\beta_j - \bar{\beta} \right), \qquad (4.2.7)$$

for $j = 1, \ldots, k$, where $c_{jN} \to c_j$ and $\bar{\beta} = \sum \lambda_i \beta_i$.

Proof. Initially suppose that X_{ij} has cdf $F(x - \theta_j)$, $i = 1, \ldots, n_j$, $j = 1, \ldots, k$. Since $R_{.j} = \sum_{i=1}^{n_j} R_{ij}$, where R_{ij} is the rank of X_{ij}, we have

$$ER_{.j} = E\sum_{i \neq j} W_{ij} + \frac{n_j(n_j + 1)}{2},$$

where $W_{ij} = \#(X_{ui} \leqslant X_{vj})$, $u = 1, \ldots, n_i$; $v = 1, \ldots, n_j$. Further,

$$EW_{ij} = n_i n_j P(X_{ui} \leqslant X_{vj})$$

$$= n_i n_j \int F(t - \theta_i) f(t - \theta_j) \, dt$$

$$= n_i n_j \int F(t + (\theta_j - \theta_i)) f(t) \, dt.$$

Expanding EW_{ij} as a function of (θ_i, θ_j) about $(0,0)$ we have

$$EW_{ij} \doteq n_i n_j \left\{ \frac{1}{2} + \theta_j \int f^2(x) \, dx - \theta_i \int f^2(x) \, dx \right\},$$

where the error of the approximation is of small order in θ_i and θ_j. This yields

$$ER_{\cdot j} \doteq n_j (N - n_j) \left\{ \frac{1}{2} + \theta_j \int f^2(x) \, dx \right\} - n_j \sum_{i \neq j} n_i \theta_i \int f^2(x) \, dx + \frac{n_j(n_j + 1)}{2}$$

and

$$ET_j^* \doteq c_{jN} \sqrt{\frac{12 n_j}{(N - n_j)(N + 1)}} \; N \left(\theta_j - \sum_{i=1}^{k} \frac{n_i}{N} \theta_i \right) \int f^2(x) \, dx.$$

Now substitute $\theta_j = \alpha + \beta_j / N^{1/2}$ and take the limit as $N \to \infty$, then

$$E_{\theta_N} T_j^* \to c_j \sqrt{\frac{12 \lambda_j}{(1 - \lambda_j)}} \left(\beta_j - \sum_{i=1}^{k} \lambda_i \beta_i \right) \int f^2(x) \, dx.$$

This completes the proof.

Since we are interested in the Kruskal–Wallis statistic H^*, let $c_{jN} = [(N - n_j)/N]^{1/2} \to (1 - \lambda_j)^{1/2}$. Then the extension of Theorem 2.6.1 states that $\mathbf{T}^{*'} = (T_1^*, \ldots, T_k^*)$ is asymptotically $MVN(\boldsymbol{\mu}, \mathbf{B}^*)$, where $b_{ii}^* = 1 - \lambda_i$ if $i = j$, and $b_{ij}^* = -(\lambda_i \lambda_j)^{1/2}$ if $i \neq j$, and $\mu_j = 12^{1/2} \int f^2(x) \, dx \, \lambda_j^{1/2} (\beta_j - \bar{\beta})$, with $\bar{\beta} = \sum \lambda_i \beta_i$.

Finally, recall that $H^* = \sum_1^k (T_j^*)^2$. This statistic continues to have an asymptotic chi-square distribution. However, it is a noncentral chi-square distribution with $k - 1$ degrees of freedom. The noncentrality parameter is found by substituting the means into H^* and is given by

$$\delta_{H^*} = \sum_{j=1}^{k} \left[ET_j^* \right]^2$$

$$= 12 \left(\int f^2(x) \, dx \right)^2 \sum_{j=1}^{k} \lambda_j \left(\beta_j - \bar{\beta} \right)^2. \qquad (4.2.8)$$

The magnitude of the asymptotic power, along the sequence of local alternatives, is determined by the noncentrality parameter. Since the expected value of a noncentral chi-square distribution, with r degrees of freedom and noncentrality parameter δ, is $r + \delta$, larger values of δ correspond to larger values of asymptotic power. See Andrews (1954) for further discussion. This prompts the definition of asymptotic efficiency as the ratio of noncentrality parameters; see Hannan (1956).

Let $(k-1)F$ denote $(k-1)$ times the usual one-way F test. Under regularity conditions, discussed by Arnold (1981, Chapter 10), $(k-1)F$ has an asymptotic noncentral chi-square distribution when $\theta'_N = (\alpha + \beta_1/N^{1/2}, \ldots, \alpha + \beta_k/N^{1/2})$. The noncentrality parameter is found by substituting the mean into the equation for F and is given by

$$\delta_F = \frac{1}{\sigma^2} \sum \lambda_j \left(\beta_j - \bar{\beta} \right)^2 \qquad (4.2.9)$$

where σ^2 is the variance of F and $\bar{\beta} = \sum \lambda_j \beta_j$; see Arnold (1981, p. 93). Hence we have

$$e(H^*, F) = \delta_{H^*}/\delta_F = 12\sigma^2 \left(\int f^2(x) \, dx \right)^2, \qquad (4.2.10)$$

which is the Pitman efficiency of the Wilcoxon signed rank test realtive to the one-sample t test and the Pitman efficiency of the Mann–Whitney–Wilcoxon test relative to the two-sample t test. Thus the Kruskal–Wallis test shares the efficiency properties of the one- and two-sample rank tests.

It may well be that the omnibus alternative hypothesis of unequal locations is not appropriate for the experiment under consideration. The researcher may wish to detect an increasing (or decreasing) experimental effect. This is similar to the one-sided alternative in the one- and two-sample problems. The Kruskal–Wallis test is not appropriate becaue it is designed to detect any departure from equal locations. It is possible to tailor a test which has more power to detect an increasing alternative.

Suppose the data consists in k samples as described at the beginning of this section. Suppose we wish to test $H_0 : \theta_1 = \cdots = \theta_k$ versus $H_A : \theta_1$

$\leqslant \cdots \leqslant \theta_k$ with at least one strict inequality. We construct a statistic which assesses the degree of agreement between the observed average ranks, $\overline{R}_{\cdot j}, j = 1, \ldots, k$, and the hypothesized ordering. Let

$$L = \frac{1}{\sqrt{N}} \sum_{j=1}^{k} \left(j - \frac{k+1}{2} \right)\left(\overline{R}_{\cdot j} - \frac{N+1}{2} \right), \qquad (4.2.11)$$

then large values of L support the alternative hypothesis. Under the null hypothesis, $EL = 0$ and

$$\operatorname{Var} L = \frac{N+1}{12} \sum_{j=1}^{k} \frac{1}{n_j} \left(j - \frac{k+1}{2} \right)^2; \qquad (4.2.12)$$

see Exercise 4.5.3. When the sample sizes are all equal to n, $\operatorname{Var} L = (k^2 - 1)(nk + 1)/(144n)$. Theorem 4.2.2, with $c_{jN} = j - (k+1)/2, j = 1, \ldots, k$, and Theorem A2(b) imply that L is asymptotically $n(0, \sigma^2)$ with σ^2 given by (4.2.12). Hence, to test $H_0 : \theta_1 = \cdots = \theta_k$ versus $H_A : \theta_1 \leqslant \cdots \leqslant \theta_k$ with at least one strict inequality, reject H_0 at approximate level α if $L \geqslant Z_\alpha (\operatorname{Var} L)^{1/2}$ where Z_α is the upper α percentile of the standard normal distribution.

The problem has the flavor of regression. When the ordering specified by the alternative hypothesis is quantitative, rather than qualitative, the regression methods in the next chapter are appropriate. With only a qualitative ordering (ordinal scale) to work with, the test based on L can be quite useful. A significant loss of power can result when the ordered alternative is appropriate, but the Kruskal–Wallis test is used.

Example 4.2.3. In the Stanford heart transplant study various quantitative and semiquantitative measurements were taken on the patients. One measure, the mismatch score, indicates the degree to which the donor and the recipient are mismatched for tissue type. It could be hypothesized that survival time will tend to increase with lower mismatch scores. The survival times, presented by Mosteller and Tukey (1977, p. 571), are given in Table 4.3. Mismatch scores are classified as low (0–1), medium (1–2), and high (2–). If θ_L, θ_M, and θ_H denote the population median survival times corresponding to these three groups, then we wish to test $H_0 : \theta_L = \theta_M = \theta_H$ versus $H_A : \theta_L \geqslant \theta_M \geqslant \theta_H$ with a least one strict inequality. If we take $\alpha = 0.05$, then we reject H_0 if $L \leqslant - Z_\alpha (\operatorname{Var} L)^{1/2}$. Now $n_1 = 14$, $n_2 = 13$, $n_3 = 12$, $N = 39$, $\operatorname{Var} L = .52$, $- Z_\alpha = - 1.645$ and we reject H_0 if $L \leqslant - 1.19$. For $k = 3$, $L = [1/(39)^{1/2}] (\overline{R}_{\cdot 3} - \overline{R}_{\cdot 1}) = - 0.8$, and we fail to reject H_0. Hence the data does not support, at $\alpha = 0.05$, the hypothesis that survival time tends to increase with decreasing mismatch scores. There are

Table 4.3.   Survival Times

	Mismatch Category		
	Low	Medium	High
	44 (11)[a]	15 (5)	3 (2)
	551 (33)	280 (30)	136 (27)
	127 (26)	1024 (38)	65 (22.5)
	1 (1)	253 (29)	25 (7)
	297 (31)	66 (24)	64 (21)
	46 (12)	29 (9)	322 (32)
	60 (19)	161 (28)	23 (6)
	65 (22.5)	624 (34)	54 (18)
	12 (4)	39 (10)	63 (20)
	1350 (39)	51 (16.5)	50 (15)
	730 (35)	68 (25)	10 (3)
	47 (13)	836 (36)	48 (14)
	994 (37)	51 (16.5)	
	26 (8)		
$\bar{R}_{\cdot j}$	21	23	16

[a] Number in parentheses is the rank in the combined sample.

other variables, such as age or waiting time for the donor, or the general physical condition of the patient, that may have a stronger impact on survival time.

Terpstra (1952) and Jonckheere (1954) independently proposed a test for the ordered alternative which is based on the pairwise Mann–Whitney–Wilcoxon statistics. This approach has the attractive feature that the comparison of samples i and j does not depend on the rest of the combined data. For testing $H_0 : \theta_1 = \cdots = \theta_k$ versus $H_A : \theta_1 \leqslant \cdots \leqslant \theta_k$, with at least one strict inequality, let

$$J = \sum_{i<j} \sum W_{ij} \qquad (4.2.13)$$

where $W_{ij} = \#(X_{vj} > X_{ui})$, $v = 1, \ldots, n_j$; $u = 1, \ldots, n_i$, discussed in Section 3.2. We reject H_0 for large values of J.

Let $\mathbf{W}' = (W_{12}, \ldots, W_{1k}, W_{23}, \ldots, W_{2k}, \ldots, W_{k-1k})$ the $k(k-1)/2$-component vector of Mann–Whitney–Wilcoxon statistics and let $N = \sum_1^k n_i$, the combined sample size. Suppose that $n_j/N \to \lambda_j$, $0 < \lambda_j < 1$. Under H_0, the limiting multivariate normality of $N^{-3/2}(\mathbf{W} - E\mathbf{W})$ follows from the componentwise limiting normality of the projections, Theorem 3.2.4, and the convergence of the variance-covariance matrix, Theorem A13. In order to make the test operational we need the covariances.

Under H_0, from (3.2.3), $EW_{ij} = n_i n_j / 2$, $\operatorname{Var} W_{ij} = n_i n_j (n_i + n_j + 1)/12$. Further, $\operatorname{Cov}(W_{st}, W_{uv}) = 0$ when s, t are both different from u, v. Exercise 4.5.4 provides the other covariances.

In order to find the variance of J we define

$$W_i = \sum_{u=1}^{i-1} W_{ui}, \tag{4.2.14}$$

so that $J = \sum_{i=2}^{k} W_i$. Further,

$$
\begin{aligned}
\operatorname{Cov}(W_s, W_t) &= \operatorname{Cov}\left(\sum_{u=1}^{s-1} W_{us}, \sum_{v=1}^{t-1} W_{vt} \right) \\
&= \sum_{u=1}^{s-1} \operatorname{Cov}\left(W_{us}, \sum_{v=1}^{t-1} W_{vt} \right) \\
&= \sum_{u=1}^{s-1} \operatorname{Cov}(W_{us}, W_{ut} + W_{st}).
\end{aligned}
$$

But, from Exercise 4.5.4,

$$\operatorname{Cov}(W_{us}, W_{ut} + W_{st}) = \operatorname{Cov}(W_{us}, W_{ut}) + \operatorname{Cov}(W_{us}, W_{st}) = 0.$$

Hence $\operatorname{Cov}(W_s, W_t) = 0$ and

$$\operatorname{Var} J = \sum_{i=2}^{k} \operatorname{Var} W_i. \tag{4.2.15}$$

Now W_i is the Mann–Whitney–Wilcoxon statistic computed on the ith sample versus the combined data in the first $i - 1$ samples. If we let $N_i = \sum_{j=1}^{i} n_j$, where $N_1 = n_1$ and $N_k = N$, then, from (3.2.3), $\operatorname{Var} W_i = n_i N_{i-1}(N_i + 1)/12$. Hence

$$\operatorname{Var} J = \frac{1}{12} \sum_{i=2}^{k} n_i N_{i-1}(N_i + 1). \tag{4.2.16}$$

Jonckheere (1954) developed the cumulant generating function and from this provided the following alternate equation for (4.2.16)

$$\operatorname{Var} J = \frac{1}{72} \left\{ N^2(2N+3) - \sum_{j=1}^{k} n_j^2(2n_j + 3) \right\}. \qquad (4.2.17)$$

From (4.2.13)

$$EJ = \sum\sum_{i<j} n_i n_j / 2$$

$$= N^2 - \sum_{j=1}^{k} n_j^2 / 4, \qquad (4.2.18)$$

where the second inequality follows from $(\sum n_i)^2 = \sum n_i^2 + 2\sum\sum_{i<j} n_i n_j$.

The Jonckheere–Terpstra test rejects $H_0 : \theta_1 = \cdots = \theta_k$ in favor of $H_A : \theta_1 \leqslant \cdots \leqslant \theta_k$ with at least one strict inequality, at approximate level α, when $J \geqslant EJ + Z_\alpha(\operatorname{Var} J)^{1/2}$, where EJ and $\operatorname{Var} J$ are given by (4.2.18) and (4.2.16) or (4.2.17), and Z_α is the upper α percentile from the standard normal distribution.

The property that $W_2, \ldots, W_k$, defined in (4.2.14) are uncorrelated can be strengthened. They are, in fact, independent. The independence can be used to develop an alternative argument for the asymptotic normality of J.

Exercise 4.5.5 describes a relationship between L, (4.2.11), and statistics based on pairwise Mann–Whitney–Wilcoxon statistics. Tryon and Hettmansperger (1974) consider weighted linear combinations of Mann–Whitney–Wilcoxon statistics. Other approaches can be found in Chacko (1963) and Johnson and Mehrotra (1971). The monograph by Barlow et al. (1972) provides an excellent overview of the area of statistical inference under order restrictions.

4.3.   THE TWO-WAY LAYOUT: THE FRIEDMAN TEST

In the last section we considered the comparison of k samples to detect significant differences among the sampled populations. In the randomization model, N subjects would be randomly assigned to k treatments. Existing differences among the k treatments may be obscured by relatively large variablility of subjects within the samples. Often this problem can be alleviated by dividing the subjects into more homogeneous subgroups or blocks. Comparisons are then carried out within the blocks.

We restrict attention to the complete randomized block design with one observation per cell. Hence we have $N = nk$ subjects divided into n blocks and subjects are assigned to the k treatments at random.

Repeated measures on subjects also provides an important example. In this application, a single subject forms a block and k measurements are made on the subject. The order in which the k measurements are taken is often randomized and we wish to detect a consistent pattern of measurement differences among the subjects. For example, n judges ranking k items could be analyzed using this design.

The sampling model can be defined in two ways: (1) by X_{ij}, $i = 1, \ldots, n, j = 1, \ldots, k$, with cdf $F_i(x - \theta_j)$, $F_i \in \Omega_0$, $i = 1, \ldots, n$. Hence F_i is the distribution of the observations in the ith block, and, within the ith block, θ_j is the median corresponding to the jth treatment. All observations are independent. (2) by $(X_{i1}, \ldots, X_{ik})$ with joint cdf $F_i(x_1 - \theta_1, \ldots, x_k - \theta_k)$, $i = 1, \ldots, n$ where $F_i(x_1, \ldots, x_k) = F_i(y_1, \ldots, y_k)$ and $(y_1, \ldots, y_k)$ is a permutation of $(x_1, \ldots, x_k)$. Hence the random variables $X_{i1}, \ldots, X_{ik}$ are said to be exchangeable. This is the appropriate model for repeated measures where it is not appropriate to assume independence within a block.

Example 4.3.1. Todd et al. (1980) attempt to isolate a formal mathematical transformation that a human subject would perceive as descriptive of growth. The idea that a geometric transformation might be helpful in describing morphological change can be traced back to the work of D'Arcy Wentworth Thompson in the early 1900s. Most of the work had been of a qualitative nature until Todd et al. studied the effects of several specific mathematical transformations on infant facial profiles. They used five different transformations: cardioidal strain (CS), spiral strain (SS), affine shear (AS), reflected shear (RS), and rotation (R). The strain transforms tend to change circles into heart-shaped figures, and the shear transforms tend to change circles into diagonally oriented ellipses. The article contains illustrations.

After preliminary research, the authors hypothesized that the effects of a cardioidal strain are perceptually equivalent to the morphological changes produced by normal growth of a human head. Subjects were shown different sequences of five facial profiles. The sequences were designed so that the perceived age would increase from left to right. One set of sequences consisted of actual growth (AG) profiles traced from X-rays. Profiles generated by the mathematical transformations all began with an actual growth profile. There was also another group of control (C) sequences in which all five profiles were identical. Subjects were asked to rate each sequence from 0 to 4 on the basis of its resemblance to actual growth.

We have a two-way layout with $k = 7$ "treatments": AG, CS, SS, AS, RS, R, and C; and the subjects, producing repeated measurements, constitute the blocks. The data, presented in Example 4.3.2, hopefully will reject the null hypothesis of no difference among these treatments and reveal an association of the strain transformations, CS and SS, with actual growth.

We wish to test $H_0 : \theta_1 = \cdots = \theta_k$ versus H_A : not all θ_i's are equal. Let R_{ij} be the rank of X_{ij} among $X_{i1}, \ldots, X_{ik}$, the observations in the ith block. Let $R_{.j} = \sum_{i=1}^{n} R_{ij}$, the sum of ranks corresponding to the jth treatment. The ranks can be displayed in a two-way array as follows:

	Treatments			
Blocks	1	2	$\cdots$	k
1	R_{11}	R_{12}	$\cdots$	R_{1k}
2	R_{21}	R_{22}	$\cdots$	R_{2k}
$\vdots$	$\vdots$			
n	R_{n1}	R_{n2}	$\cdots$	R_{nk}
	$R_{.1}$	$R_{.2}$	$\cdots$	$R_{.k}$

Under the null hypothesis, $R_{i1}, \ldots, R_{ik}$ are distributed according to the results in Theorem 3.2.1, even under the exchangeability model. Hence $ER_{ij} = (k + 1)/2$, $\operatorname{Var} R_{ij} = (k^2 - 1)/12$, and

$$ER_{.j} = n(k + 1)/2,$$

$$\operatorname{Var} R_{.j} = n(k^2 - 1)/12, \tag{4.3.1}$$

$$\operatorname{Cov}(R_{.i}, R_{.j}) = -n(k + 1)/12;$$

see Exercise 4.5.6.

Friedman (1937) proposed a statistic of the form:

$$K = \sum_{j=1}^{k} c_{jN}^2 \left\{ \frac{R_{.j} - ER_{.j}}{\sqrt{\operatorname{Var} R_{.j}}} \right\}^2, \tag{4.3.2}$$

where the weighting constants are chosen so that K has an asymptotic $\chi^2(k - 1)$ distribution.

Let $\mathbf{T}' = (T_1, \ldots, T_k)$ where

$$T_j = c_{jN} \frac{1}{\sqrt{n}} \left(R_{.j} - \frac{n(k + 1)}{2} \right). \tag{4.3.3}$$

Then the asymptotic distribution of $\mathbf{T}$ is $MVN(\mathbf{0}, \mathbf{B})$ where

$$b_{ij} = \begin{cases} c_i^2(k^2 - 1)/12 & i = j \\ -c_i c_j(k + 1)/12 & i \neq j \end{cases} \tag{4.3.4}$$

and $c_{jN} \to c_j, j = 1, \ldots, k$. The proof of this result is outlined in Exercise 4.5.7. The result forms the basis for the asymptotic distribution theory needed in the two-way layout, similar to the one-way layout.

Further, Exercise 4.5.7 shows that Friedman's statistic,

$$K^* = \sum_{j=1}^{k} \left(1 - \frac{1}{k}\right) \left\{ \frac{R_{.j} - n(k + 1)/2}{\sqrt{n(k^2 - 1)/12}} \right\}^2$$

$$= \frac{12}{nk(k + 1)} \sum_{j=1}^{k} \left[R_{.j} - n(k + 1)/2 \right]^2$$

$$= \left[\frac{12}{nk(k + 1)} \sum_{j=1}^{k} R_{.j}^2 \right] - 3n(k + 1), \tag{4.3.5}$$

rejects the null hypothesis $H_0 : \theta_1 = \cdots = \theta_k$ at approximate level α, if $K^* \geq \chi_\alpha^2(k - 1)$, where $\chi_\alpha^2(k - 1)$ is the upper α percentile for a chi-square distribution with $k - 1$ degrees of freedom.

When the Friedman test rejects H_0, we can construct pairwise multiple comparisons based on the rank sums: $R_{.1}, \ldots, R_{.k}$. From Exercise 4.5.7 and Theorem A2(b), $(R_{.j} - R_{.i})/[\text{Var}(R_{.j} - R_{.i})]^{1/2}$ has an asymptotic $n(0, 1)$ distribution under H_0. Using (4.3.1) it is easy to see that $\text{Var}(R_{.j} - R_{.i}) = nk(k + 1)/6$. Hence, similar to (4.2.6) in the one-way layout, declare θ_i and θ_j significantly different, at overall level α, if

$$|R_{.j} - R_{.i}| \geq Z_{\alpha'/2}\sqrt{nk(k + 1)/6}, \tag{4.3.6}$$

where $\alpha' = 2\alpha/k(k - 1)$ and $1 - \Phi(Z_{\alpha'/2}) = \alpha'/2$. By the same argument as in the proof of Theorem 4.2.5, the probability of committing at least one error, under H_0, is bounded above by α.

Example 4.3.2. This example continues Example 4.3.1. Five subjects are presented with "growth" sequences. They score each sequence from 0 to 4 with zero representing no perception of growth. The different sequences are presented in random order and each subject sees 5 sequences for each of

Table 4.4.   Ratings of "Growth" Sequences (Artificial Data)

Subject	AG	CS	SS	AS	RS	R	C
1	3.9 (7)[a]	3.5 (6)	2.8 (5)	1.5 (4)	0.5 (2)	0.6 (3)	0.2 (1)
2	3.4 (6.5)	3.4 (6.5)	2.5 (5)	1.0 (4)	0.8 (3)	0.1 (1)	0.2 (2)
3	3.8 (7)	3.0 (5)	3.1 (6)	0.9 (4)	0.6 (3)	0.2 (1)	0.4 (2)
4	3.2 (6)	3.4 (7)	3.0 (5)	1.2 (4)	0.4 (3)	0.2 (1)	0.3 (2)
5	3.7 (7)	3.2 (6)	2.7 (5)	1.0 (4)	0.2 (2)	0.3 (3)	0.1 (1)
$R_{.j}$	33.5	30.5	26	20	13	9	8

[a] Number in parentheses is rank within the row.

the 7 types. In Table 4.4 the mean of the 5 scores for each type of sequence is reported along with its rank among the 7 types. To test $H_0 : \theta_1 = \cdots = \theta_k$ versus H_A : not all equal at $\alpha = .05$, we reject H_0 if $K^* \geqslant \chi^2_{.05}(6) = 12.6$. In (4.3.5), using $n = 5$ and $k = 7$, we have $K^* = 27.5$, and hence we reject H_0 and claim, at approximately $\alpha = .05$, that there is a difference among the treatments.

There are $k(k-1)/2 = 21$ pairwise comparisons. If we take the overall error rate $\alpha = .21$, then the comparison error rate $\alpha' = 2\alpha/k(k-1) = .01$ and $Z_{\alpha'/2} = 2.576$. From (4.3.6), a pair will be declared significantly different if $|R_{.j} - R_{.i}| \geqslant 17.6$. This simple analysis shows that AG and CS are significantly far from RS, R, and C, and supports the hypothesis that the cardioidal strain is perceived as growth while eliminating reflected shear and rotation.

Exercise 4.5.8 describes the behavior of K^* for a sequence of alternatives that converges to the null hypothesis. From Arnold (1981, p. 87) $(k-1)F$, which is $(k-1)$ times the usual F statistic for testing H_0, is asymptotically noncentral chi-square with $k-1$ degrees of freedom and noncentrality parameter

$$\delta_F = \frac{k}{\sigma^2} \sum_1^k (\beta_j - \bar{\beta})^2. \tag{4.3.7}$$

We have assumed the same model as in Exercise 4.5.8.

Hence, as in the discussion of (4.2.10), the efficiency of K^* relative to F is

$$e(K^*, F) = \delta_{K^*}/\delta_F$$

$$= 12\sigma^2 \left(\int f^2(x)\, dx \right)^2 \frac{k}{k+1}. \tag{4.3.8}$$

The striking point is that K^* does not inherit the efficiency of the Wilcoxon one- and two-sample tests relative to the t tests, as was the case with the Kruskal–Wallis test, H^*. When the underlying distribution is normal,

$$e(K^*, F) = 3k / [(k + 1)\pi].$$ (4.3.9)

Hence, when $k = 2$, $e(K^*, F) = 2/\pi = 0.64$, the efficiency of the sign test relative to the t test. See Exercise 4.5.9. This reflects the loss of information incurred by ranking within blocks, especially for a small number of treatments. This efficiency loss vanishes as k increases, and $e(K^*, F) \to 3/\pi = 0.955$ as $k \to \infty$. Following are some values of $(k, e(K^*, F))$: (2, 0.64), (3, 0.72), (4, 0.76), (5, 0.80), (10, 0.87), (20, 0.91).

Methods, to take the interblock information into account, in the general linear model setting, are presented in the next chapter. In the randomized block design, the lost efficiency can be recovered. These methods, which generally involve constructing a rank statistic in the residuals after the block effect has been removed, are more complicated and less easy to apply than Friedman's test. The Friedman test, despite its lower efficiency for small k, is a versatile technique, useful for both the randomized block design and the repeated measures design.

Friedman's test can be extended to the case of several observations per cell. The general case of unequal cell-sample sizes is treated by Bernard and van Elteren (1953). When the cells each have m observations, the statistic with its asymptotic distribution, under H_0, is given in Exercise 4.5.10.

As in the case of a one-way layout, the omnibus alternative may not be appropriate. We may wish to test $H_0 : \theta_1 = \cdots = \theta_k$ versus $H_A : \theta_1 \leqslant \cdots \leqslant \theta_k$ with at least one strict inequality. The test statistic, in the two-way layout with one observation per cell, proposed by Page (1963), is

$$Q = \frac{1}{\sqrt{n}} \sum_{j=1}^{k} \left(j - \frac{k+1}{2} \right) \left(R_j - \frac{n(k+1)}{2} \right),$$ (4.3.10)

similar to (4.2.11). Under H_0, from Exercise 4.5.11, $EQ = 0$ and

$$\text{Var } Q = k^2(k^2 - 1)(k + 1)/144.$$ (4.3.11)

Further $Q/(\text{Var } Q)^{1/2}$ is asymptotically $n(0, 1)$. Hence Page's test rejects H_0, at approximate level α, if $Q \geqslant Z_\alpha (\text{Var } Q)^{1/2}$, where $1 - \Phi(Z_\alpha) = \alpha$.

An analog of the Jonckheere–Terpstra test J, (4.2.13), in the two-way layout consists in computing J for each block and then combining the statistics across the blocks. Skillings and Wolfe (1978) consider this statistic. Their statistic allows for unequal numbers of observations per cell and for differential weighting of the different J statistics. Let J_i denote J computed on the ith block and

$$J^* = \sum_1^n J_i .$$ (4.3.12)

Then under the null hypothesis the mean, variance, and asymptotic normality of J^* are given in Exercise 4.5.12.

Another analog of the Jonckheere–Terpstra test, (4.2.13), in the two-way layout would be the statistic $A = \sum\sum_{i<j} T_{ij}$, where T_{ij} is the Wilcoxon signed rank statistic computed on the ith and jth paired samples. These statistics have been studied by Hollander (1967) and Puri and Sen (1968). It might be hoped that the A test does not suffer the efficiency loss (similar to the Friedman test) for small k since interblock information is taken into account. Page's test relies strictly on ranking within blocks.

Pirie (1974) made an extensive comparison of tests based on A and Q. Pirie considers asymptotic efficiency for both k fixed, $n \to \infty$, and n fixed, $k \to \infty$. He shows that superior test performance depends on the underlying distribution, and the values of k and n. Hence A is not necessarily more efficient than Q. In addition, the statistic A is not distribution free under H_0 and is not easy to implement. For these reasons we recommend the test based on Q and we do not present here any of the details concerning the test based on A.

One final variation, the balanced incomplete block design, is discussed in Exercise 4.5.13. Durbin's (1951) test, with its asymptotic null distribution is presented there. The efficiency of Durbin's test relative to the F test was computed by van Elteren and Noether (1959). When there are t treatments and there are $k \leqslant t$ treatments ranked within each block, the efficiency is identical to $e(K^*, F)$, (4.3.8). Hence, for example, in paired comparisons in which n judges compare t objects, pairwise, we have $k = 2$, and the efficiency is once again that of the sign test relative to the t test.

4.4.   RANK CORRELATION AND ASSOCIATION

In this section we introduce the ideas of rank correlation and association as measures of agreement between two sets of rankings. We consider, in detail, a bivariate model in which the data is ranked separately within each component. The distribution theory will also cover the case in which the original data is a set of ranks. This case commonly arises when judges are asked to express their preferences by assigning ranks directly to a set of objects.

We also interpret the various tests proposed in the earlier sections of this chapter in the light of how they assess the degree of agreement between the observed rank sums and the alternative hypotheses. This interpretation provides additional insight into how the tests work and illustrates the connection between correlation and the sums of squares. Fisher (1970) used this connection to great advantage in developing the analysis of variance out of previous correlation approaches.

The sampling model consists in a sample $(X_1, Y_1), \ldots, (X_n, Y_n)$ from $F(x, y)$ where $F(\cdot, \cdot)$ is absolutely continuous with absolutely continuous marginal cdfs $F_x(\cdot)$ and $F_y(\cdot)$. We think of the data arranged in two rows:

$$X_1 X_2 \ldots X_n$$
$$Y_1 Y_2 \ldots Y_n \qquad (4.4.1)$$

and, without loss of generality, we suppose that $X_1 < \cdots < X_n$. Let $R_1, \ldots, R_n$ be the corresponding ranks of $Y_1, \ldots, Y_n$, then we have the corresponding rank array:

$$
\begin{array}{cccc}
1 & 2 & \ldots & n \\
R_1 & R_2 & \ldots & R_n
\end{array} \qquad (4.4.2)
$$

We now present the two major methods for assessing the degree of agreement between the two sets of rankings in (4.2.2). See Kruskal (1958) for a historical overview of measures of association. Spearman's (1904) measure is simply the product-moment correlation coefficient computed on the ranks:

$$
r_s = \frac{\sum_{i=1}^{n} \left(i - \frac{n+1}{2}\right)\left(R_i - \frac{n+1}{2}\right)}{\sqrt{\sum_{i=1}^{n} \left(i - \frac{n+1}{2}\right)^2 \sum_{i=1}^{n} \left(R_i - \frac{n+1}{2}\right)^2}}. \qquad (4.4.3)
$$

Since the ranks are a rearrangement of the integers from 1 to n, the denominator is $\sum_1^n (i - (n+1)/2)^2 = n(n^2 - 1)/12$ from Theorem A21. Further, since $\sum[i - (n+1)/2] = 0$,

$$
r_s = \frac{12}{n(n^2 - 1)} \sum_{i=1}^{n} \left[i - (n+1)/2\right] R_i
$$

$$
= 1 - \frac{6}{n(n^2 - 1)} \sum_{i=1}^{n} (i - R_i)^2. \qquad (4.4.4)
$$

The last equality is the most convenient computationally; see Exercise 4.5.14. Since r_s is a correlation coefficient, it has the usual property that $-1 \leqslant r_s \leqslant 1$. It is easy to check that the extremes are attainable. If we have independence, so that the two rankings are independent, the joint distribution of $R_1, \ldots, R_n$ is uniform on the $n!$ permutations. Hence, if there is no association, $Er_s = 0$, from (4.4.4). Thus, r_s should be between -1 and $+1$, and around 0 in the case of independence.

The second measure of association was introduced by Kendall (1938). We say that the pairs (X_i, Y_i) and (X_j, Y_j) are concordant if $X_i > X_j$ and $Y_i > Y_j$ or if $X_i < X_j$ and $Y_i < Y_j$. If $\mathrm{sgn}(x) = 1, 0, -1$ as $x > 0, 0, < 0$, then the pairs are concordant if $\mathrm{sgn}(X_j - X_i)\mathrm{sgn}(Y_j - Y_i) = 1$. In a similar fashion we call the pairs discordant if $\mathrm{sgn}(X_j - X_i)\mathrm{sgn}(Y_j - Y_i) = -1$. Let P and Q denote the number of concordant and discordant pairs, respectively. Then the excess of concordance over discordance is

$$S = P - Q$$

$$= \sum\sum_{i<j} \mathrm{sgn}(X_j - X_i)\mathrm{sgn}(Y_j - Y_i). \qquad (4.4.5)$$

The possible values of S range from $-n(n-1)/2$ to $n(n-1)/2$. For example, $\max S = n(n-1)/2$ occurs when there is perfect agreement in the order of $X_1, \ldots, X_n$ and $Y_1, \ldots, Y_n$; that is, perfect agreement in their rankings. Kendall (1938) suggested the coefficient:

$$\tau = \frac{S}{\max S}$$

$$= \frac{2(P - Q)}{n(n - 1)}$$

$$= 1 - \frac{4}{n(n - 1)} Q \qquad (4.4.6)$$

since $P + Q = n(n - 1)/2$.

Note that the Y ordering can be transformed into the X ordering by successively interchanging neighboring pairs of Y values. Then Q is the needed number of interchanges or inversions that will bring the Y's into the same order as the X's. Hence Q (or τ) can be thought of as measuring the disarray of the Y's relative to the X's.

It is easy to check that $-1 \leqslant \tau \leqslant 1$, and the extremes are attainable. Further, if we have independence of X and Y, then $E\,\mathrm{sgn}(Y_j - Y_i) = 0$ and $E\tau = 0$, similar to r_s.

Note that since $X_1 < \cdots < X_n$, $Q = \sum\sum_{i<j} s(Y_i - Y_j) = \sum\sum_{i<j} s(R_i - R_j)$ where $s(x) = 1$ if $x > 0$ and 0 otherwise. This shows that Q can be computed either from the raw data or the ranks. A simple modification of Q can be developed by weighting the inversions proportional to the distance apart of the ranks. Hence, for example, $s(1 - 2)$ will get less weight in the measure of disarray than $s(1 - 5)$. If we take $j - i$ for the weight of $s(R_i - R_j)$, then the new measure is

$$Q^* = \sum\sum_{i<j} (j - i)s(R_i - R_j).$$

In the next theorem we show that Spearman's r_s is a function of Q^*. This will offer some insight into the relationship between r_s and τ.

Theorem 4.4.1.

$$r_s = 1 - \frac{12}{n(n^2 - 1)} \sum\sum_{i<j} (j - i)s(R_i - R_j)$$

$$\tau = 1 - \frac{4}{n(n - 1)} \sum\sum_{i<j} s(R_i - R_j).$$

Proof. We have already pointed out the equation for τ in (4.4.6). From (4.4.4), it is sufficient to show $Q^* = \{\sum(i - R_i)^2\}/2$.

Note that by interchanging the i, j notation we can write $\sum\sum_{j<i} js(R_i - R_j) = \sum\sum_{i<j} is(R_j - R_i)$. Now,

$$Q^* = \sum\sum_{i<j} js(R_i - R_j) - \sum\sum_{i<j} is(R_i - R_j)$$

$$+ \sum\sum_{j<i} js(R_i - R_j) - \sum\sum_{j<i} js(R_i - R_j)$$

$$= \sum\sum_{i,j} js(R_i - R_j) - \sum\sum_{i<j} i\{s(R_i - R_j) + s(R_j - R_i)\}$$

$$= \sum_{j=1}^{n} j \sum_{i=1}^{n} s(R_i - R_j) - \sum_{i=1}^{n-1} i \sum_{j=i+1}^{n} 1$$

$$= \sum_{j=1}^{n} j(n - R_j) - \sum_{i=1}^{n-1} i(n - i).$$

The last equality follows since $\sum_{i=1}^{n} s(R_i - R_j) = \#(R_i > R_j)$, $i = 1, \ldots,$ n, which is $n - R_j$. Replace $n - 1$ by n in the upper limit of the second sum and combine the sums to get

$$Q^* = \sum_{1}^{n} i^2 - \sum_{1}^{n} jR_j .$$

Next, note that

$$\sum_{1}^{n} (i - R_i)^2 = 2\sum_{1}^{n} i^2 - 2\sum_{1}^{n} iR_i$$

$$= 2Q^*,$$

and the formula for r_s follows from (4.4.4).

The theorem shows that, except in the extreme cases, r_s and τ will not generally be equal. Since r_s gives greater weight to inversions of ranks that are farther part, generally r_s is larger in absolute value than τ. This means that r_s seems to indicate there is more agreement (or disagreement) between rankings than τ, but this difference is an artifact of their construction. Kendall and Stuart (1973) point out that, when X and Y are independent, r_s and τ are highly correlated. In fact their correlation declines from 1 at $n = 2$ to .98 at $n = 5$ and tends to 1 as n increases. Hence, for testing independence they are asymptotically equivalent under the null hypothesis.

The major reference on the properties of r_s and τ is Kendall (1970). In his book on rank correlation, Kendall claims that from many practical and most theoretical points of view τ is preferable to r_s; see Kendall (1970, Section 1.24). As an indication of the theoretical difficulties encountered by r_s, consider the population characteristics estimated by τ and r_s. Under the sampling model: $(X_1, Y_1), \ldots, (X_n, Y_n)$ i.i.d. $F(x, y)$, introduced at the beginning of this section, and from (4.4.5) and (4.4.6),

$$E\tau = \frac{2}{n(n-1)} \sum\sum_{i<j} E\left\{ \mathrm{sgn}(X_j - X_i)\mathrm{sgn}(Y_j - Y_i) \right\}$$

$$= E\left\{ \mathrm{sgn}\left[(X_2 - X_1)(Y_2 - Y_1) \right] \right\}$$

$$= P\left\{ (X_2 - X_1)(Y_2 - Y_1) > 0 \right\} - P\left\{ (X_2 - X_1)(Y_2 - Y_1) < 0 \right\}$$

$$= 1 - 2P\left\{ (X_2 - X_1)(Y_2 - Y_1) < 0 \right\}. \tag{4.4.7}$$

Thus, the parameter of interest is the probability of a discordance. Clearly, when X and Y are independent, the probability of discordance is $1/2$ and $E\tau = 0$. The situation for r_s is much more complicated and there is no simple population characteristic for r_s. Kendall (1970, Chapter 9) shows

$$E r_s = \frac{3}{n+1} \left\{ E\tau + (n-2)(2\gamma - 1) \right\} \qquad (4.4.8)$$

where $\gamma = P[(X_2 - X_1)(Y_3 - Y_1) > 0]$, which is called a type 2 concordance. For large n, $E r_s \doteq 6(\gamma - 1/2)$ which is not so easy to interpret as the probability of a simple concordance. As a result, interpretation of r_s is mostly confined to the observation that r_s is a correlation coefficient computed on the two rankings. However, as will be seen later in this section, this property makes r_s useful in the motivation and interpretation of the rank tests in the one- and two-way layouts.

Under the null hypotehsis that X and Y are independent, the Y ranks are uniformly distributed over the first n integers. The results in Exercise 4.5.15 show that $E r_s = 0$, $\operatorname{Var} r_s = 1/(n-1)$ and $(n-1)^{1/2} r_s$ is asymptotically $n(0, 1)$. Hence it is simple to construct an hypothesis test based on r_s.

The asymptotic normality of τ, under the null hypothesis, is outlined in Exercise 2.10.35(b). You are asked to construct the projection and argue the limiting normality. The $\operatorname{Var} \tau = 2(2n+5)/[9n(n-1)]$ and it is pointed out that it takes a rather tedious counting argument to derive it. In the next theorem, due to Jirina (1976), we present a very clever argument which yields the $\operatorname{Var} \tau$ and the asymptotic normality. The argument rests on the development of a recursion formula for the distribution of τ, similar to that of the Mann–Whitney–Wilcoxon statistic in Theorem 3.2.3, and the recognition of it as a convolution.

Theorem 4.4.2. Suppose X and Y are independent. Let $P(S = s) = p_n(s)$ where S in (4.4.5) is based on a sample of size n. Then

$$p_n(s) = \frac{1}{n} \sum_{j=1}^{n} p_{n-1}(s - 2j + n + 1)$$

for $n \geqslant 3$ and $s = -n(n-1)/2, \ldots, n(n-1)/2$, and $p_2(s) = 1/2$, $s = -1, 1$.

Further, $ES = 0$, $\operatorname{Var} S = (n-1)n(2n+5)/18$, and $S/(\operatorname{Var} S)^{1/2}$ is asymptotically $n(0, 1)$.

Proof. Suppose $X_1 < \cdots < X_n$ so that we need only consider $R_1, \ldots, R_n$, the ranks of $Y_1, \ldots, Y_n$. Let $\bar{p}_n(s)$ be the number of the permutations of $1, \ldots, n$ such that $S = s$. These permutations can be built from

$1, \ldots, n - 1$ by inserting the integer n. Hence

$$\bar{p}_n(s) = \bar{p}_{n-1}(s + (n - 1)) + \bar{p}_{n-1}(s - 1 + (n - 2)) + \cdots$$

$$+ \bar{p}_{n-1}(s - (n - 2) + 1) + \bar{p}_{n-1}(s - (n - 1)).$$

Recall $S = P - Q$ and consider $\bar{p}_{n-1}(s + (n - 1))$. This term arises from $n, R_1, \ldots, R_{n-1}$. The n in the first position adds 0 to P and $n - 1$ to Q. Hence S computed on $R_1, \ldots, R_{n-1}$ is reduced by $n - 1$ when we include n at the beginning. In the second term we consider $R_1, n, R_2, \ldots, R_{n-1}$. Hence P is increased by 1 and Q by $n - 2$.

We can now write

$$\bar{p}_n(s) = \sum_{j=1}^{n} \bar{p}_{n-1}\left[s + (n - j) - (j - 1)\right]$$

$$= \sum_{j=1}^{n} \bar{p}_{n-1}\left[s - 2j + n + 1\right].$$

Since the $n!$ permutations are equally likely under independence,

$$p_n(s) = \frac{1}{n!}\, \bar{p}_n(s)$$

$$= \frac{1}{n} \sum_{j=1}^{n} p_{n-1}\left[s - 2j + n + 1\right].$$

It is obvious that $p_2(s) = 1/2$, $s = -1, 1$.

Next, let $k = 2j - n - 1$ and rewrite $p_n(s)$ as

$$p_n(s) = {\sum_{k}}'\, p_{n-1}(s - k)\frac{1}{n}\,.$$

The $'$ on the summation indicates that k takes the values $k = -n + 1$, $-n + 3, \ldots, n - 3, n - 1$. Define $q_n(k) = 1/n$ if $k = -n + 1, -n + 3$, $\ldots, n - 3, n - 1$, and 0 otherwise. Then

$$p_n(s) = \sum_{k} p_{n-1}(s - k)q_n(k)$$

and $p_n(s)$ is the convolution of the two discrete mass functions p_{n-1} and q_n; see Definition A3 in the Appendix. We write $p_n = q_n * p_{n-1}$. Repeating this argument shows

$$p_n = q_n * \cdots * q_3 * p_2\,.$$

But $p_2 = q_2$, hence $p_n = q_n * \cdots * q_2$, and this means that S has the same distribution as $\sum_2^n Z_i$, where $Z_2, \ldots, Z_n$ are independent and $P(Z_i = k) = 1/i$ for $k = -i + 1, -i + 3, \ldots, i - 3, i - 1$. The variance is the sum of the variances:

$$\operatorname{Var} S = \sum_{i=2}^{n} \operatorname{Var} Z_i$$

$$= \sum_{i=2}^{n} \frac{1}{3}(i^2 - 1)$$

$$= (n - 1)n(2n + 5)/18.$$

Note that $EZ_i = 0$ so $\operatorname{Var} Z_i = EZ_i^2 = (i^2 - 1)/3$. Since $Z_1, \ldots, Z_n$ are independent but not identically distributed we apply Theorem A6 to establish the asymptotic normality of $S/(\operatorname{Var} S)^{1/2}$. In the notation of Theorem A6, $B_n^2 = \sum_2^n \operatorname{Var} Z_i \sim n^3$. Since $|Z_i| \leqslant i - 1$,

$$E\left[Z_i^2 I(|Z_i| > \epsilon B_n) \right] \leqslant (i - 1)^2 P(|Z_i| > \epsilon B_n)$$

$$\leqslant (i - 1)^2 (EZ_i^2)/\epsilon^2 B_n^2.$$

Hence, the Lindeberg condition becomes

$$\frac{1}{B_n^2} \sum_1^n E\left[Z_i^2 I(|Z_i| > \epsilon B_n) \right] \leqslant \frac{1}{\epsilon^2 B_n^4} \sum_1^n \frac{(i - 1)^2(i^2 - 1)}{3}$$

$$\sim \frac{1}{\epsilon^2 n^6} n^5 \to 0,$$

since by Theorem A21, $\sum_1^n i^4 = n(n + 1)(2n + 1)(3n^2 + 3n + 1)/30 \sim n^5$. This completes the proof.

We now present an example to compare r_s and τ, and to illustrate the various calculations.

Example 4.4.1. From the *World Almanac 1982*, we list the olympic times, in seconds, of the men's 400-meter dash, 1500-meter run, and marathon (Table 4.5). Ties are assigned the average rank. If there are extensive ties in a data set, the reader should consult Kendall (1970, Chapter 3) since the various formulas for r_s (and for τ) are no longer computationally equivalent. We have two ties in the 400-meter data, in 1932 and 1948, but they do not effect the computations very much. In Table 4.6 we display r_s, (4.4.3),

Table 4.5.   Olympic Times in Seconds with Ranks

Year	1896	1900	1904	1906	1908
400 m	54.2 (20)	49.4 (16)	49.2 (15)	53.2 (19)	50.0 (18)
1500 m	373.2 (20)	246.0 (18)	245.4 (17)	252.0 (19)	243.4 (16)
Marathon[a]	3530 (18)	3585 (19)	5333 (20)	3084 (16)	3318 (17)

Year	1912	1920	1924	1928	1932
400 m	48.2 (14)	49.6 (17)	47.6 (12)	47.8 (13)	46.2 (8.5)
1500 m	236.8 (14)	241.8 (15)	233.6 (13)	233.2 (12)	231.2 (11)
Marathon[a]	2215 (14)	1956 (11)	2483 (15)	1977 (12)	1896 (10)

Year	1936	1948	1952	1956	1960
400 m	46.5 (10)	46.2 (8.5)	45.9 (7)	46.7 (11)	44.9 (5)
1500m	227.8 (9)	229.8 (10)	225.2 (8)	221.2 (7)	215.6 (2)
Marathon[a]	1759 (9)	2092 (13)	1383 (7)	1500 (8)	916 (5)

Year	1964	1968	1972	1976	1980
400 m	45.1 (6)	43.8 (1)	44.7 (4)	44.3 (2)	44.6 (3)
1500 m	218.1 (4)	214.9 (1)	216.3 (3)	219.2 (6)	218.4 (5)
Marathon[a]	731 (3)	1226 (6)	740 (4)	595 (1)	663 (2)

[a] Actual marathon times are 2 hours + entry.

and τ from (4.4.5). The numbers in parentheses are $r_s/(\mathrm{Var}\, r_s)^{1/2}$ and $\tau/(\mathrm{Var}\,\tau)^{1/2}$. The table illustrates the tendency of r_s to be numerically more extreme than τ. However, in terms of standard deviations, τ is further from 0 than r_s.

Kendall (1970, Chapter 8) has shown that τ (unlike r_s) can be extended to the case of partial correlation. Kendall points out that it is remarkable (but apparently only a coincidence) that the partial τ has the same structural form as the partial product-moment correlation. Hence he shows

$$\tau_{XY\cdot Z} = \frac{\tau_{XY} - \tau_{XZ}\tau_{YZ}}{\sqrt{\left(1 - \tau_{XZ}^2\right)\left(1 - \tau_{YZ}^2\right)}} \qquad (4.4.9)$$

is a partial rank-correlation between X and Y with Z held fixed. In

Table 4.6.   Rank Correlations

Test	Event		
	400, 1500 m	1500, Marathon	400, Marathon
r_s	.940(4.10)	.905(3.95)	.878(3.83)
τ	.800(4.94)	.695(4.29)	.695(4.29)

the foregoing example, $\tau_{1500,M} = .695$. Further, it is not surprising that olympic times decrease with the year. Computations from Table 4.5 show $\tau_{1500,\text{year}} = \tau_{M,\text{year}} = -.832$. Then we have $\tau_{1500,M\cdot\text{year}} = .009$. Hence, once we account for the trend with time, there is not much association left between the 1500-meter run and the marathon. At present there are no tests for the significance of the partial τ.

We now turn to a discussion of the relationship between rank correlation and the tests in the one- and two-way layouts. Some of the connections are obvious. For example, Page's test, (4.3.10), for an ordered alternative in the two-way layout can be written as follows

$$Q = \frac{1}{\sqrt{n}} \sum_{j=1}^{k} \left(j - \frac{k+1}{2}\right)\left(R_{\cdot j} - \frac{n(k+1)}{2}\right)$$

$$= \frac{1}{\sqrt{n}} \sum_{i=1}^{n} \left\{ \sum_{j=1}^{k} \left(j - \frac{k+1}{2}\right)\left(R_{ij} - \frac{k+1}{2}\right)\right\}. \qquad (4.4.10)$$

The expression in the braces is the numerator of Spearman's r_s computed between the ith row and the hypothesized ordering. From (4.4.3), multiplying and dividing by $\sum_1^k[j - (k+1)/2]^2 = k(k^2-1)/12$, we have

$$Q = \frac{12}{\sqrt{n}\,k(k^2-1)} \sum_{i=1}^{n} r_i \qquad (4.4.11)$$

where r_i is Spearman's r_s between the ith row and the hypothesized ordering. Hence Q assesses the degree of agreement among the rows (or blocks) with respect to the hypothesized ordering.

If there is no specified ordering with which to correlate the rows, we can consider the average rank correlation among all $n(n-1)/2$ pairs of rows in a two-way layout. The average rank correlation then measures the degree of agreement among the rows, but does not specify what they should agree upon. We write the average Spearman correlation as

$$r_{\text{ave}} = \frac{2}{n(n-1)} \sum_{i<j}\sum \left\{ \sum_{t=1}^{k} \frac{[R_{it}-(k+1)/2][R_{jt}-(k+1)/2]}{k(k^2-1)/12} \right\}$$

$$= \frac{12}{n(n-1)k(k^2-1)}$$

$$\times \sum_{t=1}^{k} \left\{ \sum\sum_{i\neq j}[R_{it}-(k+1)/2][R_{jt}-(k+1)/2]\right\}. \qquad (4.4.12)$$

Note that

$$\left\{ \sum_i \left[R_{it} - (k+1)/2 \right] \right\}^2 = \sum_i \left[R_{it} - (k+1)/2 \right]^2$$

$$+ \sum\sum_{i \neq j} \left[R_{it} - (k+1)/2 \right]\left[R_{jt} - (k+1)/2 \right].$$

$$(4.4.13)$$

Then, since $\sum_t\sum_i[R_{it} - (k+1)/2]^2 = nk(k^2 - 1)/12$ and the left side of (4.4.13) is $[R_{\cdot t} - n(k+1)/2]^2$, (4.4.12) becomes

$$r_{\text{ave}} = \frac{12}{n(n-1)k(k^2-1)} \left\{ \sum_{t=1}^{k} \left[R_{\cdot t} - n(k+1)/2 \right]^2 - nk(k^2-1)/12 \right\}.$$

$$(4.4.14)$$

This equation makes r_{ave} much more practical since, rather than compute $n(n-1)/2$ correlations, we only need the k column rank sums. We can also see the connection with Freidman's K^*, (4.3.5). We have

$$r_{\text{ave}} = \frac{1}{(n-1)(k-1)} \left[K^* - (k-1) \right]$$

and, conversely,

$$K^* = (n-1)(k-1)r_{\text{ave}} + (k-1).$$

$$(4.4.15)$$

Hence not only is r_{ave} easy to compute, but Friedman's K^* is a linear function of r_{ave}. This means that K^* is a measure of the amount of agreement among the rows (or blocks). When there is a high degree of agreement, K^* will be large and reject the null hypothesis of no treatment effect.

This same two-way layout arises when n judges are asked to rank k objects. Kendall (1970) calls this the problem of n-rankings. Then r_{ave} is a measure of the concordance or agreement among the judges. Kendall introduces a coefficient of concordance

$$W = \frac{12}{n^2 k(k^2-1)} \sum_{j=1}^{k} (R_{\cdot j} - n(k+1)/2)^2$$

$$(4.4.16)$$

where $R_{\cdot j}$ is the sum of ranks for the jth object. In Exercise 4.5.16 you are asked to show that $0 \leqslant W \leqslant 1$ and

$$W = \frac{n-1}{n} r_{\text{ave}} + \frac{1}{n}$$

$$W = \frac{1}{n(k-1)} K^*.$$

If the null hypothesis of no concordance among the judges is interpreted to mean that the judges act, as a group, as if they are assigning ranks at random, then we can reject the null hypothesis, at approximate level α, if $W \geqslant [\chi_\alpha^2(k-1)]/(n(k-1))$, where $\chi_\alpha^2(k-1)$ is the chi-square critical value with $k-1$ degrees of freedom. In fact, W, r_{ave}, and K^* are all linearly related, so significance tests are identical for them all and may as well be carried out using K^*, which is directly comparable to the chi-square critical value.

Schucany and Frawley (1973) extend the measure of concordance to the problem of two-group concordance. Suppose two groups of judges are asked to rank k objects. We might wish to know if there is agreement within the two groups and agreement between the two groups.

Let R_{ij}, $i = 1, \ldots, m$; $j = 1, \ldots, k$, be the ranks assigned by m judges in Group 1 and R'_{ij}, $i = 1, \ldots, n$; $j = 1, \ldots, k$, be the ranks assigned by n judges in Group 2. Let $R_{\cdot j}$ and $R'_{\cdot j}$, $j = 1, \ldots, k$ be the respective rank sums for each object. The Schucany–Frawley statistic is

$$L^* = \sum_{j=1}^{k} R_{\cdot j} R'_{\cdot j}. \qquad (4.4.17)$$

Under the null hypothesis that all rankings are uniformly distributed, the mean and variance are given in Exercise 4.5.17, along with $\max L^*$ and $\min L^*$.

A generalized coefficient of two-group concordance, which ranges from -1 to $+1$, is defined by

$$W^* = \frac{L^* - EL^*}{\max L^* - EL^*}. \qquad (4.4.18)$$

The coefficient W^* simply centers and rescales L^* to the closed interval $[-1, +1]$. The intuitive appeal of W^* (or L^*) is provided in Exercise 4.5.18; you are asked to show that

$$W^* = \frac{1}{mn} \sum_{j=1}^{n} \sum_{i=1}^{m} r_{ij} \qquad (4.4.19)$$

where r_{ij} is Spearman's r_s computed on the ith judge in Group 1 and the jth judge in Group 2. The coefficient W^* has the following interpretation: large values approaching $+1$ mean high agreement within both groups and high agreement across the groups; small values approaching -1 mean high agreement within both groups and strong disagreement across the groups; and values around 0 indicate either disagreement within the groups or agreement within the groups but no association across the groups.

The hypothesis testing problem is a bit unusual since rejection of the null hypothesis that all row permutations are equally likely entails two very different decisions. The limiting distribution is given in the next theorem and is also unusual. For fixed k, the limiting distribution, as $m, n \to \infty$, is not normal. However, for moderate values of k, a normal approximation will work. See Li and Schucany (1975) for a discussion.

Theorem 4.4.3. Under the null hypothesis that all row permutations are equally likely, for fixed k and as $m, n \to \infty$,

$$\frac{L^* - EL^*}{\sqrt{\operatorname{Var} L^*}} \xrightarrow{D} \frac{1}{\sqrt{k-1}} \sum_{i=1}^{k-1} V_i W_i$$

where $V_1, \ldots, V_{k-1}, W_1, \ldots, W_{k-1}$ are independent $n(0,1)$ random variables.

Proof. First note, by (4.3.3) with $c_{jN} = 1$,

$$\frac{1}{\sqrt{mn}}(L^* - EL^*) = \frac{1}{\sqrt{m}}(\mathbf{S} - \mathbf{J}m(k+1)/2)' \frac{1}{\sqrt{n}}(\mathbf{U} - \mathbf{J}n(k+1)/2)$$

where, $\mathbf{S}' = (R_{.1}, \ldots, R_{.k})$, $\mathbf{U}' = (R'_{.1}, \ldots, R'_{.k})$, $\mathbf{J}' = (1, \ldots, 1)$, $ER_{.j} = m(k+1)/2$, and $ER'_{.j} = n(k+1)/2$. The result in Exercise 4.5.7, along with Theorem A2, implies that

$$\frac{1}{\sqrt{mn}}(L^* - EL^*) \xrightarrow{D} \mathbf{Z}_1'\mathbf{Z}_2$$

where $\mathbf{Z}_i$ is $MVN(\mathbf{0}, \mathbf{B})$, $i = 1, 2$, and $\mathbf{B}$ is defined by (4.3.4) with $c_j = 1$. The vectors $\mathbf{Z}_1$ and $\mathbf{Z}_2$ are independent. Further, from (4.3.4),

$$\mathbf{B} = \frac{k(k+1)}{12}\left\{\mathbf{I} - \frac{1}{k}\mathbf{J}\mathbf{J}'\right\}.$$

Now, for $i = 1, 2$, let the vector $\mathbf{Y}_i = \{12/[k(k+1)]\}^{1/2}\mathbf{Z}_i$. Then $\mathbf{Y}_i$ is distributed as $MVN(\mathbf{0}, \mathbf{I} - (1/k)\mathbf{J}\mathbf{J}')$ and the covariance matrix is idempotent with rank $k - 1$. Hence there exists an orthogonal matrix Γ such that $U_i = \Gamma'Y_i$, $i = 1, 2$, are distributed as $MVN(0, D)$ where D is a diagonal matrix with $k - 1$ ones and 1 zero on the diagonal. This means that the components of U_1 and U_2 are $k - 1$ i.i.d. $n(0, 1)$ random variables and one 0 with probability one.

We now apply these tranformations to rewrite $Z_1'Z_2$ as

$$Z_1'Z_2 = \frac{k(k + 1)}{12} Y_1'Y_2$$

$$= \frac{k(k + 1)}{12} U_1'\Gamma'\Gamma U_2$$

$$= \frac{k(k + 1)}{12} \sum_{j=1}^{k-1} V_i W_i,$$

with $V_1, \ldots, V_{k-1}, W_1, \ldots, W_{k-1}$ given in the statement of the theorem.

From Exercise 4.5.17, $\operatorname{Var} L^* = mn(k - 1)k^2(k + 1)^2/144$. Hence

$$\frac{1}{\sqrt{mn} \sqrt{\dfrac{(k - 1)k^2(k + 1)^2}{144}}} (L^* - EL^*) \overset{D}{\to} \frac{1}{\sqrt{k - 1}} \sum_{j=1}^{k-1} V_i W_i$$

and the proof is complete.

In the next theorem we construct the moment-generating function of $\sum_1^{k-1} V_i W_i$ and discuss the limiting distribution.

Theorem 4.4.4. Suppose $V_1, \ldots, V_{k-1}, W_1, \ldots, W_{k-1}$ are i.i.d. $n(0, 1)$. Then the moment-generating function of $\sum_1^{k-1} V_i W_i = V'W$ is given by

$$M(t) = \left(1 - t^2\right)^{-(k-1)/2}$$

Proof. Let $V' = (V_1, \ldots, V_{k-1})$ and $W' = (W_1, \ldots, W_{k-1})$. We use a conditional expectation to find

$$M(t) = E(e^{tV'W})$$

$$= E\{E(e^{tV'W} | V = v)\}.$$

But, the conditional expectation is the moment-generating function for W

which is distributed as $MVN(\mathbf{0}, \mathbf{I})$. Hence

$$M(t) = E\left\{e^{(t^2/2)\mathbf{V}'\mathbf{V}}\right\}$$

$$= (1 - t^2)^{-(k-1)/2}$$

since $\mathbf{V}'\mathbf{V}$ is $\chi^2(k-1)$. This completes the proof.

Now note that

$$\frac{L^* - EL^*}{\sqrt{\operatorname{Var} L^*}} \xrightarrow{D} \frac{1}{\sqrt{k-1}} \mathbf{V}'\mathbf{W}.$$

Hence the moment-generating function of the limiting distribution is

$$M(t) = \left(1 - \frac{t^2}{2(k-1)/2}\right)^{-(k-1)/2}$$

$$\doteq e^{-t^2/2}.$$

The approximation is valid as $k \to \infty$, and $e^{-t^2/2}$ is the moment-generating function for $n(0,1)$. Hence L^*, when standardized, is approximately normally distributed only for large m, n, and k. Li and Schucany (1975) suggest that $k \geqslant 6$ provides an adequate normal approximation for large m and n. Oddly, if m or n is small, then k need not be so large.

Finally, as an example of a nonnormal limiting distribution, we consider what happens if $k = 3$. In this case $M(t) = 1/(1 - t^2)$. But this is the moment-generating function for the double exponential distribution, $f(x) = e^{-|x|}/2$ for $-\infty < x < \infty$. Thus, for $k = 3$ and $m, n \to \infty$,

$$\sqrt{k-1} \; \frac{(L^* - EL^*)}{\sqrt{\operatorname{Var} L^*}} \xrightarrow{D} Z$$

where Z has a double exponential distribution.

Hollander and Sethuraman (1978) argue that Schucany and Frawley (1973) do not consider the correct null hypothesis. They reformulate the problem and propose a conditionally distribution-free test. Their paper also contains comments by Schucany.

We have shown that tests of Friedman and Page in the two-way layout are intimately connected to Spearman's rank correlation. We now turn to the one-way layout and explore the connection between rank correlation and the Kruskal–Wallis statistic. We must replace the familiar product-moment correlation coefficient with the intraclass correlation coefficient; see Fisher (1970, Chapter 7). The need for the intraclass correlation coefficient arises when there is no natural way to order X and Y in the pair (X, Y). For example, if we wish to measure the correlation in the IQ of twins, there is no way to say which twin should be listed first.

Given $(X_1, Y_1), \ldots, (X_k, Y_k)$, create k additional pairs $(Y_1, X_1), \ldots, (Y_k, X_k)$ and compute the product-moment correlation coefficient on the $2k$ pairs. This removes the order effect within the pairs. The result is

$$r_I = \frac{2 \sum_1^k (X_i - M)(Y_i - M)}{\left\{ \sum_1^k (X_i - M)^2 + \sum_1^k (Y_i - M)^2 \right\}}$$

where $M = (\bar{X} + \bar{Y})/2$. Hence, r_I, the intraclass correlation coefficient, replaces the individual sample means and sample sums of squares by their averages.

Next consider the one-way layout with n observations under each of k treatments. The preceding discussion is relevant for a one-way layout with $n = 2$. In general, there are $n(n-1)/2$ possible pairs under each treatment and if we add the reversed pair in each case we have $n(n-1)$ pairs from each treatment. The intraclass correlation immediately generalizes to

$$r_I = \frac{2 \sum_{t=1}^k \sum_{i<j} (X_{it} - M)(X_{jt} - M)}{(n-1) \sum_{t=1}^k \sum_{i=1}^n (X_{it} - M)^2} \tag{4.4.20}$$

where M is the grand mean.

The intraclass rank correlation in the one-way layout is computed by applying r_I to the nk ranks of the combined data. In this case $M = (kn+1)/2$ and $\sum\sum(R_{it} - M)^2 = nk(n^2k^2 - 1)/12$, so that

$$R_I = \frac{24 \sum_{t=1}^k \sum_{i<j} (R_{it} - (kn+1)/2)(R_{jt} - (kn+1)/2)}{(n-1)nk(n^2k^2 - 1)}$$

$$= \frac{12}{(n-1)nk(n^2k^2 - 1)} \left\{ \sum_{j=1}^k [R_{\cdot j} - n(nk+1)/2]^2 - \frac{nk(n^2k^2 - 1)}{12} \right\}.$$

$$\tag{4.4.21}$$

See Exercise 4.5.20 for the second equality.

When the sample sizes are equal, $N = nk$, and from Theorem 4.2.4, we have

$$\sum_{j=1}^{k} \left[R_j - n(N+1)/2 \right]^2 = \frac{n(nk)(nk+1)}{12} H^*$$

where H^* is the Kruskal–Wallis statistic. Hence, the intraclass rank correlation coefficient is

$$R_I = \frac{nH^*}{(n-1)(nk-1)} - \frac{1}{n-1} \qquad (4.4.22)$$

and, also,

$$H^* = \frac{(nk-1)}{n} \left[(n-1)R_I + 1 \right].$$

Thus, when the intraclass rank correlation is large, there is a high degree of agreement among the ranks under each treatment. In this case, the Kruskal–Wallis statistic will be large and reject the null hypothesis of no treatment effect. We also note that H^* can be used to test for significant intraclass rank correlation provided that n is large so that Theorem 4.2.4 can be applied. Since $H^* \geqslant 0$, (4.4.22) shows that $-1/(n-1) \leqslant R_I \leqslant 1$. This asymmetry requires that some care be exercised in the interpretation of R_I. There is a similar relationship between the one-way F statistic and r_I.

When the sample sizes are unequal, the intraclass correlations can still be computed, but they are no longer linearly related to the one-way layout test statistics.

4.5. EXERCISES

4.5.1. In the proof of Theorem 4.2.2 show that $\text{Cov}(V_i, V_j) \rightarrow -c_i c_j / 12$.

4.5.2. Verify that H^* reduces to the other equations in the statement of Theorem 4.2.4.

4.5.3. Verify (4.2.12) for the variance of L. Hint: Note that $\sum_{j=1}^{k}[j - (k+1)/2]^2 = k(k^2-1)/12$ and

$$2 \sum_{i=1}^{k-1} \sum_{j=i+1}^{k} \left[i - (k+1)/2 \right]\left[j - (k+1)/2 \right]$$

$$= - \sum_{t=1}^{k} \left[t - (k+1)/2 \right]^2.$$

4.5.4. Let $J = \sum\sum_{i<j} W_{ij}$ be given by (4.2.13). Under H_0, show that:

$$\text{Cov}(W_{uv}, W_{ut}) = \text{Cov}(W_{vu}, W_{tu}) = n_u n_v n_t / 12$$

$$\text{Cov}(W_{uv}, W_{tu}) = \text{Cov}(W_{vu}, W_{ut}) = -n_u n_v n_t / 12.$$

Hint: Note that $W_{uv} + W_{ut}$ is the Mann–Whitney–Wilcoxon statistic computed on the uth sample versus the combined vth and tth samples and $\text{Var}(W_{uv} + W_{ut}) = \text{Var } W_{uv} + \text{Var } W_{ut} + 2\text{Cov}(W_{uv}, W_{ut})$.

4.5.5. Suppose we have k samples, each of size n; hence $N = nk$. Show L, (4.2.11), can be expressed as

$$L = \frac{1}{\sqrt{N}}\left\{ \frac{1}{n} \sum\sum_{i<j} (j-i) W_{ij} - \frac{nk(k^2-1)}{4} \right\}.$$

Hence L is equivalent to a weighted sum of pairwise Mann–Whitney–Wilcoxon statistics.

4.5.6. Verify the moments for $R_{.1}, \ldots, R_{.k}$ in the two-way layout, under the null hypothesis, given in (4.3.1).

4.5.7. a. Let the vector $\mathbf{T}' = (T_1, \ldots, T_k)$ have jth component T_j given by (4.3.3). Under the null hypothesis, show that $\mathbf{T}$ has a $MVN(\mathbf{0}, \mathbf{B})$ limiting distribution, where $\mathbf{B}$ is defined by (4.3.4). Hint: Note that the rows of ranks in the two-way layout are i.i.d. vectors so that the multivariate central limit theorem, discussed below Theorem A8, can be directly applied.

   b. Show that if $c_{jN}^2 = (k-1)/k$, $j = 1, \ldots, k$, then K^*, (4.3.5), has an asymptotic chi-square distribution with $k-1$ degrees of freedom. Hint: Apply Theorem 4.2.3.

4.5.8. In the two-way layout, suppose X_{ij} are independent and have cdf $F(x - \mu - \alpha_i - \beta_j / N^{1/2})$, $i = 1, \ldots, n$, $j = 1, \ldots, k$. Let $\boldsymbol{\theta}_N' = (\mu + \alpha_1 + \beta_1/N^{1/2}, \ldots, \mu + \alpha_k + \beta_k/N^{1/2})$ and

$$T_j^* = c_{jN}\left\{ \frac{R_{.j} - ER_{.j}}{\sqrt{\text{Var } R_{.j}}} \right\}.$$

a. Similar to Theorem 4.2.6, argue that, when $c_{jN} \to c_j$,

$$E_{\theta_N} T_j^* \to c_j \sqrt{12k/(k^2-1)} \int f^2(x)\,dx \left(\beta_j - \bar{\beta} \right),$$

where $\bar{\beta} = \sum_1^k \beta_j / k$.

b. Let $c_{jN} = (k-1)/k$ and argue that the Friedman statistic K^*, (4.3.5), is asymptotically noncentral chi-square with $k-1$ degrees of freedom and noncentrality parameter

$$\delta_{K^*} = 12 \left(\int f^2(x)\,dx \right)^2 \frac{1}{k+1} \sum_1^k \left(\beta_j - \bar{\beta} \right)^2.$$

4.5.9.   Show that Friedman's statistic reduces to the sign statistic when $k = 2$.

4.5.10.   The two-way layout with m observations per cell. Consider a model in which the independent random variables X_{ijt}, $i = 1, \ldots, n$, $j = 1, \ldots, k$, $t = 1, \ldots, m$, have cdf $F(x - \mu - \alpha_i - \beta_j)$, $F \in \Omega_0$. We wish to test $H_0 : \beta_1 = \cdots = \beta_k$ versus H_A : $\beta_1, \ldots, \beta_k$ not all equal. Rank the data within the ith block, $i = 1, \ldots, n$; hence rank from 1 to mk. Let $R_{.j.}$ be the sum of ranks for the jth treatment, $j = 1, \ldots, k$. Under H_0, show

$$ER_{.j.} = nm(mk+1)/2$$

$$\operatorname{Var} R_{.j.} = nm^2(mk+1)(k-1)/12$$

$$\operatorname{Cov}(R_{.u.}, R_{.v.}) = -nm^2(mk+1)/12.$$

Further, argue that

$$K_m^* = \sum_{j=1}^k \left(\frac{k-1}{k} \right) \left\{ \frac{R_{.j.} - ER_{.j.}}{\sqrt{\operatorname{Var} R_{.j.}}} \right\}^2$$

$$= \left[\frac{12}{nkm^2(mk+1)} \sum_{j=1}^k R_{.j.}^2 \right] - 3n(mk+1)$$

is asymptotically chi-square with $k-1$ degrees of freedom. See Exercise 5.5.8 for an application with data.

4.5.11. Show that Page's statistic Q, (4.3.10), has mean 0 and variance given by (4.3.11). Further, show that $Q/(\mathrm{Var}\, Q)^{1/2}$ is asymptotically $n(0, 1)$.

4.5.12. The statistic J^*, (4.3.12), can be used to test for an ordered alternative in a two-way layout with several observations per cell. Suppose there are k treatments, n blocks, and n_{ij} observations in the (i, j) cell. Suppose the null hypothesis $H_0: \theta_1 = \cdots = \theta_k$ is true. Let $n_{i.} = \sum_{j=1}^{k} n_{ij}$ and show that:

$$EJ^* = \sum_{i=1}^{n} \left[n_{i.}^2 - \sum_{j=1}^{k} n_{ij}^2 \right] \Big/ 4$$

$$\mathrm{Var}\, J^* = \sum_{i=1}^{n} \left\{ n_{i.}^2 (2n_{i.} + 3) - \sum_{j=1}^{k} n_{ij}^2 (2n_{ij} + 3) \right\} \Big/ 72.$$

If the n_{ij} remain fixed and $n \to \infty$, use Theorem A6 to show that $(J^* - EJ^*)/(\mathrm{Var}\, J^*)^{1/2}$ is asymptotically $n(0, 1)$.

4.5.13. The balanced incomplete block design: Durbin's (1951) test. In this design there are n blocks (judges) and t treatments. There are $k \leqslant t$ treatments ranked within each block, every treatment appears in $r \leqslant n$ blocks, and every treatment appears with every other treatment an equal number of times. These designs are discussed by Cochran and Cox (1957). Let $R_{.j}$ be the sum of ranks under the jth treatment, $j = 1, \ldots, t$. Under the null hypothesis of no treatment effect, show

$$ER_{.j} = r(k + 1)/2$$

$$\mathrm{Var}\, R_{.j} = r(k^2 - 1)/12$$

$$\mathrm{Cov}(R_{.k}, R_{.j}) = -r(k^2 - 1)/\left[12(t - 1) \right].$$

Argue that, under the null hypothesis,

$$D = \sum_{j=1}^{t} \left(\frac{t-1}{t} \right) \left\{ \frac{R_{.j} - ER_{.j}}{\sqrt{\mathrm{Var}\, R_{.j}}} \right\}^2$$

$$= \left[\frac{12(t-1)}{rt(k^2-1)} \sum_{j=1}^{t} R_j^2 \right] - \frac{3r(t-1)(k+1)}{k-1}$$

has an asymptotic chi-square distribution with $k - 1$ degrees of freedom.

4.5.14. Show that (4.4.4) holds. Hint: Expand

$$\sum \left(i - \frac{n+1}{2} - R_i + \frac{n+1}{2} \right)^2.$$

4.5.15. Suppose X and Y are independent and show that $\operatorname{Var} r_s = 1/(n - 1)$, for r_s given by (4.4.4). We have already seen $Er_s = 0$. Let $S^* = \sum_{j=1}^{n} j(R_j - (n+1)/2)$ in (4.4.4). Show the projection is given by

$$E[S^* \mid Y_k = y] = n[k - (n+1)/2][F_y(y) - 1/2]$$

where $F_y(y)$ is the marginal cdf of Y. Then show $(n-1)^{1/2} r_s$ is asymptotically $n(0, 1)$. Hint: It is helpful to write $R_j = 1 + \sum_{i \neq j} s(Y_j - Y_i)$ for the rank of Y_j. First compute $E[s(Y_j - Y_i) - 1/2 \mid Y_k = y]$ and then compute $E[R_j - (n+1)/2 \mid Y_k = y]$.

4.5.16. Show that Kendall's coefficient of concordance W, (4.4.16), satisfies $0 \leqslant W \leqslant 1$. Further, express W, r_{ave}, and K^* as linear functions of each other.

4.5.17. Show that, under the null hypothesis that all row permutations are equally likely, we have, for the Schucany–Frawley statistic L^*, (4.4.17),

$$EL^* = \frac{mnk(k+1)^2}{4}$$

$$\operatorname{Var} L^* = \frac{mn(k-1)k^2(k+1)^2}{144}.$$

Further,

$$\min L^* = \frac{mnk(k+1)(k+2)}{6}$$

and

$$\max L^* = \frac{mnk(k+1)(2k+1)}{6}.$$

4.5.18. Verify (4.4.19), which connects W^* to the average of the rank correlation coefficients between the two groups.

4.5.19. Show that, under the null hypothesis that all row permutations are equally likely, the Schucany–Frawley statistic L^* is uncorrelated with K_1^* and K_2^*, the Friedman statistics on the individual groups.

4.5.20. Using an argument similar to the one used to establish (4.4.14), establish (4.4.21).

4.5.21. Suppose we have two samples of size n. Let U, (3.2.2), be the sum of ranks of the first sample, in the combined sample. Let R_I be the intraclass rank correlation coefficient and show that

$$R_I = \frac{12}{n(n-1)(4n^2-1)} \left[U - \frac{n(2n+1)}{2} \right]^2 - \frac{1}{n-1}.$$

This provides the connection between the intraclass correlation coefficient and the Mann–Whitney–Wilcoxon statistic.

CHAPTER 5

The Linear Model

5.1. INTRODUCTION AND SIMPLE REGRESSION

We now turn to the linear model. In this book we have concentrated on models described by a set of independent observations $Y_1, \ldots, Y_N$ with respective cdfs $F(y - \theta_1), \ldots, F(y - \theta_N)$, $F \in \Omega_0$. Major problems were formulated in terms of hypothesis tests and estimation of the location parameters $\theta_1, \ldots, \theta_N$. In the one-sample location model the unknown locations are the same: $\theta_1 = \cdots = \theta_N = \theta$. In the two-sample location model, $\theta_1 = \cdots = \theta_m = \mu_1$ and $\theta_{m+1} = \cdots = \theta_N = \mu_2$, and attention was focused on $\Delta = \mu_1 - \mu_2$. The one-way layout extends the two-sample model to several samples. In the two-way layout it is more convenient to use double subscripts on the observations to indicate the two factors, for example, treatment and block. Hence, we have Y_{ij}, $i = 1, \ldots, n$; $j = 1, \ldots, k$, with cdf $F_i(y - \theta_j)$. In this chapter we will be interested in the particular case $F_i(y - \theta_j) = F(y - \theta_{ij})$ in which the nk observations are independent. The data could be displayed in a two-way table in which i labels the n rows (blocks) and j labels the k columns (treatments). Then the additive two-way model is specified by $\theta_{ij} = \mu + \alpha_i + \beta_j$, $i = 1, \ldots, n$; $j = 1, \ldots, k$. Hypotheses are formulated in terms of the treatment effects $\beta_1, \ldots, \beta_k$ with μ and $\alpha_1, \ldots, \alpha_n$ treated as nuisance parameters. In the last chapter, in the discussion of the two-way layout, we avoided estimating the nuisance parameters by ranking within the blocks. Equation (4.3.9) indicated a loss of efficiency due to ranking in this way. The methods developed in the present chapter recover the lost efficiency, but they are only asymptotically nonparametric.

In the general linear model we have a vector $\mathbf{Y}' = (Y_1, \ldots, Y_N)$ of independent observations with cdfs $F(y - \theta_1), \ldots, F(y - \theta_N)$, $F \in \Omega_0$. The prime denotes the transpose of the vector. The linearity is imposed on

θ_i by supposing that it is a linear function of p given, independent variables: $x_{i1}, \ldots, x_{ip}$. Hence

$$\theta_i = x_{i1}\beta_1 + \cdots + x_{ip}\beta_p. \tag{5.1.1}$$

The coefficients $\beta_1, \ldots, \beta_p$ are the unknown parameters. If we denote the relevant vectors by $\mathbf{x}_i' = (x_{i1}, \ldots, x_{ip})$ and $\boldsymbol{\beta}' = (\beta_1, \ldots, \beta_p)$, then we can write

$$\theta_i = \mathbf{x}_i'\boldsymbol{\beta}. \tag{5.1.2}$$

We could also write $Y_i = \theta_i + e_i = \mathbf{x}_i'\boldsymbol{\beta} + e_i$ where $\mathbf{e}' = (e_1, \ldots, e_N)$ is a vector of i.i.d. random variables with cdf $F \in \Omega_0$.

Let $\mathbf{X}$ be the $N \times p$ matrix in which $\mathbf{x}_i'$ is the ith row; then, in matrix notation, the linear model becomes

$$\mathbf{Y} = \mathbf{X}\boldsymbol{\beta} + \mathbf{e}. \tag{5.1.3}$$

This general linear model covers regression, analysis of variance and analysis of covariance designs; see Section 5.4 for examples with data.

If we let $\mathbf{X}$ be an $N \times 1$ vector of ones, then we have the one-sample location model. If we let

$$\mathbf{X} = \begin{bmatrix} \mathbf{1} & \begin{matrix} 1 \\ 0 \end{matrix} \end{bmatrix}$$

where the second column contains n ones, then we have the two-sample location model with n and $N - n$ observations in the respective samples. In this case $\theta_i = \beta_1 + \beta_2$ for $i = 1, \ldots, n$, and $\theta_i = \beta_1$ for $i = n + 1, \ldots, N$ and $\Delta = \beta_2$. Next let

$$\mathbf{X} = \begin{bmatrix} \mathbf{1} & \begin{matrix} 1 & 1 \\ 1 & -1 \\ -1 & 1 \\ -1 & -1 \end{matrix} \end{bmatrix}$$

where each smaller 1 denotes n ones. This provides the design for a two-way layout with 2 treatments, 2 blocks, and n observations per cell. In this example, $\theta_{11} = \beta_1 + \beta_2 + \beta_3$, $\theta_{12} = \beta_1 + \beta_2 - \beta_3$, $\theta_{21} = \beta_1 - \beta_2 + \beta_3$, and $\theta_{22} = \beta_1 - \beta_2 - \beta_3$, so that β_1 corresponds to the grand mean, $2\beta_2$ represents the change in passing from the first to second block, and $2\beta_3$ represents the change in levels of the treatment. The hypothesis of no treatment effect is $H_0 : \beta_3 = 0$. Draper and Smith (1981, Chapter 9) discuss multiple regression as applied to analysis of variance problems with special attention to the two-way layout. See Example 5.4.2 for a worked example with data.

Before turning to the general estimation and hypothesis testing problems in the linear model we will discuss the simple regression model. In this case $Y_i = \beta_1 + x_i\beta_2 + e_i$, $i = 1, \ldots, N$. The independent variables $x_1, \ldots, x_N$ are given numbers and we concentrate on the unknown slope parameter β_2. The standard test of $H_0: \beta_2 = 0$ versus $H_A: \beta_2 \neq 0$ is based on $\sum(x_i - \bar{x})$ Y_i. As in the earlier chapters, this suggests a rank test statistic

$$U = \sum_{i=1}^{N} (x_i - \bar{x})R_i \qquad (5.1.4)$$

where $R_1, \ldots, R_N$ are the ranks of $Y_1, \ldots, Y_N$. We reject $H_0: \beta_2 = 0$ for extreme values of U. We need the moments of U and its asymptotic distribution under $H_0: \beta_2 = 0$ in order to approximate the critical values.

Under $H_0: \beta_2 = 0$, $Y_1, \ldots, Y_N$ are i.i.d. random variables with cdf $F(y - \beta_1)$, $F \in \Omega_0$. Hence Exercise 3.7.1 provides the distribution theory for $R_1, \ldots, R_N$. Further, $EU = 0$, and the variance of U is given by

$$\text{Var } U = \sum_{i=1}^{N} (x_i - \bar{x})^2 \frac{(N^2 - 1)}{12} - \sum\sum_{i \neq j} (x_i - \bar{x})(x_j - \bar{x}) \frac{(N + 1)}{12}$$

$$= \left[\frac{(N^2 - 1)}{12} + \frac{N + 1}{12} \right] \sum_{i=1}^{N} (x_i - \bar{x})^2$$

$$= \frac{N(N + 1)}{12} \sum_{i=1}^{N} (x_i - \bar{x})^2, \qquad (5.1.5)$$

where we have used the identity $0 = [\sum(x_i - \bar{x})]^2 = \sum(x_i - \bar{x})^2 + \sum\sum_{i \neq j}(x_i - \bar{x})(x_j - \bar{x})$. In the next theorem we provide the projection of U and discuss the limiting distribution of U.

Theorem 5.1.1. In the simple regression model, suppose $H_0: \beta_2 = 0$ is true. Further, suppose that

$$\frac{1}{N} \sum_{i=1}^{N} (x_i - \bar{x})^2 \to \delta^2 > 0.$$

Then

$$U^* = \frac{1}{(N + 1)\sqrt{N}} \sum_{i=1}^{N} (x_i - \bar{x})(R_i - (N + 1)/2) \xrightarrow{D} Z \sim n(0, \delta^2/12).$$

Proof. The argument is similar to that of Theorem 3.2.4, and Exercise 4.5.15 is a special case. First, note that the rank of Y_j among $Y_1, \ldots, Y_N$ can be written as

$$R_j = 1 + \sum_{i=1}^{N} s(Y_j - Y_i),$$

where $s(x) = 1$ if $x > 0$ and 0 otherwise. Then, since

$$E\big[s(Y_j - Y_i)\mid Y_k = y\big] = \begin{cases} F(y) & k = j \\ 1 - F(y) & k = i \\ 1/2 & k \neq i \text{ or } j, \end{cases}$$

we have

$$E\big[R_j \mid Y_k = y\big] = 1 + \sum_{i=1}^{N} E\big[s(Y_j - Y_i)\mid Y_k = y\big]$$

$$= \begin{cases} 1 + (N-1)F(y) & k = j \\ 1 + (N-2)/2 + (1 - F(y)) & k \neq j. \end{cases}$$

Then, using a little algebra, we have

$$E\big[U^* \mid Y_k = y\big] = \frac{1}{(N+1)\sqrt{N}}\left\{ \sum_{j \neq k}(x_j - \bar{x})[1/2 - F(y)] \right.$$

$$\left. + (x_k - \bar{x})\big[(N-1)F(y) - (N-1)/2\big] \right\}$$

$$= \frac{\sqrt{N}}{(N+1)}(x_k - \bar{x})\big[F(y) - \tfrac{1}{2}\big].$$

Hence the projection V_p of U^* is

$$V_p = \frac{\sqrt{N}}{(N+1)} \sum_{k=1}^{N}(x_k - \bar{x})\big[F(Y_k) - 1/2\big].$$

Since $F(Y_k)$ has a uniform distribution on $(0,1)$ with mean $\tfrac{1}{2}$ and variance $1/12$, we have

$$\operatorname{Var} V_p = \frac{N}{12(N+1)^2} \sum_{k=1}^{N}(x_k - \bar{x})^2 \to \frac{1}{12}\delta^2.$$

From (5.1.5), $\operatorname{Var} U^* \to \delta^2/12$, so by the Projection Theorem 2.5.2 $E(U^* - V_p)^2 \to 0$. Thus, U^* and V_p have the same limiting distribution. Theorem A10 implies that V_p (and hence U^*) is asymptotically $n(0, \delta^2/12)$.

The result in the theorem can be restated as $U/(\operatorname{Var} U)^{1/2}$ is asymptotically $n(0, 1)$. Hence $H_0: \beta_2 = 0$ is rejected in favor of $H_A: \beta_2 \neq 0$ at approximate level α, if $|U| > Z_{\alpha/2}(\operatorname{Var} U)^{1/2}$ and $1 - \Phi(Z_{\alpha/2}) = \alpha/2$.

In order to discuss the estimation of β_2, we introduce

$$U(\beta_2) = \sum (x_i - \bar{x}) R_i(\beta_2), \qquad (5.1.6)$$

where $R_i(\beta_2)$ is the rank of $Y_i - \beta_1 - x_i\beta_2$, $i = 1, \ldots, N$. Note that the rank $R_i(\beta_2)$ is invariant under changes in β_1 and can be computed with $\beta_1 = 0$. If β_2 is the true parameter value, then $EU(\beta_2) = 0$. This implies that the Hodges–Lehmann estimate of β_2 is the value $\hat{\beta}_2$ such that $U(\hat{\beta}_2) \doteq 0$. Adichie (1967) was the first to introduce rank estimates in the simple regression model. Unlike the one- and two-sample problems, $\hat{\beta}_2$ does not have a simple representation and must be determined using numerical methods.

To see why numerical methods are necessary, consider the graph of $U(\beta_2)$ as a function of β_2. Suppose $x_1 < x_2 < \cdots < x_N$. If $\beta_2 = (Y_j - Y_i)/(x_j - x_i)$, $j > i$, then $Y_j - x_j\beta_2 = Y_i - x_i\beta_2$ and these residuals will be assigned the average rank. If, on the other hand, $\beta_2 < (Y_j - Y_i)/(x_j - x_i)$, then $Y_i - x_i\beta_2 < Y_j - x_j\beta_2$. Suppose that β_2 is also close to this slope, then the rank of $Y_j - x_j\beta_2$ is one greater than the rank of $Y_i - x_i\beta_2$. As β_2 moves across this slope so that $(Y_j - Y_i)/(x_j - x_i) < \beta_2$, the ranks of $Y_i - x_i\beta_2$ and $Y_j - x_j\beta_2$ are interchanged. This shows that $U(\beta_2)$ is a decreasing step function which steps at the $N(N - 1)/2$ pairwise slopes. When β_2 is just barely below $(Y_j - Y_i)/(x_j - x_i)$ the ranks of $Y_i - x_i\beta_2$ and $Y_j - x_j\beta_2$ are $R_i(\beta_2)$ and $R_i(\beta_2) + 1$, respectively. They appear in $U(\beta_2)$ as $(x_i - \bar{x})R_i(\beta_2)$ and $(x_j - \bar{x})[R_i(\beta_2) + 1]$. The change in $U(\beta_2)$ as β_2 crosses this slope is $(x_i - \bar{x})R_i(\beta_2) + (x_j - \bar{x})[R_i(\beta_2) + 1] - \{(x_i - \bar{x})[R_i(\beta_2) + 1] + (x_j - \bar{x})R_i(\beta_2)\} = x_j - x_i$. Hence $U(\beta_2)$ steps down by the amount $x_j - x_i$ at the slope $(Y_j - Y_i)/(x_j - x_i)$. This is in contrast to the one- and two-sample location models in which the steps at Walsh averages and pairwise differences, respectively, are constant. These variable step sizes result in a weighted median as the estimate of β_2; see Jaeckel (1972). A naive computational approach would be to order the slopes and carry along the $x_j - x_i$ step sizes. Then, beginning with $\max U(\beta_2) = \sum(x_i - \bar{x})i$, accumulate the steps until $U(\beta_2)$ crosses zero or steps onto an interval of zeros; see Example 5.1.1. Compare the discussions in Section 1.5, 2.3, and 3.2. Section 5.4 has further discussion of computation. We will discuss the statistical properties of $\hat{\beta}_2$ in the context of the general linear model in Section 5.2 where it is shown that $N^{1/2}(\hat{\beta}_2 - \beta_2)$ is asymptotically normally distributed.

At first thought, a natural estimate of β_2 based on $U(\beta_2)$ would be the median of the pairwise slopes. This, however, is not the case in general as shown by the foregoing discussion. Exercise 5.5.1 shows that this estimate corresponds to the test based on Kendall's τ. See Sen (1968) for further discussion.

Finally, we note that, because the ranks are invariant to a constant shift, the present approach does not provide an estimate or test of β_1, the intercept parameter. A natural estimate of β_1 is $\hat{\beta}_1$, the median of the residuals $Y_1 - x_1\hat{\beta}_2, \ldots, Y_N - x_N\hat{\beta}_2$ or, if we assume symmetry of F, the median of the Walsh averages of the residuals. The joint limiting distribution of $N^{1/2}(\hat{\beta}_1, \hat{\beta}_2)'$ is discussed at the end of Section 5.2.

Example 5.1.1. Hubble's law in astronomy states that the recession velocity of a galaxy is directly proportional to the distance. Hence, once the constant ratio of velocity to distance is estimated, recession velocity can be predicted from a distance estimate. In Table 5.1 we provide the distance in millions of light years and velocity in hundreds of miles per second for 11 galactic clusters from Clason (1958, p. 337). If $x =$ distance and y = velocity, then we wish to fit the model $Y = \beta x + e$. The intercept is constrained to be zero. A plot of velocity (y) versus distance (x) shows the points to be close to a straight line. The problem is quite simple, and we use the data to illustrate the behavior of $U(\beta) = \sum(x_i - \bar{x})R_i(\beta)$ where $R_i(\beta)$ is the rank of $y_i - x_i\beta$. The max $U(\beta) = \sum(x_i - \bar{x})i = 10{,}320$. There are $11(10)/2 = 55$ pairwise slopes beginning with the smallest slope 0.187

Table 5.1. Distance and Velocity Data

Cluster	Distance (x)	Velocity (y)	y/x
Virgo	22	7.5	.341
Pegasus	68	24	.353
Perseus	108	32	.296
Coma Berenices	137	47	.343
Ursa Major No. 1	255	93	.365
Leo	315	120	.381
Corona Borealis	390	134	.344
Gemini	405	144	.356
Bootes	685	245	.358
Ursa Major No. 2	700	260	.371
Hydra	1100	380	.345

determined by Leo and Corona Borealis and a jump of 75. At 0.353, the 23rd slope, the graph of $U(\beta)$ steps from 332 to -90. Hence, $\hat{\beta} = .353$, close to the average ratio of .350. The estimate of β which corresponds to Kendall's τ (see Exercise 5.5.1) is the median of the 55 slopes. This estimate is .359, the 28th slope. The least-squares estimate is also .353.

In Section 1.6 we introduced the influence curve of an estimator. Robust estimates have bounded influence curves, so that single outlying observations cannot have inordinately large effects on the estimates. Examples 1.6.4 and 2.4.2 show that the median and the median of the Walsh averages both have bounded influence curves as opposed to the linear, unbounded influence curve of the sample mean. The discussion around (1.6.1) also explains the connection between the influence curve and the asymptotic variance of the estimate.

We now discuss the influence curves for the least-squares and rank estimators of β_2, the slope parameter in the simple regression model. The results carry over to the general linear model with minor notation changes; hence the discussion will not be repeated in the general case. Cook and Weisberg (1982) describe in detail how influence curves and their finite-sample counterparts can be used in diagnostic data analysis in the linear model.

Recall the simple linear regression model is specified by $Y_i = \beta_1 + x_i\beta_2 + e_i$, $i = 1, \ldots, N$. Without loss of generality, we will suppose the independent variables have been centered so that $\bar{x} = 0$. Define

$$V(\beta_2) = N^{-1}\sum x_i(y_i - x_i\beta_2), \tag{5.1.7}$$

then the least-squares estimate $\hat{\beta}_2 = \sum x_i y_i / \sum x_i^2$ is the solution of $V(\beta_2) = 0$. Let $H_N(x, y)$ be the empirical bivariate cdf, then (5.1.7) can be written

$$V(\beta_2) = \int\int x(y - x\beta_2)\,dH_N(x, y) \tag{5.1.8}$$

since $H_N(x, y)$ assigns mass N^{-1} to the points (x_i, y_i), $i = 1, \ldots, N$.

For a fixed value of x, $F(y - \beta_1 - \beta_2 x)$ represents the conditional distribution of Y given $X = x$. We generalize the problem slightly and allow x to be a realization of a random variable X with marginal cdf $M(x)$. The joint distribution of X and Y is denoted by $H(x, y)$. Then the parameter β_2 can be defined as the solution $\beta_2 = T(H)$ of the equation

$$\int\int x[y - xT(H)]\,dH(x, y) = 0, \tag{5.1.9}$$

and when H is replaced by H_N, (5.1.8) becomes $V(\hat{\beta}_2) = 0$.

In order to compute the influence curve, we replace H by a contaminated version:

$$H_t(x, y) = (1 - t)H(x, y) + t\delta_{x_0, y_0}(x, y), \qquad (5.1.10)$$

where $\delta_{x_0, y_0}(x, y)$ puts mass one at the point (x_0, y_0), and differentiate with respect to t; see Definition 1.6.4. Inserting H_t into (5.1.9) yields

$$\int \int x \left[y - xT(H_t) \right] d \left\{ H(x, y) + t(\delta_{x_0, y_0}(x, y) - H(x, y)) \right\} = 0.$$

$$(5.1.11)$$

Differentiate with respect to t and set $t = 0$ to get

$$- \left. \frac{dT(H_t)}{dt} \right|_{t=0} \int \int x^2 \, dH(x, y)$$

$$+ \int \int x \left[y - xT(H) \right] d \left\{ \delta_{x_0, y_0}(x, y) - H(x, y) \right\} = 0. \quad (5.1.12)$$

This yields the influence curve evaluated at $x = x_0$ and $y = y_0$ as

$$\Omega(x_0, y_0) = \left. \frac{d}{dt} T(H_t) \right|_{t=0} = \frac{x_0(y_0 - x_0 \beta_2)}{\int x^2 \, dM(x)}, \qquad (5.1.13)$$

where $\int \int x^2 \, dH(x, y) = \int x^2 \, dM(x)$. The influence curve could have been derived more directly from the formula for $\hat{\beta}_2$. We have presented this development because it provides an outline for the derivation of the influence curve of the rank estimate of β_2. The main point to be noted from (5.1.13) is that the least-squares estimate has unlimited influence in x and y directions. Hence the influence is unbounded, and either an extreme x value or a large residual will have a large impact on the estimate. This generalizes the unbounded influence of the sample mean discussed in Example 1.6.4.

The defining equation for the R estimate is given by (5.1.6). We introduce a factor of N^{-2} in order to anticipate the defining equation for the parameter β_2. Hence we consider

$$\frac{1}{N} \sum x_i \left[\frac{R_i(\beta_2)}{N} - \frac{1}{2} \right] = 0. \qquad (5.1.14)$$

Since $\bar{x} = 0$, the $1/2$ does not alter the equation, but provides a convenient centering for the influence curve later. Further, since the rank is not effected by β_1, without loss of generality we let $\beta_1 = 0$. Now, since $R_i(\beta_2) = NF_N(y_i - x_i\beta_2)$ where $F_N(\cdot)$ is the empirical cdf based on the residuals $y_i - x_i\beta_2$, $i = 1, \ldots, N$, we have

$$\int \int x \left[F_N(y - x\beta_2) - 1/2 \right] dH_N(x, y) = 0, \qquad (5.1.15)$$

where H_N assigns the mass N^{-1} to each point (x_i, y_i), $i = 1, \ldots, N$.

Let $\beta_2 = T(H)$ where $H(x, y)$ is the joint cdf of X and Y as described in the preceding discussion of the least-squares estimate. Then the defining equation for $\beta_2 = T(H)$ is

$$\int \int x \left[F(y - xT(H)) - 1/2 \right] dH(x, y) = 0. \qquad (5.1.16)$$

The contaminated version of H is given by (5.1.10). Note that $\delta_{x_0, y_0}(x, y) = \delta_{x_0}(x)\delta_{y_0}(y)$ so the degenerate distribution function, at x_0, y_0, factors into independent degenerate distribution functions. The contaminated version of the conditional cdf of Y given $X = x$ is

$$F_t(y - x\beta_2) = \begin{cases} F(y - x\beta_2) & \text{if} \quad x \neq x_0 \\ \delta_{y_0}(y) & \text{if} \quad x = x_0. \end{cases} \qquad (5.1.17)$$

In the following calculations we suppose that $M(x)$ does not have a jump at x_0. This is reasonable since if $M(x)$ were the design measure assigning N^{-1} to each x_i, $i = 1, \ldots, N$, no mass would be assigned to the outlying x_0 point.

We now must insert F_t and H_t into (5.1.16), differentiate with respect to t and set $t = 0$. We first have

$$(1 - t) \int \int x \left[F_t(y - xT(H_t)) - 1/2 \right] dH(x, y)$$

$$+ t \int \int x \left[F_t(y - xT(H_t)) - 1/2 \right] d\delta_{x_0, y_0}(x, y) = 0.$$

Using (5.1.17), we have

$$(1 - t) \int \int x \left[F(y - xT(H_t)) - 1/2 \right] dH(x, y)$$

$$+ tx_0 \left[F_t(y_0 - x_0 T(H_t)) - 1/2 \right] = 0.$$

Differentiation, along with (5.1.16), yields

$$-\frac{dT(H_t)}{dt}\Bigg|_{t=0}\int\int x^2 f(y - xT(H))\,dH(x, y)$$

$$+ x_0\big[F(y_0 - x_0 T(H)) - 1/2\big] = 0.$$

After replacing $T(H)$ by β_2, the influence curve, evaluated at x_0, y_0, becomes:

$$\Omega(x_0, y_0) = \frac{x_0\big[F(y_0 - x_0\beta_2) - 1/2\big]}{\int\int x^2 f(y - x\beta_2)\,dH(x, y)}. \qquad (5.1.18)$$

Recall that $H(x, y)$ is constructed from the marginal cdf of X, $M(x)$, and the conditional cdf of Y given $X = x$, $F(y - x\beta_2)$. Hence

$$\int\int x^2 f(y - x\beta_2)\,dH(x, y) = \int\int x^2 f(y - x\beta_2)\,dF(y - x\beta_2)\,dM(x)$$

$$= \int x^2 \int f^2(y - x\beta_2)\,dy\,dM(x)$$

$$= \int x^2\,dM(x)\int f^2(u)\,du.$$

The last line results when we make the change of variable $u = y - x\beta_2$ since the inside integral will no longer depend on x. Thus, (5.1.18) becomes

$$\Omega(x_0, y_0) = \frac{x_0\big[F(y_0 - x_0\beta_2) - 1/2\big]}{\int f^2(u)\,du\int x^2\,dM(x)}. \qquad (5.1.19)$$

This influence curve is similar to that for the median of the Walsh averages in the one-sample problem; see (2.4.6). The influence is bounded relative to large or extreme residuals. However, the influence is unbounded with respect to design points. Extreme design points [sometimes called high leverage points; see Huber (1981) and Cook and Weisberg (1982)] can have a large impact on the R estimator. Hence we conclude that the R estimator is robust relative to outlying residuals but not robust relative to outlying x values. The same thing happens in the general linear model so that care must be exercised when estimating parameters in the presence of points with high leverage. For a review of bounded influence regression see Huber (1983).

We see from this section that the linear model subsumes a large number of important models. In the following sections we treat these models in a unified fashion through the linear model. In the previous chapters we first introduced test statistics and then the estimates derived from them. We now reverse this approach and first treat the problem of estimation. Then we present three approaches to testing hypotheses. Finally, in Section 5.4 we illustrate the methods on data.

5.2. RANK ESTIMATES IN THE LINEAR MODEL

We refine the notation of the linear model, (5.1.3), to explicitly distinguish the intercept parameter from the regression parameters. The examples in Section 5.1 included the intercept as part of the design when needed. We now let

$$Y = \begin{bmatrix} 1 & X \end{bmatrix} \begin{pmatrix} \alpha \\ \beta \end{pmatrix} + e \qquad (5.2.1)$$

where Y is an $N \times 1$ observation vector, 1 is an $N \times 1$ column vector of ones, X is an $N \times p$ matrix of known regression constants, α is the scalar intercept parameter, β is a $p \times 1$ vector of unknown regression parameters, and e is an $N \times 1$ vector of i.i.d. errors with cdf $F \in \Omega_0$. Hence the median of the distribution of Y_i is $\alpha + x_i'\beta$ where x_i' is the ith row of X.

Center the matrix X by subtracting column means to get $X_c = X - 1(\bar{x}_1, \ldots, \bar{x}_p)$, where $\bar{x}_i$ is the mean of the ith column of X. Then (5.2.1) can be written in the form

$$Y = \begin{bmatrix} 1 & X_c \end{bmatrix} \begin{pmatrix} \alpha^* \\ \beta \end{pmatrix} + e \qquad (5.2.2)$$

where $\alpha^* = \alpha + \bar{x}'\beta$ and $\bar{x}' = (\bar{x}_1, \ldots, \bar{x}_p)$. Now the subspaces spanned by 1 and the columns of X_c are orthogonal. This results in uncorrelated estimates of α^* and β, similar to least-squares. We concentrate mainly on estimation and testing of β. Estimation of β, separately from α or α^*, is effected by minimizing a measure of dispersion of residuals. The nature of such a measure is described in the following definition.

Definition 5.2.1. Let $D(\cdot)$ be a measure of variability that satisfies the following two properties: (1) $D(Z + 1a) = D(Z)$ and (2) $D(-Z) = D(Z)$ for every $N \times 1$ vector Z and scalar a. Then $D(\cdot)$ is called an even, location-free measure of dispersion.

If $D(\cdot)$ is even, location free, then $D(\mathbf{Y} - \mathbf{1}\alpha - \mathbf{X}\boldsymbol{\beta}) = D(\mathbf{Y} - \mathbf{1}\alpha^* - \mathbf{X}_c\boldsymbol{\beta}) = D(\mathbf{Y} - \mathbf{X}_c\boldsymbol{\beta}) = D(\mathbf{Y} - \mathbf{X}\boldsymbol{\beta})$. Hence, when working with $D(\cdot)$, we can use either $\mathbf{X}$ or $\mathbf{X}_c$ without altering the results. We write $D(\mathbf{Y} - \mathbf{X}\boldsymbol{\beta})$ or $D(\mathbf{Y} - \mathbf{X}_c\boldsymbol{\beta})$ interchangeably. By minimizing $D(\mathbf{Y} - \mathbf{X}\boldsymbol{\beta})$, as a function of $\boldsymbol{\beta}$, we have an estimate of $\boldsymbol{\beta}$, generated by $D(\cdot)$. For example, if $D(\mathbf{Z}) = (\mathbf{Z} - \mathbf{1}\bar{Z})'(\mathbf{Z} - \mathbf{1}\bar{Z})$ where $\bar{Z} = \sum Z_i / N$, then $D(\mathbf{Y} - \mathbf{X}_c\boldsymbol{\beta}) = (\mathbf{Y} - \mathbf{X}_c\boldsymbol{\beta})'(\mathbf{Y} - \mathbf{X}_c\boldsymbol{\beta}) + N\bar{Y}^2$. The resulting estimate of $\boldsymbol{\beta}$ is the usual least-squares estimate.

Our goal is to define an even, location-free measure of dispersion that produces a rank estimate of $\boldsymbol{\beta}$. This estimate will be considered the extension of the Hodges–Lehmann estimate from the two-sample location model to the regression model; see Example 5.2.1.

Suppose $a_1 \leqslant \cdots \leqslant a_N$ is a nonconstant sequence of scores such that $a_k + a_{N-k+1} = 0$. They can be constructed using the score-generating functions discussed in Chapter 3. See Definition 3.4.2 for an example. Jaeckel (1972) defined the following measure of dispersion of the vector $\mathbf{Z}' = (Z_1, \ldots, Z_N)$:

$$D(\mathbf{Z}) = \sum_{i=1}^{N} a(i) Z_{(i)} \qquad (5.2.3)$$

where $Z_{(1)} \leqslant \cdots \leqslant Z_{(N)}$. Further, if $R_1, \ldots, R_N$ are the ranks of $Z_1, \ldots, Z_N$ then we can also write

$$D(\mathbf{Z}) = \sum_{i=1}^{N} a(R_i) Z_i \qquad (5.2.4)$$

where we assign the average score to tied Z values.

Definition 5.2.2. A rank estimate (R estimate) of $\boldsymbol{\beta}$ is the value $\hat{\boldsymbol{\beta}}$ which minimizes

$$D(\mathbf{Y} - \mathbf{X}\boldsymbol{\beta}) = \sum a\left[R(Y_i - \mathbf{x}_i'\boldsymbol{\beta})\right](Y_i - \mathbf{x}_i'\boldsymbol{\beta}) \qquad (5.2.5)$$

where $\mathbf{x}_i'$ is the ith row of $\mathbf{X}$ and $R(Y_i - \mathbf{x}_i'\boldsymbol{\beta})$ is the rank of $Y_i - \mathbf{x}_i'\boldsymbol{\beta}$ among $Y_1 - \mathbf{x}_1'\boldsymbol{\beta}, \ldots, Y_N - \mathbf{x}_N'\boldsymbol{\beta}$.

This even, location-free measure is a linear, rather than quadratic, function of the residuals with coefficients determined by the rank, or size, of the residuals. Hence it is hoped that the estimates generated by (5.2.5) will be more robust than least-squares estimates because the influence of outliers enters in a linear rather than quadratic fashion. The next theorem shows that $D(\mathbf{Y} - \mathbf{X}\boldsymbol{\beta})$ is a proper measure of dispersion.

Theorem 5.2.1. The function $D(\mathbf{Y} - \mathbf{X}\boldsymbol{\beta})$ is a nonnegative, continuous, and convex function of $\boldsymbol{\beta}$.

Proof. Let $Z_i = Y_i - \mathbf{x}_i'\boldsymbol{\beta}$, $i = 1, \ldots, N$. Let t be such that $a(1) \leqslant \cdots \leqslant a(t-1) \leqslant 0 \leqslant a(t) \leqslant \cdots \leqslant a(N)$. Then

$$D(\mathbf{Z}) = \sum_{i=1}^{N} a(i)\left[Z_{(i)} - Z_{(t)} \right]$$

since $\sum a(i) Z_{(t)} = Z_{(t)} \sum a(i) = 0$. But each term $a(i)[Z_{(i)} - Z_{(t)}] \geqslant 0$; hence, $D(\mathbf{Z}) = D(\mathbf{Y} - \mathbf{X}\boldsymbol{\beta}) \geqslant 0$.

Now let $p = (p(1), \ldots, p(N))$ be a permutation of $1, \ldots, N$ and define

$$D_p(\mathbf{Z}) = \sum_{i=1}^{N} a(i) Z_{p(i)}.$$

Theorem 368 of Hardy et al. (1952) states that

$$\max_p \sum_{i=1}^{N} a(i) Z_{p(i)} = \sum_{i=1}^{N} a(i) Z_{(i)}.$$

Hence

$$D(\mathbf{Y} - \mathbf{X}\boldsymbol{\beta}) = \max_p \sum_{i=1}^{N} a(i)\left(Y_{p(i)} - \mathbf{x}_{p(i)}'\boldsymbol{\beta} \right).$$

This shows that $D(\mathbf{Y} - \mathbf{X}\boldsymbol{\beta})$ is the maximum of a finite set of linear functions. It then follows that $D(\mathbf{Y} - \mathbf{X}\boldsymbol{\beta})$ is a continuous and convex function of $\boldsymbol{\beta}$.

Jaeckel (1972) further points out that if $\mathbf{X}_c$ is a full rank, then $D(\mathbf{Y} - \mathbf{X}\boldsymbol{\beta})$ attains its minimum, and the set of $\boldsymbol{\beta}$ for which this occurs is bounded. Hence the estimate in Definition 5.2.2 may be taken to be any value which minimizes $D(\mathbf{Y} - \mathbf{X}\boldsymbol{\beta})$.

The domain ($\boldsymbol{\beta}$-space) of $D(\mathbf{Y} - \mathbf{X}\boldsymbol{\beta})$ is divided into a finite number of convex polygonal subsets, on each of which $D(\mathbf{Y} - \mathbf{X}\boldsymbol{\beta})$ is a linear function of $\boldsymbol{\beta}$. The partial derivatives (gradient) of $D(\mathbf{Y} - \mathbf{X}\boldsymbol{\beta})$ exist almost everywhere and are given by

$$\frac{\partial}{\partial \beta_j} D(\mathbf{Y} - \mathbf{X}\boldsymbol{\beta}) = \sum_{i=1}^{N} a\left[R(Y_i - \mathbf{x}_i'\boldsymbol{\beta}) \right](-x_{ij})$$

$$= -\sum_{i=1}^{N} (x_{ij} - \bar{x}_j) a\left[R(Y_i - \mathbf{x}_i'\boldsymbol{\beta}) \right] \qquad (5.2.6)$$

for $j = 1, \ldots, p$. We have used $\sum a_k = 0$ in the second equality to introduce the centered design. Hence, for each $j = 1, \ldots, p$, the negative of the partial derivative of the dispersion is the regression rank statistic:

$$S_j(\mathbf{Y} - \mathbf{X}\boldsymbol{\beta}) = \sum_{i=1}^{N} (x_{ij} - \bar{x}_j) a[R(Y_i - \mathbf{x}_i'\boldsymbol{\beta})]. \qquad (5.2.7)$$

If we denote the gradient of $D(\cdot)$ by $\nabla D(\cdot)$ and let $\mathbf{S}'(\mathbf{Y} - \mathbf{X}\boldsymbol{\beta}) = (S_1(\mathbf{Y} - \mathbf{X}\boldsymbol{\beta}), \ldots, S_p(\mathbf{Y} - \mathbf{X}\boldsymbol{\beta}))$, then the analog of the linear normal equations of least squares is the set of nonlinear equations

$$-\nabla D(\mathbf{Y} - \mathbf{X}\boldsymbol{\beta}) = \mathbf{S}(\mathbf{Y} - \mathbf{X}\boldsymbol{\beta}) \doteq \mathbf{0}. \qquad (5.2.8)$$

We have approximately $\mathbf{0}$ because of the discrete nature of the regression rank statistics. This extends the discussion of simple regression at the end of Section 5.1. Hence minimizing $D(\mathbf{Y} - \mathbf{X}\boldsymbol{\beta})$ is equivalent to solving (5.2.8). But this is the extension of the Hodges–Lehmann estimate to the linear model since $E\mathbf{S}(\mathbf{Y} - \mathbf{X}\boldsymbol{\beta}) = \mathbf{0}$ when $\boldsymbol{\beta}$ is the true parameter value. The reader is asked to verify that $E\mathbf{S}(\mathbf{Y} - \mathbf{X}\boldsymbol{\beta}) = \mathbf{0}$ in Exercise 5.5.2.

As in previous chapters we are primarily concerned with the Wilcoxon scores. Let $a(i) = \phi(i/(N+1))$ where

$$\phi(u) = \sqrt{12}\,(u - 1/2). \qquad (5.2.9)$$

This standardization is convenient since $\int \phi(u)\,du = 0$ and $\int \phi^2(u)\,du = 1$.

In the simple linear regression problem, (5.2.8) corresponds to (5.1.6). The following example illustrates the preceding ideas on the two-sample location model.

Example 5.2.1. The two-sample location model can be expressed as follows

$$\mathbf{Y} = \begin{pmatrix} Y_1 \\ \vdots \\ Y_n \\ Y_{n+1} \\ \vdots \\ Y_N \end{pmatrix} = \begin{pmatrix} \mathbf{1} & \mathbf{1}_n \\ & \mathbf{0} \end{pmatrix} \begin{pmatrix} \alpha \\ \beta \end{pmatrix} + \mathbf{e}.$$

Hence, if $m = N - n$, the first n observations are from a distribution with median $\alpha + \beta$ and the last m observations are from a distribution with median α. The difference in population medians is $\Delta = \alpha + \beta - \alpha = \beta$. In this example $p = 1$ so we suppress the subscript on $S_1(Y - X\beta)$. The vector X is the second column in the full design matrix. Using the Wilcoxon score function, (5.2.9), we have

$$D(Y - X\beta) = \frac{\sqrt{12}}{N+1} \sum_{i=1}^{N} \left[R(Y_i - x_i\beta) - (N+1)/2 \right](Y_i - x_i\beta)$$

$$S(Y - X\beta) = \frac{\sqrt{12}}{N+1} \sum_{i=1}^{N} (x_i - \bar{x})\left[R(Y_i - x_i\beta) - (N+1)/2 \right]$$

$$= \frac{\sqrt{12}}{N+1} \sum_{i=1}^{n} \left[R(Y_i - \beta) - (N+1)/2 \right].$$

The third expression follows easily by noting $\sum \bar{x}[R(Y_i - x_i\beta) - (N+1)/2] = 0$. Now, if $S(Y - X\beta) \doteq 0$, then $\sum_{i=1}^{n} R(Y_i - \beta) \doteq n(N+1)/2$, and $\hat{\beta}$ is the median of the pairwise differences across the two samples since we have the rank-sum form of the Mann–Whitney–Wilcoxon statistic. [Compare (3.2.4).] For future reference we note that for this example

$$X_c = \frac{1}{N} \begin{bmatrix} m \\ \vdots \\ m \\ -n \\ \vdots \\ -n \end{bmatrix} \quad \text{and} \quad X_c'X_c = \frac{mn}{N},$$

where there are n m's and $m - n$'s.

We now develop the limiting distribution for $N^{1/2}(\hat{\beta} - \beta_0)$ where $\hat{\beta}$ is the R estimate found by minimizing (5.2.5) or solving (5.2.8) and β_0 is the true parameter value. Recall from Section 2.7 that the Wilcoxon signed rank statistic $n^{1/2}\bar{T}(\theta)$ is approximately linear in an appropriate sense, and this leads to a heuristic derivation of the asymptotic normality of $n^{1/2}(\hat{\theta} - \theta_0)$. We have a similar result for the regression rank statistic $S_j(Y - X\beta)$. The strategy for determining the limiting distribution is as follows: (1) from the linear approximation to $S(Y - X\beta) = -\nabla D(Y - X\beta)$ we construct a quadratic approximation to $D(Y - X\beta)$; (2) for the value $\tilde{\beta}$, which minimizes the quadratic approximation, it will be shown that $N^{1/2}(\tilde{\beta} - \beta_0)$ has a limiting normal distribution; and then (3) we will show that $\hat{\beta}$ is close to $\tilde{\beta}$ and $N^{1/2}(\hat{\beta} - \beta_0)$ has the same limiting distribution as $N^{1/2}(\tilde{\beta} - \beta_0)$.

We begin by stating the linearity result for the vector of regression rank statistics. Let β_0 denote the true parameter value and suppose we are using the Wilcoxon scores, (5.2.9). Then we have

$$\frac{1}{\sqrt{N}} S(Y - X\beta) = \frac{1}{\sqrt{N}} S(Y - X\beta_0) - \sqrt{12} \int f^2(x)\, dx\, \frac{1}{N} X_c' X_c N^{1/2}(\beta - \beta_0)$$

$$+ o_p(1) \tag{5.2.10}$$

where $o_p(1)$ tends to zero in probability uniformly for all vectors β such that $N^{1/2}\|\beta - \beta_0\| \leqslant C$ for any $C > 0$. Here $\|\cdot\|$ denotes Euclidean length. This result is proved by Aubuchon (1982) under the mild assumptions listed in (5.2.11) later. The proof is similar to, but more complex than, that of Theorem 2.7.2. Jureckova (1969, 1971) establishes (5.2.10) for general scores with $12^{1/2}\int f^2(x)\, dx$ replaced by $\int_0^1 \phi(u)\phi_f(u)\, du$ where $\phi(u)$ is a score-generating function such that $\int\phi(u)\, du = 0$, $\int\phi^2(u)\, du = 1$, and that satisfies Definition 3.4.2; and $\phi_f(u)$ is given in (3.5.14). Her proof requires more technical assumptions on the structure of the design matrix.

We gather together at this point the assumptions that are made throughout this chapter. We will indicate from time to time more general results for arbitrary score functions and refer the reader to the primary sources for proofs. Henceforth, we assume:

a. $\phi(u) = \sqrt{12}\,(u - 1/2)$, $a(i) = \sqrt{12}\,[i/(N+1) - 1/2]$.

b. $[1X]$ has full column rank, $p + 1$.

c. $N^{-1}[1X]'[1X]$ converges to a positive definite matrix which implies that $N^{-1}X_c'X_c \to \Sigma$, positive definite. (5.2.11)
(See the following lemma.)

d. $F \in \Omega_0$ and $f(\cdot)$ has finite Fisher information (Definition 2.9.1).

Two implications of these assumptions are isolated in the following lemma for future reference. The proof is due to Aubuchon (1982).

Lemma.   Condition (c) implies that $N^{-1}X_c'X_c \to \Sigma$, positive definite and condition (d) implies that $\int_{-\infty}^{\infty} f^2(x)\, dx < \infty$.

Proof.   Note first that

$$N^{-1}[1X]'[1X] = \begin{bmatrix} 1 & \bar{x}' \\ \bar{x} & N^{-1}X'X \end{bmatrix} \to \begin{bmatrix} 1 & \mu' \\ \mu & \Delta \end{bmatrix} = \Lambda$$

where $\bar{\mathbf{x}}' = (\bar{x}_1, \ldots, \bar{x}_p)$ and the limit is positive definite. Now define the following convergent matrix

$$\mathbf{A}_N = \begin{bmatrix} 1 & -\bar{\mathbf{x}}' \\ 0 & \mathbf{I} \end{bmatrix} \rightarrow \begin{bmatrix} 1 & -\boldsymbol{\mu}' \\ 0 & \mathbf{I} \end{bmatrix} = \mathbf{A}.$$

Then

$$\mathbf{A}_N' \begin{bmatrix} 1 & \bar{\mathbf{x}}' \\ \bar{\mathbf{x}} & N^{-1}\mathbf{X}'\mathbf{X} \end{bmatrix} \mathbf{A}_N = \begin{bmatrix} 1 & 0 \\ 0 & N^{-1}\mathbf{X}_c'\mathbf{X}_c \end{bmatrix}$$

converges to $\mathbf{A}'\boldsymbol{\Lambda}\mathbf{A}$ which is positive definite, and the first part of the lemma is established.

For the second part note that if Y has cdf $F \in \Omega_0$ then $-Y$ has cdf $1 - F(-y) \in \Omega_0$ with pdf $f(-y)$. Now let X and $-Y$ be independent with pdfs $f(x)$ and $f(-y)$, respectively. Theorem A17 in the Appendix implies that the convolution density has the form

$$f^*(x) = \int_{-\infty}^{\infty} f(x - y)f(-y)\,dy$$

$$= \int_{-\infty}^{\infty} f(x + t)f(t)\,dt$$

$$= \int_{-\infty}^{\infty} \int_{-\infty}^{t} f'(x + z)f(t)\,dz\,dt$$

$$\leqslant \int_{-\infty}^{\infty} \int_{-\infty}^{\infty} |f'(x + z)|f(t)\,dz\,dt$$

$$= \int_{-\infty}^{\infty} \int_{-\infty}^{\infty} |f'(u)|f(t)\,du\,dt$$

$$= \int_{-\infty}^{\infty} |f'(u)|\,du \int_{-\infty}^{\infty} f(t)\,dt$$

$$= \int_{-\infty}^{\infty} \frac{|f'(u)|}{\sqrt{f(u)}} \sqrt{f(u)}\,du$$

$$\leqslant \left\{ \int_{-\infty}^{\infty} \frac{[f'(u)]^2}{f(u)}\,du \right\}^{1/2} = \{I(f)\}^{1/2} < \infty.$$

We have used the absolute continuity, The Cauchy–Schwarz inequality (Theorem A19), and Definition 2.9.1. Now $f^*(0) = \int_{-\infty}^{\infty} f^2(x)\,dx < \infty$.

The result in (5.2.10), valid under the assumptions in (5.2.11), provides a linear approximation to the gradient $\nabla D(\mathbf{Y} - \mathbf{X}\boldsymbol{\beta})$. This suggests how to construct a quadratic approximation to $D(\mathbf{Y} - \mathbf{X}\boldsymbol{\beta})$. Suppose $N^{-1}\mathbf{X}_c'\mathbf{X}_c$ converges to $\boldsymbol{\Sigma}$, a positive definite matrix and suppose $\boldsymbol{\beta}_0$ is the true parameter value. Then define

$$Q(\mathbf{Y} - \mathbf{X}\boldsymbol{\beta}) = D(\mathbf{Y} - \mathbf{X}\boldsymbol{\beta}_0) - (\boldsymbol{\beta} - \boldsymbol{\beta}_0)'\mathbf{S}(\mathbf{Y} - \mathbf{X}\boldsymbol{\beta}_0)$$

$$+ \frac{1}{2}\sqrt{12}\int f^2(x)\,dx\, N(\boldsymbol{\beta} - \boldsymbol{\beta}_0)'\boldsymbol{\Sigma}(\boldsymbol{\beta} - \boldsymbol{\beta}_0). \quad (5.2.12)$$

This quadratic in $\boldsymbol{\beta}$ has the property that $Q(\mathbf{Y} - \mathbf{X}\boldsymbol{\beta}_0) = D(\mathbf{Y} - \mathbf{X}\boldsymbol{\beta}_0)$, and the gradient of $Q(\mathbf{Y} - \mathbf{X}\boldsymbol{\beta})$ is the linear approximation on the right side of (5.2.10). The next theorem, proved by Jaeckel (1972), shows that $Q(\mathbf{Y} - \mathbf{X}\boldsymbol{\beta})$ provides a useful approximation to $D(\mathbf{Y} - \mathbf{X}\boldsymbol{\beta})$.

Theorem 5.2.2. For any $B > 0$ and $\epsilon > 0$, under the assumptions in (5.2.11),

$$P\left\{ \sup_{\sqrt{N}\,\|\boldsymbol{\beta} - \boldsymbol{\beta}_0\| \leqslant B} |Q(\mathbf{Y} - \mathbf{X}\boldsymbol{\beta}) - D(\mathbf{Y} - \mathbf{X}\boldsymbol{\beta})| \geqslant \epsilon \right\} \to 0.$$

Proof. Let $\boldsymbol{\sigma}_j'$ be the jth row of $\boldsymbol{\Sigma}$. Then from (5.2.10), for all $j = 1, \ldots, p$,

$$P\left\{ \sup_{\sqrt{N}\,\|\boldsymbol{\beta} - \boldsymbol{\beta}_0\| \leqslant B} \frac{1}{\sqrt{N}} \left| S_j(\mathbf{Y} - \mathbf{X}\boldsymbol{\beta}) - S_j(\mathbf{Y} - \mathbf{X}\boldsymbol{\beta}_0) \right. \right.$$

$$\left. \left. + \sqrt{12}\int f^2(x)\,dx\, N\boldsymbol{\sigma}_j'(\boldsymbol{\beta} - \boldsymbol{\beta}_0) \right| > \frac{\epsilon}{B\sqrt{p}} \right\} \to 0.$$

Hence

$$P\left\{ \sup_{\sqrt{N}\,\|\boldsymbol{\beta} - \boldsymbol{\beta}_0\| \leqslant B} \left| \frac{\partial}{\partial \beta_j} Q(\mathbf{Y} - \mathbf{X}\boldsymbol{\beta}) - \frac{\partial}{\partial \beta_j} D(\mathbf{Y} - \mathbf{X}\boldsymbol{\beta}) \right| \geqslant \frac{\sqrt{N}\,\epsilon}{B\sqrt{p}} \right\} \to 0$$

$$(5.2.13)$$

Consider a point $\boldsymbol{\beta}^* \neq \boldsymbol{\beta}_0$ such that $N^{1/2}\|\boldsymbol{\beta}^* - \boldsymbol{\beta}_0\| \leqslant B$. The point $\boldsymbol{\beta}_t = t\boldsymbol{\beta}^* + (1 - t)\boldsymbol{\beta}_0 = \boldsymbol{\beta}_0 + t(\boldsymbol{\beta}^* - \boldsymbol{\beta}_0)$, for $0 \leqslant t \leqslant 1$, is on the line segment connecting β^* and β_0. Apply the chain rule to get

$$\frac{d}{dt}\left[Q(\mathbf{Y} - \mathbf{X}\beta_t) - D(\mathbf{Y} - \mathbf{X}\beta_t) \right]$$

$$= \sum_{j=1}^{p} (\beta_j^* - \beta_{0j})\left[\frac{\partial}{\partial \beta_j} Q(\mathbf{Y} - \mathbf{X}\beta_t) - \frac{\partial}{\partial \beta_j} D(\mathbf{Y} - \mathbf{X}\beta_t) \right].$$

With high probability, for large N, from (5.2.13) we can conclude that

$$\left| \frac{d}{dt}\left[Q(\mathbf{Y} - \mathbf{X}\beta_t) - D(\mathbf{Y} - \mathbf{X}\beta_t) \right] \right| < \frac{\sqrt{N}\,\epsilon}{B\sqrt{p}} \sum_{j=1}^{p} |\beta_j^* - \beta_{0j}|$$

$$\leqslant \frac{\sqrt{N}\,\epsilon}{B\sqrt{p}} \sqrt{p} \sqrt{\sum |\beta_j^* - \beta_{0j}|^2}$$

$$\leqslant \frac{\sqrt{N}\,\epsilon}{B} \| \beta^* - \beta_0 \|$$

$$\leqslant \epsilon.$$

Now define $h(t) = Q(\mathbf{Y} - \mathbf{X}\beta_t) - D(\mathbf{Y} - \mathbf{X}\beta_t)$ and recall $\beta_t = t\beta^* + (1 - t)\beta_0$. Note $h(t)$ is differentiable almost everywhere and

$$|h(1) - h(0)| \leqslant \int_0^1 |h'(t)|\, dt.$$

Furthermore, $h(0) = Q(\mathbf{Y} - \mathbf{X}\beta_0) - D(\mathbf{Y} - \mathbf{X}\beta_0) = 0$ and $|h'(t)| \leqslant \epsilon$ for $0 < t < 1$. Hence $|h(1)| < \epsilon$, and we have $| Q(\mathbf{Y} - \mathbf{X}\beta^*) - D(\mathbf{Y} - \mathbf{X}\beta^*)| < \epsilon$ for β^* such that $N^{1/2}\| \beta^* - \beta_0 \| \leqslant B$, with high probability. Since $Q(\mathbf{Y} - \mathbf{X}\beta) - D(\mathbf{Y} - \mathbf{X}\beta)$ is continuous,

$$\sup_{N^{1/2}\| \beta - \beta_0 \| \leqslant B} | Q(\mathbf{Y} - \mathbf{X}\beta) - D(\mathbf{Y} - \mathbf{X}\beta)| < \epsilon$$

with high probability, and this completes the proof.

Recall $\hat{\beta}$ minimizes $D(\mathbf{Y} - \mathbf{X}\beta)$ and solves $\mathbf{S}(\mathbf{Y} - \mathbf{X}\beta) \doteq 0$. Define

$$\tilde{\beta} = \beta_0 + \frac{1}{\sqrt{12}\int f^2(x)\,dx} \Sigma^{-1} \frac{1}{N} \mathbf{S}(\mathbf{Y} - \mathbf{X}\beta_0) \qquad (5.2.14)$$

where $\boldsymbol{\beta}_0$ is the true, unknown parameter value. Then $\tilde{\boldsymbol{\beta}}$ minimizes the quadratic approximation $Q(\mathbf{Y} - \mathbf{X}\boldsymbol{\beta})$ and solves the right side of (5.2.10), the linear approximation. The reader is asked to find the limiting distribution of $N^{1/2}(\tilde{\boldsymbol{\beta}} - \boldsymbol{\beta}_0)$ in Exercise 5.5.3(b). We are now ready to show that $\hat{\boldsymbol{\beta}}$ behaves asymptotically like $\tilde{\boldsymbol{\beta}}$. Jaeckel (1972) proves the result for general scores.

Theorem 5.2.3. Let $\hat{\boldsymbol{\beta}}$ be any point that minimizes $D(\mathbf{Y} - \mathbf{X}\boldsymbol{\beta})$. Then, if $\boldsymbol{\beta}_0$ is the true parameter value, under assumptions in (5.2.11),

$$\sqrt{N}\,(\hat{\boldsymbol{\beta}} - \boldsymbol{\beta}_0) \xrightarrow{D} \mathbf{Z} \sim MVN\left(\mathbf{0}, \frac{1}{12\left(\int f^2(x)\,dx\right)^2}\boldsymbol{\Sigma}^{-1}\right).$$

Proof. Choose $\epsilon > 0$ and $\delta > 0$. The asymptotic normality of $N^{1/2}(\tilde{\boldsymbol{\beta}} - \boldsymbol{\beta}_0)$, Exercise 5.5.3, implies that there exists C_0 such that

$$P\left\{\|\tilde{\boldsymbol{\beta}} - \boldsymbol{\beta}_0\| \geqslant C_0/\sqrt{N}\right\} < \delta/2 \qquad (5.2.15)$$

for sufficiently large N.

Define (see Exercise 5.5.12)

$$T = \min\left\{Q(\mathbf{Y} - \mathbf{X}\boldsymbol{\beta}) : \|\boldsymbol{\beta} - \tilde{\boldsymbol{\beta}}\| = \epsilon/\sqrt{N}\right\} - Q(\mathbf{Y} - \mathbf{X}\tilde{\boldsymbol{\beta}}).$$

Note that $T > 0$ since $\tilde{\boldsymbol{\beta}}$ is the unique minimum of $Q(\mathbf{Y} - \mathbf{X}\boldsymbol{\beta})$. Then, by Theorem 5.2.2 we have

$$P\left\{\sup_{\|\boldsymbol{\beta} - \boldsymbol{\beta}_0\| \leqslant (C_0+\epsilon)/\sqrt{N}} |Q(\mathbf{Y} - \mathbf{X}\boldsymbol{\beta}) - D(\mathbf{Y} - \mathbf{X}\boldsymbol{\beta})| \geqslant T/2\right\} \leqslant \delta/2.$$

$$(5.2.16)$$

Now, for large N, with probability greater than $1 - \delta$, (5.2.15) and (5.2.16) imply $\|\tilde{\boldsymbol{\beta}} - \boldsymbol{\beta}_0\| \leqslant C_0/N^{1/2}$ and

$$D(\mathbf{Y} - \mathbf{X}\tilde{\boldsymbol{\beta}}) < Q(\mathbf{Y} - \mathbf{X}\tilde{\boldsymbol{\beta}}) + T/2.$$

For all $\boldsymbol{\beta}$ such that $\|\boldsymbol{\beta} - \tilde{\boldsymbol{\beta}}\| = \epsilon/N^{1/2}$, $\|\boldsymbol{\beta} - \boldsymbol{\beta}_0\| \leqslant \|\boldsymbol{\beta} - \tilde{\boldsymbol{\beta}}\| + \|\tilde{\boldsymbol{\beta}} - \boldsymbol{\beta}_0\| \leqslant \epsilon/N^{1/2} + C_0/N^{1/2} = (C_0 + \epsilon)/N^{1/2}$. Hence, by (5.2.16), $|Q(\mathbf{Y} - \mathbf{X}\boldsymbol{\beta}) - D(\mathbf{Y} - \mathbf{X}\boldsymbol{\beta})| < T/2$ implies that $D(\mathbf{Y} - \mathbf{X}\boldsymbol{\beta}) > Q(\mathbf{Y} - \mathbf{X}\boldsymbol{\beta}) - T/2$.

Now,

$$D(\mathbf{Y} - \mathbf{X}\boldsymbol{\beta}) > Q(\mathbf{Y} - \mathbf{X}\boldsymbol{\beta}) - T/2$$

$$\geqslant \min\left\{ Q(\mathbf{Y} - \mathbf{X}\boldsymbol{\beta}) : \|\boldsymbol{\beta} - \tilde{\boldsymbol{\beta}}\| = \epsilon/\sqrt{N} \right\} - T/2$$

$$= T + Q(\mathbf{Y} - \mathbf{X}\tilde{\boldsymbol{\beta}}) - T/2$$

$$= T/2 + Q(\mathbf{Y} - \mathbf{X}\tilde{\boldsymbol{\beta}}) > D(\mathbf{Y} - \mathbf{X}\tilde{\boldsymbol{\beta}}).$$

Hence $D(\mathbf{Y} - \mathbf{X}\boldsymbol{\beta}) > D(\mathbf{Y} - \mathbf{X}\tilde{\boldsymbol{\beta}})$ for all $\boldsymbol{\beta}$ such that $\|\boldsymbol{\beta} - \tilde{\boldsymbol{\beta}}\| = \epsilon/N^{1/2}$.
The convexity of $D(\mathbf{Y} - \mathbf{X}\boldsymbol{\beta})$, Theorem 5.2.1, implies that $D(\mathbf{Y} - \mathbf{X}\boldsymbol{\beta})$ $> D(\mathbf{Y} - \mathbf{X}\tilde{\boldsymbol{\beta}})$ for all $\boldsymbol{\beta}$ such that $\|\boldsymbol{\beta} - \tilde{\boldsymbol{\beta}}\| \geqslant \epsilon/N^{1/2}$. Thus $D(\mathbf{Y} - \mathbf{X}\tilde{\boldsymbol{\beta}})$ $\geqslant \min D(\mathbf{Y} - \mathbf{X}\boldsymbol{\beta}) = D(\mathbf{Y} - \mathbf{X}\hat{\boldsymbol{\beta}})$ which implies $\|\hat{\boldsymbol{\beta}} - \tilde{\boldsymbol{\beta}}\| < \epsilon/N^{1/2}$. But this says $N^{1/2}\|\hat{\boldsymbol{\beta}} - \tilde{\boldsymbol{\beta}}\| < \epsilon$ for large N, with high probability.
Hence $N^{1/2}(\hat{\boldsymbol{\beta}} - \boldsymbol{\beta}_0) - N^{1/2}(\tilde{\boldsymbol{\beta}} - \boldsymbol{\beta}_0)$ converges to $\mathbf{0}$ in probability and they have the same limiting distribution, given in Exercise 5.5.3. This completes the proof.

Jaeckel (1972) proves that $B_N = \{ \boldsymbol{\beta} : D(\mathbf{Y} - \mathbf{X}\boldsymbol{\beta}) = \min \}$ is bounded. Since $D(\mathbf{Y} - \mathbf{X}\boldsymbol{\beta})$ is continuous and constant on B_N, B_N is also closed. These facts, along with the last theorem, show that $N^{1/2}$ (diameter of B_N) converges to 0 in probability. Hence, for moderate to large N, the set of points that minimizes $D(\mathbf{Y} - \mathbf{X}\boldsymbol{\beta})$ should be quite small.
As a simple application recall the two-sample location problem discussed in Example 5.2.1. It was pointed out that $\mathbf{X}_c'\mathbf{X}_c = mn/N$ where $m + n = N$. If we suppose that $m/N \rightarrow \lambda$, $0 < \lambda < 1$, then $N^{-1}\mathbf{X}_c'\mathbf{X}_c \rightarrow \lambda(1 - \lambda)$. The estimate $\hat{\beta}$ is the median of the mn differences and the last theorem implies that $N^{1/2}(\hat{\beta} - \beta_0)$ is asymptotically $n(0, 1/[\lambda(1 - \lambda)12(\int f^2(x)\,dx)^2])$. This result was first mentioned after (3.5.10), the efficacy of the Mann–Whitney–Wilcoxon test. In the simple regression model discussed in the previous section, $N^{1/2}(\hat{\beta} - \beta_0)$ has a limiting normal distribution with mean 0 and variance $1/[\delta^2 12(\int f^2(x)\,dx)^2]$ where $N^{-1}\sum(x_i - \bar{x})^2 \rightarrow \delta^2$.
Wilks (1962, p. 547) defines the generalized variance of a multivarate estimator as the determinant of its variance-covariance matrix. We then define the asymptotic relative efficiency of two asymptotically multivariate normal estimators as the $(1/p)$th root of the reciprocal ratio of their asymptotic generalized variances. See Bickel (1964, Section 4). In the Appendix, following Theorem A13, it is shown that if $\boldsymbol{\beta}^*$ is the least-squares estimator then $N^{1/2}(\boldsymbol{\beta}^* - \boldsymbol{\beta}_0)$ is asymptotically $MVN(\mathbf{0}, \sigma^2\boldsymbol{\Sigma}^{-1})$ where σ^2 is the variance of F. Hence the asymptotic efficiency of the rank

estimator $\hat{\beta}$ relative to the least-squares estimator β^* is

$$
e(\hat{\beta}, \beta^*) = \left\{ \frac{|\sigma^2 \Sigma^{-1}|}{\left| \dfrac{1}{12\left(\int f^2(x)\,dx\right)^2} \Sigma^{-1} \right|} \right\}^{1/p} = 12\sigma^2\left(\int f^2(x)\,dx\right)^2. \quad (5.2.17)
$$

But this is the efficiency of the Wilcoxon methods relative to the least-squares methods and has been discussed extensively throughout the book. See Section 2.6.

The estimator $\hat{\beta}$, except in the location model, is not simple to compute. Search methods such as steepest descent can be used to find the minimum of $D(\mathbf{Y} - \mathbf{X}\beta)$ since it is a convex surface. The Minitab computing system (see Ryan et al., 1981) contains a RANK REGRESSION command that provides $\hat{\beta}$ in its output.

The linear approximation, (5.2.10), and its solution $\tilde{\beta}$, (5.2.14), can be used to construct a one-step R estimator. Let $\hat{\beta}_0$ denote an initial estimate of β, perhaps the least-squares estimate. Suppose that $\hat{\tau}$ is a consistent estimate of

$$
\tau = \frac{1}{\sqrt{12}\,\int f^2(x)\,dx} . \qquad (5.2.18)
$$

Then, replacing Σ^{-1} by $N(\mathbf{X}_c'\mathbf{X}_c)^{-1}$ in (5.2.14), define

$$
\hat{\beta}_1 = \hat{\beta}_0 + \hat{\tau}(\mathbf{X}_c'\mathbf{X}_c)^{-1}\mathbf{S}(\mathbf{Y} - \mathbf{X}\hat{\beta}_0). \qquad (5.2.19)
$$

The estimate $\hat{\beta}_1$ is a one-step estimator. By iterating this formula a k-step estimator can be quickly constructed.

In the one-sample location model, in which we assume a symmetric underlying distribution, $F \in \Omega_s$, it was pointed out in (2.7.8) that the length of the Wilcoxon confidence interval, when properly standardized, provides a consistent estimate of τ. If we compute this estimate on the residuals $Y_1 - \mathbf{x}_1'\hat{\beta}, \ldots, Y_N - \mathbf{x}_N'\hat{\beta}$ then McKean and Hettmansperger (1976) showed that this estimate is also consistent in the linear model. Hence, if we are willing to suppose $F \in \Omega_s$, we can compute $\hat{\beta}_1$, (5.2.19), or by iteration $\hat{\beta}_k$, the k-step R estimate.

Exercise 5.5.4 outlines the surprising result that $N^{1/2}(\hat{\beta} - \beta_0)$ has the same limiting distribution as $N^{1/2}(\hat{\beta} - \beta_0)$. Hence, the one-step estimate is asymptotically as efficient as the full R estimate found by minimizing

$D(\mathbf{Y} - \mathbf{X}\boldsymbol{\beta})$; see McKean and Hettmansperger (1978). It should be emphasized, however, that the limiting distribution developed in Theorem 5.2.3 for $N^{1/2}(\hat{\boldsymbol{\beta}} - \boldsymbol{\beta}_0)$ does not require symmetry of the underlying distribution. On the other hand, if $\hat{\tau}$, in the construction of $\hat{\boldsymbol{\beta}}_1$, is based on the Wilcoxon confidence interval computed from the residuals, then symmetry must be assumed for the consistency of $\hat{\tau}$.

The parameter $\tau = 1/[12^{1/2}\int f^2(x)\,dx]$ has appeared throughout the book. For example τ^{-1} is the essential part of the Pitman efficacy of rank tests based on Wilcoxon scores and appears in the asymptotic linear approximation of the rank statistics. Further, τ^2 is in the asymptotic variance or covariance matrix of the Wilcoxon scores R estimates and is the stochastic limit of the standardized squared length of the Wilcoxon confidence interval. It appears in the quadratic approximation of $D(\mathbf{Y} - \mathbf{X}\boldsymbol{\beta})$ and will appear later as a standardizing parameter in rank tests based on residuals in the linear model. Hence it is important to have a consistent estimate of τ without assuming symmetry of the underlying distribution.

We now develop a different estimate of τ which does not require symmetry. The method is based on an estimate of the density function. We consider the i.i.d case where $Y_1, \ldots, Y_N$ is a random sample from $F \in \Omega_0$ with pdf $f(x)$. The estimate of $f(x)$, called a kernel or window estimate, was proposed by Rosenblatt (1956) and studied further by Parzen (1962). Wegman (1972) and Bean and Tsokos (1980) provide extensive surveys of the area of density estimation.

Definition 5.2.3.   Suppose $F \in \Omega_0$ with pdf $f(x)$. Suppose $w(x)$ is a square integrable density, symmetric about 0, and h_N a sequence of constants such that

$$h_N \to 0 \quad \text{and} \quad Nh_N \to \infty.$$

Then

$$f_N(y) = \frac{1}{Nh_N} \sum_{i=1}^{N} w\left(\frac{y - Y_i}{h_N}\right)$$

is called a window estimator of $f(y)$. The function $w(x)$ is called the window or kernel, and h_N is called the window width or bandwidth.

We restrict attention to a uniform window:

$$w(x) = 1 \quad \text{for} \quad -1/2 < x \leqslant 1/2$$
$$0 \quad \text{otherwise.} \tag{5.2.20}$$

We do this because a uniform window is easy to work with, and Bean and Tsokos (1980, p. 277) indicate that the choice of window is not nearly so crucial as the choice of window width.

Using the indicator function $I(E) = 1$ if the event E occurs and 0 otherwise, with (5.2.20), we have

$$f_N(y) = \frac{1}{Nh_N}\left[\sum_{i=1}^{N} I\left(Y_i \leqslant y + \frac{h_N}{2}\right) - \sum_{i=1}^{N} I\left(Y_i \leqslant y - \frac{h_N}{2}\right)\right]$$

$$= \frac{1}{h_N}\left[F_N(y + h_N/2) - F_N(y - h_N/2)\right] \qquad (5.2.21)$$

where $F_N(\cdot)$ is the empirical cdf. Hence, $f_N(y)$ in (5.2.21) is an empirical derivative of the empirical cdf. In Exercise 5.5.5 you are asked to show that $f_N(y)$ converges in probability to $f(y)$.

Based on $f_N(y)$ we can now construct an estimator of τ, (5.2.18). We consider the parameter

$$\gamma = \int f^2(y)\,dy = \int f(y)\,dF(y). \qquad (5.2.22)$$

The estimator is given by

$$\hat{\gamma} = \int f_N(y)\,dF_N(y) = \frac{1}{N^2 h_N}\sum_{i=1}^{N}\sum_{j=1}^{N} w\left(\frac{Y_i - Y_j}{h_N}\right)$$

$$= \frac{1}{N^2 h_N}\sum_{i=1}^{N}\sum_{j=1}^{N} I_{ij} \qquad (5.2.23)$$

where $I_{ij} = 1$ if $-h_N/2 < Y_i - Y_j < h_N/2$ and 0 otherwise, from (5.2.20).

The estimate $\hat{\gamma}$ (5.2.23), was first proposed by Schuster (1974) and studied extensively, for the case of a random sample, by Schweder (1975). Schweder (1975) discusses the asymptotic normality of $\hat{\gamma}$ based on general windows and general scores. He further discusses several practical considerations in the implementation of the estimator. Bhattacharyya and Roussas (1969) study $\int f_N^2(x)\,dx$, and Cheng and Serfling (1981) extend this estimate to the general scores case. It is possible to show when using the Wilcoxon scores that the Bhattacharyya–Roussas estimate is a special case of the estimate $\hat{\gamma}$; see Exercise 5.5.6. Hence we restrict our attention to $\hat{\gamma}$, (5.2.23).

Theorem 5.2.4. Suppose $Y_1, \ldots, Y_N$ are i.i.d. $F \in \Omega_0$ with pdf $f(\cdot)$. Suppose $\int f^2(x)\,dx < \infty$. Then $\hat{\gamma}$, (5.2.23), converges in probability to $\gamma = \int f^2(x)\,dx$.

Proof.   We first compute the expectation:

$$E\hat{\gamma} = \frac{1}{N^2 h_N} \sum_{i=1}^{N} \sum_{j=1}^{N} EI_{ij} = \frac{1}{N^2 h_N} \left[N + N(N-1) P(|Y_1 - Y_2| < h_N/2) \right].$$

Let $Z = Y_1 - Y_2$, then, from Theorem A17,

$$F^*(z) = P(Z \leqslant z) = \int_{-\infty}^{\infty} F(z+y) f(y)\, dy$$

with pdf

$$f^*(z) = \int_{-\infty}^{\infty} f(z+y) f(y)\, dy.$$

Hence

$$E\hat{\gamma} \sim \frac{1}{h_N} \left[F^*(h_N/2) - F^*(-h_N/2) \right]$$

and

$$E\hat{\gamma} \to f^*(0) = \int f^2(y)\, dy.$$

Now,

$$\operatorname{Var}\hat{\gamma} = \frac{1}{N^4 h_N^2} \operatorname{Var}\left(\sum_{i=1}^{N} \sum_{j=1}^{N} I_{ij} \right) = \frac{N^2}{N^4 h_N^2} \operatorname{Var} I_{12}$$

$$+ \frac{1}{N^4 h_N^2} E\left\{ \sum \sum \sum \sum (I_{ij} I_{kt} - EI_{ij} EI_{kt}) \right\}.$$

Recall $Nh_N \to \infty$, then using the same argument as in Theorem 2.5.1 and Theorem 2.7.2, we have

$$\operatorname{Var}\hat{\gamma} \sim \frac{1}{Nh_N^2} \left\{ EI_{12} I_{13} - EI_{12} EI_{13} \right\}.$$

Now,

$$|EI_{12} I_{13} - EI_{12} EI_{13}| \leqslant EI_{12} + (EI_{12})^2$$

and

$$\frac{1}{Nh_N}\left[\frac{1}{h_N}EI_{12}\right]\left[1+EI_{12}\right]\to 0$$

since $Nh_N\to\infty$, $1/(Nh_N)\to 0$, $h_N^{-1}EI_{12}\to f^*(0)$ and $1+EI_{12}$ is bounded above by 2.

Thus we have $E\hat\gamma\to\gamma$ and $\operatorname{Var}\hat\gamma\to 0$, so $\hat\gamma$ is a consistent estimate of γ by Theorem A5.

We now discuss some practical issues that are necessary for the implementation of $\hat\gamma$. Note that the $i=j$ terms in (5.2.23) are nonrandom and $\hat\gamma$ can be written

$$\hat\gamma=\frac{1}{Nh_N}w(0)+\frac{1}{N^2h_N}\sum_{i\ne j}\sum w\left(\frac{Y_i-Y_j}{h_N}\right).$$

We will consider the following modification of $\hat\gamma$,

$$\gamma^*=\frac{1}{NK}+\frac{1}{N(N-1)h_N}\sum_{i\ne j}\sum w\left(\frac{Y_i-Y_j}{h_N}\right),\qquad(5.2.24)$$

where K is a constant. Obviously, Theorem 5.2.4 holds also for γ^*, and γ^* is a consistent estimator of γ. Now if $f^*(z)=\int f(y+z)f(y)\,dy$ is the pdf of Y_1-Y_2, then

$$E\gamma^*=\frac{1}{NK}+\frac{1}{h_N}\int_{-\infty}^{\infty}w(z/h_N)f^*(z)\,dz=\frac{1}{NK}+\int_{-\infty}^{\infty}w(u)f^*(h_Nu)\,du$$

$$=\frac{1}{NK}+\int_{-\infty}^{\infty}w(u)\left[f^*(0)+h_Nuf^{*\prime}(0)+h_N^2u^2f^{*\prime\prime}(0)/2\right]du+h_N^2o(1),$$

$$(5.2.25)$$

provided $f^{*\prime\prime}(y)$ is continuous at 0. Now since $w(u)$ is symmetric about 0 and $\gamma=f^*(0)$, we have

$$\operatorname{Bias}(\gamma^*)=\frac{1}{NK}+\frac{1}{2}h_N^2f^{*\prime\prime}(0)\int_{-\infty}^{\infty}u^2w(u)\,du+o(h_N^2).\quad(5.2.26)$$

After differentiating $f^*(z)$ twice, plus an integration by parts, we have

$$f^{*\prime\prime}(0)=-\int_{-\infty}^{\infty}\left[f'(x)\right]^2dx.\qquad(5.2.27)$$

The estimator γ^* will be used to standardize test statistics in the next section and we are concerned more about the bias of γ^* than efficiency. Hence we wish to let h_N tend to zero as fast as possible. Theory, developed by Aubuchon (1982, Section 4.3) for the use of γ^* based on dependent residuals in the linear model with random h_N, shows that h_N can go to zero no faster than $N^{-1/2}$. Hence we take

$$h_N = KN^{-1/2},$$

then the bias, ignoring the $o(h_N^2)$ term, becomes

$$\text{Bias}(\gamma^*) \doteq \frac{1}{NK} + \frac{K^2}{2N} f^{*\prime\prime}(0) \int_{-\infty}^{\infty} u^2 w(u)\, du. \tag{5.2.28}$$

The bias is thus of the order $1/N$, and setting (5.2.28) equal to zero and solving for K yields

$$K = \left[-\frac{f^{*\prime\prime}(0)}{2} \int_{-\infty}^{\infty} u^2 w(u)\, du \right]^{-1/3}.$$

Let $f(y) = \delta^{-1} f_1(\delta^{-1} y)$, then from (5.2.27),

$$-f^{*\prime\prime}(0) = \delta^{-3} \int_{-\infty}^{\infty} \left[f_1'(x) \right]^2 dx,$$

and

$$K = \delta \left[\frac{1}{2} \int_{-\infty}^{\infty} \left[f_1'(x) \right]^2 dx \int_{-\infty}^{\infty} u^2 w(u)\, du \right]^{-1/3}. \tag{5.2.29}$$

Thus, the estimate γ^*, (5.2.24), with $h_N = K/N^{1/2}$ where K is any fixed constant, is a consistent estimate of γ. The bias of γ^* tends to zero at the rate $1/N$, and we have shown previously that if we know the form of $f(\cdot)$ we could choose K, (5.2.29), to remove the $1/N$ bias terms. Of course, we do not know $f(\cdot)$. We have separated out two components in (5.2.29): a scale parameter δ and the standardized form of $f(\cdot)$, namely $f_1(\cdot)$. Schweder (1975) suggests using the data to estimate both δ and $\int [f_1'(x)]^2 dx$. A simpler approach is to take δ to be a robust scale parameter such as the mean absolute deviation from the mean or the interquartile range and estimate it from the data while assuming a distribution, such as normal, for computing $\int [f_1'(x)]^2 dx$. The bias is still of the order $1/N$ and we think of the calculations leading to (5.2.29) as providing a practical guide in the problem of selecting K and hence selecting the crucial window width. See Aubuchon and Hettmansperger (1982b) for additional discussion of the properties of γ^*.

As a simple example we illustrate the calculations for $\delta = F^{-1}(3/4) - F^{-1}(1/4)$, the interquartile range, and $f(\cdot)$, the normal density. The standardized density $f_1(\cdot)$ is that normal density with $\delta = 1$. Since $\delta = \sigma[\Phi^{-1}(3/4) - \Phi^{-1}(1/4)] = \sigma(1.348)$ where σ is the standard deviation, $f_1(\cdot)$ is the normal density with mean 0 and variance .5503. A simple calculation then shows $\int [f_1'(x)]^2 dx = .3455$. We have been using the rectangular window, (5.2.20), for which $\int u^2 w(u) du = 1/12$; however, the equations are valid for any smooth, symmetric window. Hence, from (5.2.29) we have $K = 4.11\delta$ and δ is estimated by the interquartile range of the data.

Using a rectangular window for $w(\cdot)$ and the sample interquartile range to estimate δ, the estimate of γ, (5.2.22), is given by

$$\gamma^* = \frac{1}{4.11N\hat{\delta}} + \frac{2}{\sqrt{N}(N-1)4.11\hat{\delta}} \sum\sum_{i<j} w\left(\frac{Y_i - Y_j}{4.11\hat{\delta}/\sqrt{N}}\right)$$

$$= \frac{1}{4.11N\hat{\delta}} + \frac{\sqrt{N}}{4.11\hat{\delta}}\left\{F_N^*\left(\frac{4.11\hat{\delta}}{2\sqrt{N}}\right) - F_N^*\left(\frac{-4.11\hat{\delta}}{2\sqrt{N}}\right)\right\} \quad (5.2.30)$$

where $F_N^*(\cdot)$ is the empirical cdf computed from the $N(N-1)/2$ differences $Y_i - Y_j$, $1 \leqslant i < j \leqslant N$.

Recall that our purpose for introducing the estimator $\hat{\gamma}$ (or γ^*) was to provide an estimate of $\gamma = \int f^2(x) dx$ or $\tau = 1/[12^{1/2}\int f^2(x) dx]$ without assuming that $f(\cdot)$ is symmetric. Then, for example, the one-step estimator $\hat{\beta}_1$, (5.2.19), can be computed without assuming symmetry. In the next section, tests that depend on an estimate of γ (or τ) can be constructed without the symmetry assumption. Thus far the discussion has assumed a random sample $Y_1, \ldots, Y_N$. We now turn again to the linear model, (5.2.1) or (5.2.2), with observations $Y_1, \ldots, Y_N$.

Suppose $F \in \Omega_0$ and the assumptions of (5.2.11) hold. Suppose $f_N(y)$ satisfies Definition 5.2.3 with $\int u^2 w(u) du < \infty$, $h_N = K/N^{1/2}$, and $N^{1/2}(\hat{K} - K)$ is bounded in probability. Then Aubuchon (1982) shows that Theorem 5.2.4 still holds when $\hat{\gamma}$ is computed from the dependent residuals $Y_1 - x_1'\hat{\beta}, \ldots, Y_N - x_N'\hat{\beta}$. Thus, $\hat{\gamma}$ or γ^* provides a consistent estimate of $\gamma = \int f^2(x) dx$ without the symmetry assumption. For the general scores case we would need to estimate $\int \phi'(F(x))f^2(x) dx$. An obvious candidate is $\int \phi'(F_N(x))f_N(x) dF_N(x)$. The consistency of this estimator is discussed by Schweder (1975) for the case of a random sample. His results have not yet been extended to the estimator based on residuals in a linear model.

Comparing (5.2.30) to (5.2.21), we see that γ^* is an estimate of the density (or derivative) of F^* (the cdf of the difference) at 0. Sievers and McKean (1983) suggest a slightly different estimate. They consider $\gamma^{**} = \hat{H}_N(\hat{t}/N^{1/2})/(2\hat{t}/N^{1/2})$ where $\hat{H}_N(\cdot)$ is the empirical cdf of the absolute values of differences of residuals and $\hat{t}$ is the α percentile of $\hat{H}_N(\cdot)$, $0 < \alpha < 1$. This estimate is also consistent without the symmetry of $f(\cdot)$. Note that if X_1, X_2 are i.i.d. $F \in \Omega_0$ then $X_1 - X_2$ has cdf $G \in \Omega_s$. In a designed experiment with several observations per cell, Draper (1983) develops an estimate of $\int f^2(x)\,dx$ without the symmetry of $f(\cdot)$ by combining the lengths of two-sample Hodges–Lehmann confidence intervals formed across the pairs of cells. See (3.2.5) for a description of the confidence interval. There has been no comparative study of the various approaches to the estimation of $\int f^2(x)\,dx$.

To fix the ideas involved in using γ^*, we outline the calculations needed to produce an approximate 95% confidence interval for β in Example 5.1.1. From that example, $\hat{\beta} = .353$, and the residuals $r_i = Y_i - 0.353x_i$, $i = 1, \ldots, 11$, are -8.3, -6.124, -3.670, -1.364, -0.266, -0.004, 1.035, 2.985, 3.195, 8.805, and 12.9. The quartiles are -3.67 and 3.195; the interquartile range is $\hat{\delta} = 6.865$. To compute γ^*, (5.2.30), we must form the 55 differences $r_i - r_j$, $1 \le i < j \le 11$. Then, from (5.2.20),

$$w\left(\frac{r_i - r_j}{4.11\hat{\delta}/\sqrt{11}}\right) = 1 \qquad \text{if} \quad |r_i - r_j| \le 4.254$$

and $\gamma^* = .003 + .034 = .037$. Then $\tau^* = 1/(12^{1/2}\gamma^*) = 7.734$. This is comparable to the least squares $s = 6.48$, based on the residual sum of squares. Now Σ^{-1}, in Theorem 5.2.3, is replaced by $(N^{-1}\sum(x_i - \bar{x})^2)^{-1} = .00001$. Then $N^{1/2}(\hat{\beta} - \beta)$ is approximately normally distributed with variance estimated by $\tau^{*2}\Sigma^{-1} = (59.81)(.00001) = .0006$. Finally, the approximate 95% confidence interval is $.353 \pm 2(.0006)^{1/2}/11^{1/2}$ or $(.338, .368)$.

We complete this section with a discussion of the intercept parameter α. This parameter is estimated by applying one-sample methods to the residuals $Y_1 - \mathbf{x}_1'\hat{\boldsymbol{\beta}}, \ldots, Y_N - \mathbf{x}_N'\hat{\boldsymbol{\beta}}$, where $\hat{\boldsymbol{\beta}}$ is the rank estimator developed previously. If we assume symmetry ($F \in \Omega_s$) and use Wilcoxon scores, then the estimate of α would be the median of the Walsh averages of the residuals; see (2.3.2). Suppose assumptions (5.2.11) hold. Suppose α_0 and $\boldsymbol{\beta}_0$ denote the true parameter values in the linear model, (5.2.1), and $N^{-1}[\mathbf{1}\ \mathbf{X}]'[\mathbf{1}\ \mathbf{X}] \to \Lambda$, full rank. Then Theorem 5.2.3 can be extended immediately to show

$$\sqrt{N}\left[\begin{pmatrix} \hat{\alpha} \\ \hat{\boldsymbol{\beta}} \end{pmatrix} - \begin{pmatrix} \alpha_0 \\ \boldsymbol{\beta}_0 \end{pmatrix}\right] \xrightarrow{D} \mathbf{Z} \sim MVN\left(\mathbf{0}, \frac{1}{12\left(\int f^2(x)\,dx\right)^2}\Lambda^{-1}\right). \quad (5.2.31)$$

This result also holds for one- (or k) step rank estimators, (5.2.19). See McKean and Hettmansperger (1978) for a further discussion.

When we do not wish to assume symmetry ($F \in \Omega_0$), we have to decide what location parameter the intercept represents. We take α to be the median of the distribution of $Y_i - \mathbf{x}_i'\boldsymbol{\beta}$. Hence, $\hat{\alpha}$ is the median of $Y_1 - \mathbf{x}_1'\hat{\boldsymbol{\beta}}, \ldots, Y_N - \mathbf{x}_N'\hat{\boldsymbol{\beta}}$, while $\hat{\boldsymbol{\beta}}$ is constructed using the Wilcoxon scores. Under the assumptions in (5.2.11), Aubuchon (1982) shows that

$$\sqrt{N}\left[\begin{pmatrix} \hat{\alpha} \\ \hat{\boldsymbol{\beta}} \end{pmatrix} - \begin{pmatrix} \alpha_0 \\ \boldsymbol{\beta}_0 \end{pmatrix}\right] \overset{D}{\to} \mathbf{Z} \sim MVN(\mathbf{0}, \mathbf{V}) \tag{5.2.32}$$

where

$$\mathbf{V} = \begin{bmatrix} \left[4f^2(0)\right]^{-1} + \tau^2 \boldsymbol{\mu}' \boldsymbol{\Sigma}^{-1} \boldsymbol{\mu} & -\tau^2 \boldsymbol{\mu}' \boldsymbol{\Sigma}^{-1} \\ -\tau^2 \boldsymbol{\Sigma}^{-1} \boldsymbol{\mu} & \tau^2 \boldsymbol{\Sigma}^{-1} \end{bmatrix}$$

and $\tau = 1/[12^{1/2} \int f^2(x)\,dx]$, (5.2.18). The asymptotic variance-covariance matrix is developed from projections using sign scores for the intercept and Wilcoxon scores for the regression parameters.

5.3.   RANK TESTS IN THE LINEAR MODEL

We now turn to the problem of hypothesis testing in the linear model. The basic model is given by (5.2.1) or (5.2.2). Again, as in the previous section, we restrict our attention to the $p \times 1$ vector of regression parameters $\boldsymbol{\beta}$. Partition the vector $\boldsymbol{\beta}$ as

$$\boldsymbol{\beta} = \begin{pmatrix} \boldsymbol{\beta}_1 \\ \boldsymbol{\beta}_2 \end{pmatrix} \tag{5.3.1}$$

where $\boldsymbol{\beta}_1$ and $\boldsymbol{\beta}_2$ are $(p - q) \times 1$ and $q \times 1$ vectors, respectively. Then the linear model can be written as

$$\mathbf{Y} = \mathbf{1}\alpha^* + \mathbf{X}_{1c}\boldsymbol{\beta}_1 + \mathbf{X}_{2c}\boldsymbol{\beta}_2 + \mathbf{e} \tag{5.3.2}$$

where $\mathbf{X}_{1c}$ and $\mathbf{X}_{2c}$ are $N \times (p - q)$ and $N \times q$ matrices with $[\mathbf{X}_{1c}, \mathbf{X}_{2c}] = \mathbf{X}_c$. Our goal is to construct tests of the hypotheses:

$$H_0 : \boldsymbol{\beta}_2 = \mathbf{0}, \quad \boldsymbol{\beta}_1 \text{ unspecified} \quad \text{versus}$$
$$H_A : \boldsymbol{\beta}_2 \neq \mathbf{0}, \quad \boldsymbol{\beta}_1 \text{ unspecified.} \tag{5.3.3}$$

An alternative formulation is based on specifying a collection of q linearly independent, estimable functions $\mathbf{H}\boldsymbol{\beta}$ where $\mathbf{H}$ is given $q \times p$

matrix. For example, if $\mathbf{H} = (\mathbf{O}, \mathbf{I})$, where $\mathbf{I}$ is the $q \times q$ identity, then $\mathbf{H}\boldsymbol{\beta} = \boldsymbol{\beta}_2$.

The vector $\boldsymbol{\beta}_2$ specifies the parameters under test, whereas $\boldsymbol{\beta}_1$ denotes the nuisance parameters. For example, in the examples following (5.1.3) consider the two-way layout with two treatments and two blocks. The test of $H_0 : \beta_3 = 0$ versus $H_A : \beta_3 \neq 0$ with β_2 unspecified is an example with $\boldsymbol{\beta}' = (\beta_2, \beta_3)$ and in our present notation "$\boldsymbol{\beta}_2$" $= \beta_3$ and "$\boldsymbol{\beta}_1$" $= \beta_2$. This example could easily be expanded to a test for interaction in which the main effects determine the nuisance parameters. This is illustrated in the next section with data.

The framework is quite general. The methods to be developed cover tests for main effects and interactions in multiway layouts, for the significance of covariates in analysis of covariance models, and for coefficients in regression models.

5.3.1. Tests Based on $D(\mathbf{Y} - \mathbf{X}\boldsymbol{\beta})$

In the last section we used $D(\mathbf{Y} - \mathbf{X}\boldsymbol{\beta})$, (5.2.5), to define an estimate of $\boldsymbol{\beta}$. Hence $D(\mathbf{Y} - \mathbf{X}\boldsymbol{\beta})$ is used as a criterion for fitting a linear model to data. In fact $D(\mathbf{Y} - \mathbf{X}\hat{\boldsymbol{\beta}})$ represents the minimum distance, as measured by $D(\mathbf{Y} - \mathbf{X}\boldsymbol{\beta})$, from the data vector to the subspace determined by the linear model. Let $\boldsymbol{\beta}_0$ denote the $p \times 1$ vector specified by $H_0 : \boldsymbol{\beta}_2 = \mathbf{0}$. Thus, we have $\boldsymbol{\beta}_0' = (\boldsymbol{\beta}_1', \mathbf{0}')$ and we can rewrite (5.3.3) as

$$H_0 : \boldsymbol{\beta} = \boldsymbol{\beta}_0 \quad \text{versus} \quad H_A : \boldsymbol{\beta} \neq \boldsymbol{\beta}_0 . \tag{5.3.4}$$

Further, let $\hat{\boldsymbol{\beta}}_0$ denote the R estimate of $\boldsymbol{\beta}_0$, that is, the value that minimizes $D(\mathbf{Y} - \mathbf{X}\boldsymbol{\beta}_0) = D(\mathbf{Y} - \mathbf{X}_1\boldsymbol{\beta}_1)$. The estimate $\hat{\boldsymbol{\beta}}_0$ is referred to as the reduced model estimate. The full-model estimate is denoted by $\hat{\boldsymbol{\beta}}$ and it minimizes $D(\mathbf{Y} - \mathbf{X}\boldsymbol{\beta})$.

To test the hypothesis $H_0 : \boldsymbol{\beta} = \boldsymbol{\beta}_0$ versus $H_A : \boldsymbol{\beta} \neq \boldsymbol{\beta}_0$, we will compare $D(\mathbf{Y} - \mathbf{X}\hat{\boldsymbol{\beta}}_0)$ to $D(\mathbf{Y} - \mathbf{X}\hat{\boldsymbol{\beta}})$ and reject $H_0 : \boldsymbol{\beta} = \boldsymbol{\beta}_0$ for large values of $D(\mathbf{Y} - \mathbf{X}\hat{\boldsymbol{\beta}}_0) - D(\mathbf{Y} - \mathbf{X}\hat{\boldsymbol{\beta}})$. This is the same strategy as that used to develop F tests based on the reduction in sum of squares due to fitting the reduced and full models. To make the test operational we at least need the limiting distribution under the null hypothesis. It should be noted that the test is not distribution free for finite sample sizes. The next theorem shows that the test is, however, asymptotically distribution free under the null hypothesis. The result in the next theorem shows that

$$\frac{D(\mathbf{Y} - \mathbf{X}\hat{\boldsymbol{\beta}}_0) - D(\mathbf{Y} - \mathbf{X}\hat{\boldsymbol{\beta}})}{(\tau/2)} ,$$

where $\tau = 1/(12^{1/2} \int f^2(x)\,dx)$, is asymptotically chi-square with q degrees of freedom. We propose to use

$$\tau^* = 1/(\sqrt{12}\,\gamma^*) \tag{5.3.5}$$

where

$$\gamma^* = \frac{1}{N\hat{K}} + \frac{1}{N(N-1)\hat{K}} \sum_{i\neq j}\sum w\left(\frac{r_i - r_j}{h_N}\right),$$

$r_i = Y_i - \mathbf{x}_i'\hat{\boldsymbol{\beta}}$, $i = 1, \ldots, N$, $\hat{\boldsymbol{\beta}}$ is the full-model R estimate of $\boldsymbol{\beta}$, $\hat{K}$ is an estimate of K and $h_N = \hat{K}/N^{1/2}$. See (5.2.30) with $\hat{K} = 4.11\hat{\delta}$ where $\hat{\delta}$ is the interquartile range of the residuals. The hypothesis $H_0: \boldsymbol{\beta} = \boldsymbol{\beta}_0$ (or $H_0: \boldsymbol{\beta}_2 = \mathbf{0}$) is rejected at approximate level α if

$$D^* = \frac{D(\mathbf{Y} - \mathbf{X}\hat{\boldsymbol{\beta}}_0) - D(\mathbf{Y} - \mathbf{X}\hat{\boldsymbol{\beta}})}{(\tau^*/2)} \geqslant \chi_\alpha^2(q) \tag{5.3.6}$$

where $\chi_\alpha^2(q)$ is the chi-square critical value with q degrees of freedom. This test is illustrated on data in the next section.

Theorem 5.3.1.   Given the linear model, (5.3.2), suppose the null hypothesis $H_0: \boldsymbol{\beta}_2 = \mathbf{0}$ holds. Under the assumptions (5.2.11),

$$D = \frac{D(\mathbf{Y} - \mathbf{X}\hat{\boldsymbol{\beta}}_0) - D(\mathbf{Y} - \mathbf{X}\hat{\boldsymbol{\beta}})}{(\tau/2)}$$

has an asymptotic $\chi^2(q)$ distribution, where $\hat{\boldsymbol{\beta}}_0$, $\hat{\boldsymbol{\beta}}$ are the reduced- and full-model R estimates and $\tau = 1/(12^{1/2} \int f^2(x)\,dx)$.

Proof.   The argument proceeds by approximating $D(\cdot)$ with $Q(\cdot)$, (5.2.12), and by approximating $\hat{\boldsymbol{\beta}}$, the R estimate by $\tilde{\boldsymbol{\beta}}$, (5.2.14), the value that minimizes $Q(\mathbf{Y} - \mathbf{X}\boldsymbol{\beta})$. We then show that the asymptotic behavior of $D(\mathbf{Y} - \mathbf{X}\hat{\boldsymbol{\beta}}_0) - D(\mathbf{Y} - \mathbf{X}\hat{\boldsymbol{\beta}})$ is determined by that of $Q(\mathbf{Y} - \mathbf{X}\tilde{\boldsymbol{\beta}}_0) - Q(\mathbf{Y} - \mathbf{X}\tilde{\boldsymbol{\beta}})$. Finally, the argument is completed by showing that $Q(\mathbf{Y} - \mathbf{X}\tilde{\boldsymbol{\beta}}_0) - Q(\mathbf{Y} - \mathbf{X}\tilde{\boldsymbol{\beta}})$, when properly standardized, has an asymptotic chi-square distribution. Throughout this discussion we simplify the notation by writing $D(\hat{\boldsymbol{\beta}})$ rather than $D(\mathbf{Y} - \mathbf{X}\hat{\boldsymbol{\beta}})$ and similarly for $Q(\hat{\boldsymbol{\beta}})$.

We begin by writing

$$D(\hat{\boldsymbol{\beta}}_0) - D(\hat{\boldsymbol{\beta}}) = D(\hat{\boldsymbol{\beta}}_0) - Q(\hat{\boldsymbol{\beta}}_0) + Q(\hat{\boldsymbol{\beta}}_0) - Q(\tilde{\boldsymbol{\beta}}_0) + Q(\tilde{\boldsymbol{\beta}}_0) - Q(\tilde{\boldsymbol{\beta}})$$

$$+ Q(\tilde{\boldsymbol{\beta}}) - Q(\hat{\boldsymbol{\beta}}) + Q(\hat{\boldsymbol{\beta}}) - D(\hat{\boldsymbol{\beta}}). \tag{5.3.7}$$

Under the null hypothesis, by Theorem 5.2.3, both $N^{1/2}(\tilde{\boldsymbol{\beta}}_0 - \boldsymbol{\beta}_0)$ and $N^{1/2}(\hat{\boldsymbol{\beta}} - \boldsymbol{\beta}_0)$ are asymptotically normally distributed and hence are bounded in probability. Theorem 5.2.2 can now be applied to assert that the first and fifth difference on the right side of (5.3.7) tend to zero in probability, that is, they are $o_p(1)$.

Next, substituting $\tilde{\boldsymbol{\beta}}$ and $\hat{\boldsymbol{\beta}}$ into $Q(\boldsymbol{\beta})$, (5.2.12), we have

$$Q(\tilde{\boldsymbol{\beta}}) - Q(\hat{\boldsymbol{\beta}}) = (\hat{\boldsymbol{\beta}} - \tilde{\boldsymbol{\beta}})'S(\boldsymbol{\beta}_0) + \frac{\sqrt{12}\left(\int f^2(x)\,dx\right)N}{2}$$

$$\times\left[(\tilde{\boldsymbol{\beta}} - \boldsymbol{\beta}_0)'\Sigma(\tilde{\boldsymbol{\beta}} - \boldsymbol{\beta}_0) - (\hat{\boldsymbol{\beta}} - \boldsymbol{\beta}_0)'\Sigma(\hat{\boldsymbol{\beta}} - \boldsymbol{\beta}_0)\right].$$

If the expression in square brackets in the second term is multiplied out and rearranged, it can be written as

$$(\tilde{\boldsymbol{\beta}} - \hat{\boldsymbol{\beta}})\Sigma(\tilde{\boldsymbol{\beta}} - \boldsymbol{\beta}_0 + \hat{\boldsymbol{\beta}} - \boldsymbol{\beta}_0).$$

Hence

$$Q(\tilde{\boldsymbol{\beta}}) - Q(\hat{\boldsymbol{\beta}}) = \sqrt{N}\,(\hat{\boldsymbol{\beta}} - \tilde{\boldsymbol{\beta}})'\left[\frac{1}{\sqrt{N}}\,S(\boldsymbol{\beta}_0) - \frac{\sqrt{12}\left(\int f^2(x)\,dx\right)}{2}\right.$$

$$\left. \times \Sigma\sqrt{N}\,(\tilde{\boldsymbol{\beta}} - \boldsymbol{\beta}_0 + \hat{\boldsymbol{\beta}} - \boldsymbol{\beta}_0)\right]. \quad (5.3.8)$$

From the proof of Theorem 5.2.3, $N^{1/2}(\hat{\boldsymbol{\beta}} - \tilde{\boldsymbol{\beta}}) = o_p(1)$, and $N^{1/2}(\hat{\boldsymbol{\beta}} - \boldsymbol{\beta}_0)$ and $N^{1/2}(\tilde{\boldsymbol{\beta}} - \boldsymbol{\beta}_0)$ are asymptotically normally distributed. Since $N^{-1/2}S(\boldsymbol{\beta}_0)$ is also asymptotically normally distributed (Exercise 5.5.3), the second factor in (5.3.8) is bounded in probability. Hence we conclude that the fourth term in (5.3.7) is $o_p(1)$. A similar argument shows that the second term in (5.3.7) is also $o_p(1)$.

We can now write (5.3.7) as

$$D(\hat{\boldsymbol{\beta}}_0) - D(\hat{\boldsymbol{\beta}}) = Q(\tilde{\boldsymbol{\beta}}_0) - Q(\tilde{\boldsymbol{\beta}}) + o_p(1).$$

Recall $N^{-1}(X_c'X_c) \to \Sigma$, a positive definite matrix; see assumptions in (5.2.11). Partition Σ to correspond to the partition of $\boldsymbol{\beta}$:

$$\Sigma = \begin{pmatrix} \Sigma_{11} & \Sigma_{12} \\ \Sigma_{21} & \Sigma_{22} \end{pmatrix} \quad (5.3.9)$$

where Σ_{11} is $(p - q) \times (p - q)$, Σ_{22} is $q \times q$ and $\Sigma_{12} = \Sigma_{21}'$.

Substitute $\tilde{\beta}$, given in (5.2.14), into $Q(\beta)$ and simplify to get

$$Q(\tilde{\beta}) = D(\beta_0) - \frac{1}{2N\sqrt{12}\int f^2(x)\,dx} S'(\beta_0)\Sigma^{-1}S(\beta_0).$$

Likewise, using the partition, (5.3.9),

$$Q(\tilde{\beta}_0) = D(\beta_0) - \frac{1}{2N\sqrt{12}\int f^2(x)\,dx} S'(\beta_0)\begin{pmatrix}\Sigma_{11}^{-1} & 0 \\ 0 & 0\end{pmatrix}S(\beta_0).$$

Hence

$$Q(\tilde{\beta}_0) - Q(\tilde{\beta}) = \frac{1}{2N\sqrt{12}\int f^2(x)\,dx} S'(\beta_0)\left[\Sigma^{-1} - \begin{pmatrix}\Sigma_{11}^{-1} & 0 \\ 0 & 0\end{pmatrix}\right]S(\beta_0),$$

and

$$D = \frac{D(\hat{\beta}_0) - D(\hat{\beta})}{(\tau/2)}$$

$$= \frac{1}{\sqrt{N}} S'(\beta_0)\left[\Sigma^{-1} - \begin{pmatrix}\Sigma_{11}^{-1} & 0 \\ 0 & 0\end{pmatrix}\right]\frac{1}{\sqrt{N}} S(\beta_0) + o_p(1). \quad (5.3.10)$$

Using a result on the inverse of a partitioned matrix (see Arnold, 1981, p. 450) we have

$$\Sigma^{-1} - \begin{pmatrix}\Sigma_{11}^{-1} & 0 \\ 0 & 0\end{pmatrix} = \begin{pmatrix}-\Sigma_{11}^{-1}\Sigma_{12} \\ I\end{pmatrix}(\Sigma_{22} - \Sigma_{21}\Sigma_{11}^{-1}\Sigma_{12})^{-1}(-\Sigma_{21}\Sigma_{11}^{-1}, I).$$

$$(5.3.11)$$

where (AB, C) denotes the matrix partitioned into the product AB and C. Then the right side of (5.3.10) becomes

$$\frac{1}{\sqrt{N}}\left[(-\Sigma_{21}\Sigma_{11}^{-1}, I)S(\beta_0)\right]'(\Sigma_{22} - \Sigma_{21}\Sigma_{11}^{-1}\Sigma_{12})^{-1}$$

$$\times \frac{1}{\sqrt{N}}\left[(-\Sigma_{21}\Sigma_{11}^{-1}, I)S(\beta_0)\right]. \quad (5.3.12)$$

From Exercise 5.5.3,

$$\frac{1}{\sqrt{N}} \mathbf{S}(\boldsymbol{\beta}_0) \xrightarrow{D} \mathbf{U} \sim MVN(\mathbf{0}, \boldsymbol{\Sigma}), \qquad (5.3.13)$$

so that

$$\left(-\boldsymbol{\Sigma}_{21}\boldsymbol{\Sigma}_{11}^{-1}, \mathbf{I}\right) \frac{1}{\sqrt{N}} \mathbf{S}(\boldsymbol{\beta}_0) \xrightarrow{D} \mathbf{V} \sim MVN(\mathbf{0}, \mathbf{A}) \qquad (5.3.14)$$

where

$$\mathbf{A} = \left(-\boldsymbol{\Sigma}_{21}\boldsymbol{\Sigma}_{11}^{-1}, \mathbf{I}\right) \begin{pmatrix} \boldsymbol{\Sigma}_{11} & \boldsymbol{\Sigma}_{12} \\ \boldsymbol{\Sigma}_{21} & \boldsymbol{\Sigma}_{22} \end{pmatrix} \begin{pmatrix} -\boldsymbol{\Sigma}_{11}^{-1}\boldsymbol{\Sigma}_{12} \\ \mathbf{I} \end{pmatrix}$$

$$= \boldsymbol{\Sigma}_{22} - \boldsymbol{\Sigma}_{21}\boldsymbol{\Sigma}_{11}^{-1}\boldsymbol{\Sigma}_{12} . \qquad (5.3.15)$$

Thus, from (5.3.10), D converges in distribution to $\mathbf{V}'\mathbf{A}^{-1}\mathbf{V}$. This is a quadratic form in a multivariate normal vector in which the matrix is the inverse of the covariance matrix. The distribution of such a quadratic form is chi-square with degrees of freedom equal to the rank of $\mathbf{A}$; see Arnold (1981, p. 49). But the rank of $\mathbf{A}$ is q since it is nonsingular and the proof is complete.

This theorem is proved by McKean and Hettmansperger (1976) for general score functions. In that case τ^{-1} is given by $\int \phi(u)\phi_f(u)\,du$ where $\int \phi(u)\,du = 0$, $\int \phi^2(u)\,du = 1$ and $\phi_f(u)$ is given in (3.5.14). A consistent estimate of τ is based on the length of the one-sample confidence interval constructed from the full-model residuals. This approach to the estimation of τ requires the symmetry of F, the error distribution. The consistency of a window type estimator for general τ has not yet been established rigorously. Recall the discussion following (5.2.30).

The key result in establishing the theorem is (5.3.10) which relates D to a quadratic form in the vector of rank statistics $\mathbf{S}(\boldsymbol{\beta}_0)$, where $\boldsymbol{\beta}_0' = (\boldsymbol{\beta}_1', \mathbf{0}')$, the value of $\boldsymbol{\beta}$ specified under the null hypothesis. Since $\boldsymbol{\beta}_1$, in $\boldsymbol{\beta}_0$, is generally an unknown vector of nuisance parameters, the quadratic form in $\mathbf{S}(\boldsymbol{\beta}_0)$ is not a test statistic. We now provide an example in which $\boldsymbol{\beta}_0$ is completely specified (there are no nuisance parameters) and the test based on $\mathbf{S}(\boldsymbol{\beta}_0)$ reduces to a known test.

Example 5.3.1. Consider the one-way layout. We have X_{ij}, $i = 1, \ldots, n_j$, $j = 1, \ldots, k$, independently distributed with $X_{i1} \sim F(x - \theta_1)$, $i = 1$,

$\ldots, n_1; \ldots; X_{ik} \sim F(x - \theta_k)$, $i = 1, \ldots, n_k$, $F \in \Omega_0$, and we wish to test $H_0: \theta_1 = \cdots = \theta_k$ versus $H_A: \theta_1, \ldots, \theta_k$ not all equal. In order to formulate this problem in terms of a linear model with nonsingular design matrix we let $\mathbf{Y}' = (X_{11}, \ldots, X_{n_1 1}, \ldots, X_{1k}, \ldots, X_{n_k k})$ and

$$
\mathbf{Y} = \begin{bmatrix} \mathbf{1} & \begin{matrix} \mathbf{0}_{n_1} & \mathbf{0} & & & \mathbf{0} \\ \mathbf{1}_{n_2} & \mathbf{0} & & & \mathbf{0} \\ \mathbf{0} & \mathbf{1}_{n_3} & \mathbf{0} & \cdots & \mathbf{0} \\ & & \vdots & & \\ \mathbf{0} & & & & \mathbf{1}_{n_k} \end{matrix} \end{bmatrix} \begin{bmatrix} \alpha \\ \beta_2 \\ \beta_3 \\ \vdots \\ \beta_k \end{bmatrix} + \mathbf{e},
$$

where the subscripts on the entries indicate the length of the vectors. This linear model is related to our original one-way layout as follows:

$$
\begin{aligned}
\theta_1 &= \alpha \\
\theta_2 &= \alpha + \beta_2 \qquad \beta_2 = \theta_2 - \theta_1 \\
&\vdots \qquad\qquad\qquad \vdots \\
\theta_k &= \alpha + \beta_k \qquad \beta_k = \theta_k - \theta_1.
\end{aligned}
$$

In terms of the linear model we will test $H_0: \beta_2 = \cdots = \beta_k = 0$ versus $H_A: \beta_2, \ldots, \beta_k$ not all zero, or $H_0: \boldsymbol{\beta} = \mathbf{0}$ versus $H_A: \boldsymbol{\beta} \neq \mathbf{0}$ where $\boldsymbol{\beta}' = (\beta_2, \ldots, \beta_k)$.

The $\mathbf{X}$ matrix has column means, $\bar{\mathbf{x}}' = (n_2/N, \ldots, n_k/N)$ where $N = \sum_1^k n_i$. Recall, $\mathbf{X}_c = \mathbf{X} - \mathbf{1}\bar{\mathbf{x}}'$ and hence, if $n_i/N \to \lambda_i$, $0 < \lambda_i < 1$, $i = 1, \ldots, k$,

$$
\frac{1}{N}\mathbf{X}_c'\mathbf{X}_c = \begin{bmatrix} \dfrac{n_2}{N}\left(1 - \dfrac{n_2}{N}\right) & -\dfrac{n_2 n_3}{N^2} & \cdots & -\dfrac{n_2 n_k}{N^2} \\ \vdots & & & \vdots \\ -\dfrac{n_2 n_k}{N^2} & -\dfrac{n_3 n_k}{N^2} & \cdots & \dfrac{n_k}{N}\left(1 - \dfrac{n_k}{N}\right) \end{bmatrix}
$$

$$
\to \begin{bmatrix} \lambda_2(1 - \lambda_2) & -\lambda_2\lambda_3 & \cdots & -\lambda_2\lambda_k \\ \vdots & & & \vdots \\ -\lambda_2\lambda_k & -\lambda_3\lambda_k & \cdots & \lambda_k(1 - \lambda_k) \end{bmatrix} = \boldsymbol{\Sigma}.
$$

The inverse of this matrix is given by Graybill (1969), pp. 170–171) as

$$\Sigma^{-1} = \begin{pmatrix} \lambda_2^{-1} & 0 & \cdots & 0 \\ 0 & \ddots & & \\ \vdots & & \ddots & \\ 0 & & & \lambda_k^{-1} \end{pmatrix} + \lambda_1^{-1} \begin{pmatrix} 1 & \cdots & 1 \\ \vdots & & \vdots \\ \vdots & & \vdots \\ 1 & \cdots & 1 \end{pmatrix},$$

where $\lambda_1 = 1 - \sum_2^k \lambda_j$. The vector $\mathbf{S}(\boldsymbol{\beta}_0) = \mathbf{S}(\mathbf{0})$ has jth component

$$S_j(\mathbf{0}) = \sum_{i=1}^N (x_{ij} - \bar{x}_j)\sqrt{12}\left(\frac{R_{ij}}{N+1} - \frac{1}{2}\right) = \frac{\sqrt{12}}{N+1}\sum_{i=1}^{n_j}\left(R_{ij} - \frac{N+1}{2}\right)$$

$$= \frac{\sqrt{12}\,n_j}{N+1}\left(\bar{R}_{\cdot j} - \frac{N+1}{2}\right),$$

$j = 2, \ldots, k$, where R_{ij} is the rank of X_{ij} in the combined data array and $\bar{R}_{\cdot j} = \sum_{i=1}^{n_j} R_{ij}/n_j$. We used the fact that the centered R_{ij}'s sum to zero and $x_{ij} = 1$ as i ranges over the indices in the jth sample. The quadratic form on the right side of (5.3.10) can be written as

$$\frac{1}{N}\mathbf{S}'(\mathbf{0})\Sigma^{-1}\mathbf{S}(\mathbf{0}) = \frac{1}{N}\left[\sum_{j=2}^k \frac{1}{\lambda_j}S_j^2 + \frac{1}{\lambda_1}\left(\sum_{j=2}^k S_j\right)^2\right]$$

$$= \frac{1}{N}\sum_{j=1}^k \frac{1}{\lambda_j}S_j^2$$

since $S_1 = -\sum_2^k S_j$. We now replace λ_j by n_j/N and

$$\frac{1}{N}\mathbf{S}'(\mathbf{0})\Sigma^{-1}\mathbf{S}(\mathbf{0}) \doteq \frac{1}{N}\sum_{j=1}^k \frac{N}{n_j}\frac{12n_j^2}{(N+1)^2}\left(\bar{R}_{\cdot j} - \frac{N+1}{2}\right)^2$$

$$= \frac{12}{(N+1)^2}\sum_{j=1}^k n_j\left(\bar{R}_{\cdot j} - \frac{N+1}{2}\right)^2 = \frac{N}{N+1}H^*$$

where H^* is the Kruskal–Wallis statistic given in Theorem 4.2.4. Hence, in the one-way layout, the quadratic form on the right side of (5.3.10) is essentially the Kruskal–Wallis statistic and we would not use D^*, (5.3.6), which requires the estimation of τ. However, this example shows that D^* is asymptotically equivalent to H^*.

When there are nuisance parameters present in the model and $\boldsymbol{\beta}_0$ is not fully specified under H_0, we estimate $\boldsymbol{\tau}$ and use D^*. In the next subsection we show that if estimates of the nuisance parameters are substituted into $\mathbf{S}(\boldsymbol{\beta}_0)$, then a test can be based on the quadratic form in (5.3.10) with $\mathbf{S}(\boldsymbol{\beta}_0)$ replaced by $\mathbf{S}(\hat{\boldsymbol{\beta}}_0)$. Practical aspects of implementation are discussed with the examples in Section 5.4.

5.3.2. Tests Based on $\mathbf{S}(\mathbf{Y} - \mathbf{X}\boldsymbol{\beta})$

From the proof of Theorem 5.3.1, under $H_0: \boldsymbol{\beta} = \boldsymbol{\beta}_0$ where $\boldsymbol{\beta}_0' = (\boldsymbol{\beta}_1', \mathbf{0}')$, we have

$$\frac{1}{\sqrt{N}} \mathbf{S}'(\mathbf{Y} - \mathbf{X}\boldsymbol{\beta}_0) \left[\boldsymbol{\Sigma}^{-1} - \begin{pmatrix} \boldsymbol{\Sigma}_{11}^{-1} & \mathbf{0} \\ \mathbf{0} & \mathbf{0} \end{pmatrix} \right] \frac{1}{\sqrt{N}} \mathbf{S}(\mathbf{Y} - \mathbf{X}\boldsymbol{\beta}_0)$$

$$= \frac{1}{\sqrt{N}} \left[(-\boldsymbol{\Sigma}_{21}\boldsymbol{\Sigma}_{11}^{-1}, \mathbf{I}) \mathbf{S}(\mathbf{Y} - \mathbf{X}\boldsymbol{\beta}_0) \right]' \left[\boldsymbol{\Sigma}_{22} - \boldsymbol{\Sigma}_{21}\boldsymbol{\Sigma}_{11}^{-1}\boldsymbol{\Sigma}_{12} \right]^{-1}$$

$$\times \frac{1}{\sqrt{N}} \left[(-\boldsymbol{\Sigma}_{21}\boldsymbol{\Sigma}_{11}^{-1}, \mathbf{I}) \mathbf{S}(\mathbf{Y} - \mathbf{X}\boldsymbol{\beta}_0) \right], \tag{5.3.16}$$

which has a limiting chi-square distribution with q degrees of freedom; see (5.3.10) and (5.3.12). Since $\boldsymbol{\beta}_0$ contains the unspecified $\boldsymbol{\beta}_1$ we cannot base a test on this quadratic form unless we estimate $\boldsymbol{\beta}_1$.

The reduced model in which $\boldsymbol{\beta}_2 = \mathbf{0}$ is given by

$$\mathbf{Y} = \mathbf{1}\alpha^* + \mathbf{X}_{1c}\boldsymbol{\beta}_1 + \mathbf{e}. \tag{5.3.17}$$

The reduced-model R estimate, defined by (5.2.8), is denoted by $\hat{\boldsymbol{\beta}}_{10}$ where the $(p - q) \times 1$ vector of regression rank statistics is denoted by $\mathbf{S}_1(\mathbf{Y} - \mathbf{X}_{1c}\hat{\boldsymbol{\beta}}_{10})$. Further, note that $\mathbf{S}_1(\mathbf{Y} - \mathbf{X}_{1c}\hat{\boldsymbol{\beta}}_{10}) \doteq \mathbf{0}$. Since

$$\hat{\boldsymbol{\beta}}_0 = \begin{pmatrix} \hat{\boldsymbol{\beta}}_{10} \\ \mathbf{0} \end{pmatrix} \tag{5.3.18}$$

we can also write $\mathbf{S}_1(\mathbf{Y} - \mathbf{X}_c\hat{\boldsymbol{\beta}}_0)$. We will also let $\mathbf{S}_2(\mathbf{Y} - \mathbf{X}_c\hat{\boldsymbol{\beta}}_0)$ be the remaining q components of $\mathbf{S}(\mathbf{Y} - \mathbf{X}\boldsymbol{\beta}_0)$. Hence,

$$\mathbf{S}(\mathbf{Y} - \mathbf{X}_c\hat{\boldsymbol{\beta}}_0) = \begin{pmatrix} \mathbf{S}_1(\mathbf{Y} - \mathbf{X}_c\hat{\boldsymbol{\beta}}_0) \\ \mathbf{S}_2(\mathbf{Y} - \mathbf{X}_c\hat{\boldsymbol{\beta}}_0) \end{pmatrix} \doteq \begin{pmatrix} \mathbf{0} \\ \mathbf{S}_2(\mathbf{Y} - \mathbf{X}_c\hat{\boldsymbol{\beta}}_0) \end{pmatrix}. \tag{5.3.19}$$

Now, if we insert $\hat{\boldsymbol{\beta}}_0$ into (5.3.16), we have a test statistic. For purposes of constructing the test statistics, we will replace Σ by $N^{-1}(\mathbf{X}_c'\mathbf{X}_c)$, Σ_{11} by $N^{-1}(\mathbf{X}_{1c}'\mathbf{X}_{1c})$, and Σ_{12} by $N^{-1}(\mathbf{X}_{1c}'\mathbf{X}_{2c})$. Then we have, using (5.3.19) and (5.3.16),

$$S^* = \mathbf{S}'(\mathbf{Y} - \mathbf{X}_c\hat{\boldsymbol{\beta}}_0)\left[(\mathbf{X}_c'\mathbf{X}_c)^{-1} - \begin{pmatrix} (\mathbf{X}_{1c}'\mathbf{X}_{1c})^{-1} & \mathbf{0} \\ \mathbf{0} & \mathbf{0} \end{pmatrix}\right]\mathbf{S}(\mathbf{Y} - \mathbf{X}_c\hat{\boldsymbol{\beta}}_0)$$

$$= \mathbf{S}_2'(\mathbf{Y} - \mathbf{X}_c\hat{\boldsymbol{\beta}}_0)\left[(\mathbf{X}_{2c}'\mathbf{X}_{2c}) - \mathbf{X}_{2c}'\mathbf{X}_{1c}(\mathbf{X}_{1c}'\mathbf{X}_{1c})^{-1}\mathbf{X}_{1c}'\mathbf{X}_{2c}\right]^{-1}\mathbf{S}_2(\mathbf{Y} - \mathbf{X}_c\hat{\boldsymbol{\beta}}_0).$$

$$(5.3.20)$$

It remains to show that the substitution of $\hat{\boldsymbol{\beta}}_0$, the reduced-model rank estimate for $\boldsymbol{\beta}_0$, does not alter the limiting distribution.

Theorem 5.3.2. Given the linear model, (5.3.2), suppose the null hypothesis $H_0 : \boldsymbol{\beta}_2 = \mathbf{0}$ holds. Under the assumptions in (5.2.11), S^*, (5.3.20), has an asymptotic chi-square distribution with q degrees of freedom.

Proof. We will work with the second form of S^* given in (5.3.20). Recall τ is given by (5.2.18), and (5.2.10) yields

$$\frac{1}{\sqrt{N}}\mathbf{S}(\boldsymbol{\beta}) = \frac{1}{\sqrt{N}}\mathbf{S}(\boldsymbol{\beta}_0) - \tau^{-1}\frac{1}{N}\mathbf{X}_c'\mathbf{X}_c\sqrt{N}(\boldsymbol{\beta} - \boldsymbol{\beta}_0) + o_p(1),$$

for all $\boldsymbol{\beta}$ such that $N^{1/2}\|\boldsymbol{\beta} - \boldsymbol{\beta}_0\| \leqslant C$ for $C > 0$, a constant. Writing this in partitioned form, we have

$$\frac{1}{\sqrt{N}}\begin{pmatrix} \mathbf{S}_1(\boldsymbol{\beta}) \\ \mathbf{S}_2(\boldsymbol{\beta}) \end{pmatrix} = \frac{1}{\sqrt{N}}\begin{pmatrix} \mathbf{S}_1(\boldsymbol{\beta}_0) \\ \mathbf{S}_2(\boldsymbol{\beta}_0) \end{pmatrix}$$

$$- \tau^{-1}\frac{1}{N}\begin{pmatrix} \mathbf{X}_{1c}'\mathbf{X}_{1c} & \mathbf{X}_{1c}'\mathbf{X}_{2c} \\ \mathbf{X}_{2c}'\mathbf{X}_{1c} & \mathbf{X}_{2c}'\mathbf{X}_{2c} \end{pmatrix}\sqrt{N}\begin{pmatrix} \boldsymbol{\beta}_1 - \boldsymbol{\beta}_{10} \\ \boldsymbol{\beta}_2 - \boldsymbol{\beta}_{20} \end{pmatrix} + o_p(1),$$

where $\boldsymbol{\beta}_0' = (\boldsymbol{\beta}_{10}', \mathbf{0}')$. Since $N^{1/2}(\hat{\boldsymbol{\beta}}_0 - \boldsymbol{\beta}_0)$ is bounded in probability under H_0 by Theorem 5.2.3, we may replace $\boldsymbol{\beta}$ by $\hat{\boldsymbol{\beta}}_0$ and we have

$$\frac{1}{\sqrt{N}}\mathbf{S}_2(\hat{\boldsymbol{\beta}}_0) = \frac{1}{\sqrt{N}}\mathbf{S}_2(\boldsymbol{\beta}_0) - \tau^{-1}\frac{1}{N}(\mathbf{X}_{2c}'\mathbf{X}_{1c})\sqrt{N}(\hat{\boldsymbol{\beta}}_{10} - \boldsymbol{\beta}_{10}) + o_p(1).$$

$$(5.3.21)$$

Further, applying (5.2.10) in the reduced model,

$$\frac{1}{\sqrt{N}} \mathbf{S}_1(\hat{\boldsymbol{\beta}}_{10}) \doteq \mathbf{0}$$

$$= \frac{1}{\sqrt{N}} \mathbf{S}_1(\boldsymbol{\beta}_{10}) - \tau^{-1} \frac{1}{N}(\mathbf{X}'_{1c}\mathbf{X}_{1c})\sqrt{N}\,(\hat{\boldsymbol{\beta}}_{10} - \boldsymbol{\beta}_{10}) + o_p(1),$$

and we have

$$\sqrt{N}\,(\hat{\boldsymbol{\beta}}_{10} - \boldsymbol{\beta}_{10}) = \tau \left(\frac{1}{N}\mathbf{X}'_{1c}\mathbf{X}_{1c}\right)^{-1} \frac{1}{\sqrt{N}} \mathbf{S}_1(\boldsymbol{\beta}_{10}) + o_p(1).$$

Substituting this into (5.3.21) yields

$$\frac{1}{\sqrt{N}} \mathbf{S}_2(\hat{\boldsymbol{\beta}}_0) = \frac{1}{\sqrt{N}} \mathbf{S}_2(\boldsymbol{\beta}_0) - \frac{1}{N}(\mathbf{X}'_{2c}\mathbf{X}_{1c})\left(\frac{1}{N}\mathbf{X}'_{1c}\mathbf{X}_{1c}\right)^{-1} \frac{1}{\sqrt{N}} \mathbf{S}_1(\boldsymbol{\beta}_0) + o_p(1).$$

$$(5.3.22)$$

From Exercise 5.5.3 we have

$$\frac{1}{\sqrt{N}} \begin{pmatrix} \mathbf{S}_1(\boldsymbol{\beta}_0) \\ \mathbf{S}_2(\boldsymbol{\beta}_0) \end{pmatrix} \xrightarrow{D} \mathbf{Z} \sim MVN(\mathbf{0}, \boldsymbol{\Sigma})$$

where $N^{-1}\mathbf{X}'_c\mathbf{X}_c \to \boldsymbol{\Sigma}$, positive definite. In (5.3.22), replace $N^{-1}\mathbf{X}'_{2c}\mathbf{X}_{1c}$ by $\boldsymbol{\Sigma}_{21}$ and $N^{-1}\mathbf{X}'_{1c}\mathbf{X}_{1c}$ by $\boldsymbol{\Sigma}_{11}$. Then

$$\frac{1}{\sqrt{N}} \mathbf{S}_2(\hat{\boldsymbol{\beta}}_0) \doteq (-\boldsymbol{\Sigma}_{21}\boldsymbol{\Sigma}_{11}^{-1}, \mathbf{I}) \begin{bmatrix} \dfrac{1}{\sqrt{N}} \mathbf{S}_1(\boldsymbol{\beta}_0) \\[2mm] \dfrac{1}{\sqrt{N}} \mathbf{S}_2(\boldsymbol{\beta}_0) \end{bmatrix} + o_p(1)$$

$$= (-\boldsymbol{\Sigma}_{21}\boldsymbol{\Sigma}_{11}^{-1}, \mathbf{I}) \frac{1}{\sqrt{N}} \mathbf{S}(\boldsymbol{\beta}_0) + o_p(1).$$

This is limiting $MVN(\mathbf{0}, \boldsymbol{\Sigma}_{22} - \boldsymbol{\Sigma}_{21}\boldsymbol{\Sigma}_{11}^{-1}\boldsymbol{\Sigma}_{12})$ by (5.3.14) and (5.3.15). Now S^*, given by the second equation in (5.3.20), is a quadratic form using the inverse of its covariance matrix and is thus asymptotically chi-square with the degrees of freedom equal to the rank of its matrix, namely q. This completes the proof.

The test of $H_0 : \beta_2 = 0$, β_1 unspecified versus $H_A : \beta_2 \neq 0$, β_1 unspecified, is carried out as follows: First find the reduced-model R estimate of β_1, denoted by $\hat{\beta}_{10}$ or $\hat{\beta}_0' = (\hat{\beta}_{10}', \mathbf{0}')$. Form the reduced-model residuals $Y_1 - \mathbf{x}_1'\hat{\beta}_0, \ldots, Y_N - \mathbf{x}_N'\hat{\beta}_0$ and compute the $q \times 1$ vector of rank statistics $\mathbf{S}_2(\mathbf{Y} - \mathbf{X}_1\hat{\beta}_{10}) = \mathbf{S}_2(\mathbf{Y} - \mathbf{X}\hat{\beta}_0)$. Then compute the quadratic form S^*, (5.3.20), and reject H_0 at approximate level α if $S^* \geqslant \chi_\alpha^2(q)$, where $\chi_\alpha^2(q)$ is the chi-square critical value with q degrees of freedom. Practical aspects of the computation of S^* are discussed in the next section.

The vector $\mathbf{S}_2(\mathbf{Y} - \mathbf{X}\hat{\beta}_0)$ of rank statistics is sometimes called the vector of aligned rank statistics. The original data has been aligned or adjusted by removing the effects of the unspecified nuisance parameters β_1 before constructing the test for β_2. If the null hypothesis $H_0 : \beta_2 = 0$ is true then $\mathbf{S}_2(\mathbf{Y} - \mathbf{X}\hat{\beta}_0) \doteq \mathbf{0}$. Hence the idea behind the test based on S^* is to assess the size of $\mathbf{S}_2(\mathbf{Y} - \mathbf{X}\hat{\beta}_0)$. If $\mathbf{S}_2(\mathbf{Y} - \mathbf{X}\hat{\beta}_0)$ is large then we wish to reject $H_0 : \beta_2 = 0$. The standard statistical method of assessing the size of a random vector which has a multivariate normal distribution is to construct a quadratic form in the components of the vector using the covariance matrix. This method yields statistics with chi-square distributions so that critical values for the test can be determined. This is precisely how we developed S^*.

Hodges and Lehmann (1962) introduced aligned rank tests. Adichie (1967a, b) considered aligned rank tests in a simple linear regression model. Adichie (1978) gives a proof of Theorem 5.3.2 for general score functions and does not require the reduced-model estimate to be an R estimate. He only requires that $N^{1/2}(\hat{\beta}_{10} - \beta_{10})$ be bounded in probability. Hence, for certain models, the statistic S^* could be constructed using the reduced-model least-squares residuals; this is discussed in the next section. Adichie (1978) works directly with the first form of S^* in (5.3.20). Sen and Puri (1977) studied the second form of S^* in (5.3.20). They prove a multivariate version of Theorem 5.3.2 for general score functions in which they use R estimates of the nuisance parameters. The papers of Adichie (1978) and Sen and Puri (1977) contain additional references to work on aligned rank tests.

5.3.3. Tests Based on $\hat{\beta}$

The third approach to testing $H_0 : \beta_2 = 0$, β_1 unspecified, is based directly on the full-model R estimate $\hat{\beta}$ determined by minimizing $D(\mathbf{Y} - \mathbf{X}\beta)$ or solving the nonlinear system of equations $\mathbf{S}(\mathbf{Y} - \mathbf{X}\beta) \doteq \mathbf{0}$. It is convenient to use the formulation mentioned below (5.3.3). Let the $q \times p$ matrix $\mathbf{H} = (\mathbf{0}, \mathbf{I})$, where $\mathbf{I}$ is the $q \times q$ identity matrix. Then $H_0 : \beta_2 = 0$, β_1 unspecified versus $H_A : \beta_2 \neq 0$, β_1 unspecified, may be rewritten as

$$H_0 : \mathbf{H}\beta = \mathbf{0} \quad \text{versus} \quad H_A : \mathbf{H}\beta \neq \mathbf{0}.$$

The test is based on $\mathbf{H}\hat{\boldsymbol{\beta}}$ by constructing a quadratic form that has an asymptotic chi-square distribution. Exercise 5.5.7 provides the asymptotic distribution of $N^{1/2}(\mathbf{H}\hat{\boldsymbol{\beta}} - \mathbf{H}\boldsymbol{\beta})$. Using the asymptotic covariance matrix, the quadratic form

$$\frac{\sqrt{N}\,(\mathbf{H}\hat{\boldsymbol{\beta}}\,)'\left[\mathbf{H}\boldsymbol{\Sigma}^{-1}\mathbf{H}'\right]^{-1}\sqrt{N}\,(\mathbf{H}\hat{\boldsymbol{\beta}}\,)}{\tau^2},$$

where τ is given by (5.2.18), has an asymptotic chi-square distribution with q degrees of freedom.

Replacing $\boldsymbol{\Sigma}$ by $N^{-1}\mathbf{X}_c'\mathbf{X}_c$ and using τ^*, (5.3.5), the test statistic is

$$B^* = \frac{(\mathbf{H}\hat{\boldsymbol{\beta}}\,)'\left[\mathbf{H}(\mathbf{X}_c'\mathbf{X}_c)^{-1}\mathbf{H}'\right]^{-1}(\mathbf{H}\hat{\boldsymbol{\beta}}\,)}{(\tau^*)^2}. \qquad (5.3.23)$$

Then $H_0: \boldsymbol{\beta}_2 = \mathbf{0}$, $\boldsymbol{\beta}_1$ unspecified, is rejected at approximate level α if $B^* \geqslant \chi_\alpha^2(q)$, where $\chi_\alpha^2(q)$ is the chi-square critical value. The statistic B^* is called a Wald statistic after Abraham Wald who first proposed constructing quadratic forms in estimates using their asymptotic covariance matrices.

This result is valid for general scores. The remarks following the proof of Theorem 5.3.1 are relevant for this case also.

The three test statistics D^*, (5.3.6); S^*, (5.3.20); and B^*, (5.3.23); have the same asymptotic distribution under the null hypothesis. Pitman efficiency is developed in the linear model by considering the limiting distributions, as a sequence of alternatives converges to the null hypothesis, similar to the location models considered earlier. When the statistics have asymptotic chi-square distributions, the efficiency is defined by the ratio of noncentrality parameters. The results in the papers by McKean and Hettmansperger (1976), Sen and Puri (1977), and Adichie (1978), show that for the three tests, the efficiency of any one of these tests relative to the least-squares F test is

$$e(\text{Rank}, LS\,) = 12\sigma^2\left(\int f^2(x)\,dx\right)^2. \qquad (5.3.24)$$

Thus the three tests are asymptotically equivalent in the sense of Pitman efficiency and they inherit the efficiency of the Wilcoxon signed rank test, the Mann–Whitney–Wilcoxon test, and the Kruskal–Wallis test. This, further, shows that in the two-way layout these rank tests recover the efficiency lost by the Friedman test; see the discussion surrounding (4.3.8). It should be recalled that the Friedman test does have certain attractive features such as being distribution free and not just asymptotically distribution free, being easy to compute by hand, and being useful in repeated measures designs.

As in the location models discussed earlier, the rank tests and estimates have the same efficiency relative to their least-squares counterparts. Equation 5.2.17, the efficiency of the R estimate relative to the least-squares estimate, is the same as (5.3.24).

Our preference among D^*, S^*, and B^* is for D^*. The statistic D^* and the rank estimate $\hat{\beta}$ both derive in a natural way from the data-fitting criterion $D(\mathbf{Y} - \mathbf{X}\beta)$. The even, location-free measure $D(\mathbf{Y} - \mathbf{X}\beta)$, (5.2.5), is a pseudonorm and provides a geometric interpretation of the test and estimate that is similar to least squares. See Exercise 5.5.10. Further discussion of the geometry of robust estimates and tests can be found in McKean and Schrader (1980) and Hettmansperger and McKean (1983).

The result stated in (5.3.24) also extends to general scores tests. The discussion of Example 3.5.2 carries over from the two-sample location model to the general linear model.

In the next section we discuss the implementation of these tests. Practical issues such as computation and adjustments needed for small-to-moderate sample sizes are discussed. The tests and estimates are illustrated on data arising from regression and analysis of variance models.

Before turning to the examples, we outline in some detail the implementation of the Winsorized Wilcoxon score function introduced in Example 2.8.2 for the one-sample problem and extended to the two-sample case in Exercise 3.7.8. The results at the end of Section 2.9 suggest that Winsorization is advantageous from the point of view of robustness; see Example 2.9.5. With the exception of occasional remarks, this chapter has dealt solely with the Wilcoxon score function, (5.2.9). We choose to discuss the Winsorized Wilcoxon score function because estimates and tests based on it are available through the RANK REGRESSION command in the Minitab statistical computing system.

The solution to Exercise 3.7.8 can be written in the following standardized form:

$$\phi^*(u) = \begin{cases} -(1 - \gamma)/2 & 0 < u < \gamma/2 \\ u - 1/2 & \gamma/2 \leqslant u \leqslant 1 - \gamma/2 \\ (1 - \gamma)/2 & 1 - \gamma/2 < u < 1. \end{cases} \tag{5.3.25}$$

Further,

$$\int_0^1 \left[\phi^*(u) \right]^2 du = \frac{(1 - \gamma)^2(1 + 2\gamma)}{12} .$$

Hence we define

$$\phi(u) = \sqrt{\frac{12}{(1-\gamma)^2(1+2\gamma)}} \; \phi^*(u). \qquad (5.3.26)$$

Then $\int_0^1 \phi^2(u)\,du = 1$ and $\int_0^1 \phi(u)\,du = 0$. From Example 2.9.1, the parameter $\tau^{-1} = 12^{1/2}\int f^2(x)\,dx$ is replaced by

$$\tau^{-1} = \sqrt{12} \int_a^b f^2(x)\,dx \Big/ \sqrt{(1-\gamma)^2(1+2\gamma)} \qquad (5.3.27)$$

where $a = F^{-1}(\gamma/2)$ and $b = F^{-1}(1-\gamma/2)$. Note that γ represents the fraction of sign-type scores used in the Winsorization of the Wilcoxon scores. When $\gamma = 0$ the equations reduce to the Wilcoxon case and when γ approaches 1 the equations yield the L_1 case.

Now using $a(i) = \phi[i/(N+1)]$ in (5.2.5), the dispersion function $D(\cdot)$ is defined, and the corresponding R estimate $\hat{\boldsymbol{\beta}}$ is found by minimizing $D(\mathbf{Y} - \mathbf{X}\boldsymbol{\beta})$. Theorem 5.2.3 remains valid and $N^{1/2}(\hat{\boldsymbol{\beta}} - \boldsymbol{\beta})$ is asymptotically $MVN(\mathbf{0}, \tau^2 \boldsymbol{\Sigma}^{-1})$. If we suppose that $F \in \Omega_s$, then the parameter τ will be estimated consistently by $N^{1/2}L/2Z_{\alpha/2}$ where $1 - \Phi(Z_{\alpha/2}) = 1 - \alpha$ and L is the length of the confidence interval constructed from the Winsorized Wilcoxon signed rank statistic applied to the residuals $Y_i - \mathbf{x}_i'\hat{\boldsymbol{\beta}}$, $i = 1, \ldots, N$. This approach is described in detail in Example 2.8.4. If we may only suppose that $F \in \Omega_0$ then a density estimation approach could be used. The Schweder (1975) type estimate of $\int_a^b f^2(x)\,dx$ is given by

$$\int \phi'(F_N(x)) f_N(x)\,dF_N(x) = \int_{\hat{a}}^{\hat{b}} f_N(x)\,dF_N(x)$$

where $\hat{a}, \hat{b}$ are the $\gamma/2$ and $1 - \gamma/2$ sample percentiles, F_N is the empirical cdf, and f_N is a window-type density estimate, all computed from the full-model residuals. A rigorous proof of consistency, specifying all the necessary regularity conditions, has not been given for the density-type estimate in the linear-model case for general scores.

For a specified value of γ the RANK REGRESSION command in the Minitab statistical computing system will produce the R estimates and tests based on either D^*, (5.3.6), or B^*, (5.3.23), using either the confidence interval or density estimate of τ. Except for the density estimate approach, the proofs for the results in the linear model with general scores can be found in Jaeckel (1972) and McKean and Hettmansperger (1976).

5.4. IMPLEMENTATION AND EXAMPLES

The computations involved in determining a rank estimate or test in the linear model are generally quite tedious and are carried out using a computer. From Theorem 5.2.1 we know that $D(Y - X\beta)$ is continuous and convex so that methods such as steepest descent can be used to locate the minimum. An iterative method based on the idea in the construction of k-step estimates, (5.2.19), is also useful. Since the minimizing value is not always unique, the numerical answer will depend on which solution the algorithm picks. If we have a single sample, this corresponds to taking any value between the middle two order statistics as the sample median when the sample size is even. In addition, since iterative methods are used to find the minimum, the tolerances set to determine when to stop will have some effect on the numerical answers. Discussion of the computational aspects of finding rank estimates and tests can be found in Aubuchon and Hettmansperger (1982a) and Hettmansperger and McKean (1983) and their references. For data analysis, the Minitab computing system provides a RANK REGRESSION command to compute R estimates and tests based on D^* and B^*. Wilcoxon and Winsorized Wilcoxon scores are available along with either the confidence interval approach or the density estimate approach to estimating τ.

In addition to the numerical issues, there are statistical issues that require some attention. Simulation studies by McKean and Hettmansperger (1978) and Hettmansperger and McKean (1977, 1983) indicate that the nominal significance levels of the rank tests are quite close to the true levels when the chi-square critical point $\chi_\alpha^2(q)$ is replaced by q times the F critical point $qF_\alpha(q, N - p - 1)$. Typically, the chi-square critical value, derived from the asymptotic theory, allowed too many rejections under the null hypothesis and inflated the significance level. The F critical value corrected this problem quite well in all cases studied. A bias correction is generally applied to the confidence interval estimate of τ. For example, $\hat{\tau}$ may be multiplied by $[N/(N - p - 1)]^{1/2}$, a factor suggested by the least-squares estimate of σ. Bias correction of the density estimate approach to estimating τ is discussed in the construction of γ^* in (5.2.24).

The following examples have been chosen from the literature where the reader can find different perspectives and other analyses. We provide the least-squares solution for purposes of comparison.

Example 5.4.1. This example is intended to illustrate the estimation of regression coefficients. The data, given in Table 5.2, consists of 21 observations on a response variable Y and three independent variables x_1, x_2, and x_3 relevant to the study of a manufacturing plant that oxidizes ammonia to

Table 5.2. Stack Loss Data

Run No.	Air Flow x_1	Cooling Water Inlet Temperature x_2	Acid Concentration x_3	Stack Loss Y
1	80	27	89	42
2	80	27	88	37
3	75	25	90	37
4	62	24	87	28
5	62	22	87	18
6	62	23	87	18
7	62	24	93	19
8	62	24	93	20
9	58	23	87	15
10	58	18	80	14
11	58	18	89	14
12	58	17	88	13
13	58	18	82	11
14	58	19	93	12
15	50	18	89	8
16	50	18	86	7
17	50	19	72	8
18	50	19	79	8
19	50	20	80	9
20	56	20	82	15
21	70	20	91	15

nitric acid. The variables are

Y = 10 times the percent ammonia lost up the stack in the process
x_1 = rate of the operation, measured by air flow
x_2 = temperature of the cooling water used in the process
x_3 = concentration of nitric acid in the absorbing liquid (coded by minus 50, times 10).

We will consider the equations $Y_i = \beta_0 + x_{1i}\beta_1 + x_{2i}\beta_2 + x_{3i}\beta_3 + e_i$, $i = 1, \ldots, 21$. The dependent variable Y is a measure of the inefficiency of the process.

Least-squares analyses are given by Brownlee (1965, p. 454), Draper and Smith (1981, p. 361), and Daniel and Wood (1971, Chapter 5). Daniel and Wood, after an elaborate analysis, set aside four observations—numbers 1, 3, 4, and 21—as outliers corresponding to transient states in the process. The variable x_3 was eliminated as nonsignificant. The analysis then proceeded on the remaining 17 observations.

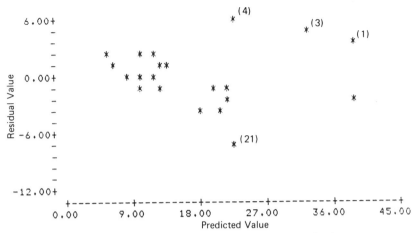

Figure 5.1. Least-squares plot of residuals versus fitted values.

The least-squares analysis is provided by inputting the data from Table 5.2 into the computer and regressing the Y column on the three variables in columns 2 through 4 using the REGRESS command in Minitab. A plot of residuals versus fitted values is provided in Fig. 5.1. Note the four extreme residuals and the downward tilt pattern in the remaining points.

Andrews (1974), using M estimation, and Hettmansperger and McKean (1977), using R estimation, while retaining the four spurious points, were able to achieve essentially the same fit as that produced by least squares without the four points.

In Table 5.3 we list the R estimates of β_0, β_1, β_2, β_3, and τ for various score functions. The target for comparison is the line corresponding to least squares without outliers. Estimates of the standard deviations of the esti-

Table 5.3. Estimates of the Parameters

	$\hat{\beta}_0$	$\hat{\beta}_1$	$\hat{\beta}_2$	$\hat{\beta}_3$	$\hat{\tau}$
Least squares	-39.9	$0.72(0.17)^a$	$1.30(0.37)$	$-0.15(0.16)$	
Least squares (w/o outliers)	-37.6	$0.80(0.07)$	$0.58(0.17)$	$-0.07(0.06)$	
Wilcoxon	-40.1	$0.82(0.13)$	$0.89(0.35)$	$-0.12(0.15)$	3.06
33% Winsorized	-38.0	$0.85(0.10)$	$0.72(0.28)$	$-0.13(0.12)$	2.49
50% Winsorized	-38.5	$0.85(0.08)$	$0.67(0.22)$	$-0.11(0.09)$	2.00
Sign (L_1)	-39.7	$0.83(0.06)$	$0.58(0.14)$	$-0.06(0.06)$	0.89
Andrews	-37.2	$0.82(0.05)$	$0.52(0.12)$	$-0.07(0.04)$	

^{a}Number in parentheses is the estimate of the standard deviation.

mates are computed from $\hat{\tau}^2(\mathbf{X}_c'\mathbf{X}_c)^{-1}$, suggested by Theorem 5.2.3. Since $\hat{\sigma}^2/\hat{\tau}^2$ is an estimate of the efficiency of the rank method relative to the least-squares method, smaller values of $\hat{\tau}$ indicate that the data is selecting a more efficient score function. This informal view of letting the sample select a score function suggests that the L_1 fit and the resulting estimates corresponding to minimizing the sum of absolute values of the residuals, is the best among those considered. The estimates corresponding to the sign scores are closest to least squares without outliers. Denby and Mallows (1977) reach the same conclusion using diagnostic plots in robust regression.

Unfortunately, we cannot expect, in general, to have a least-squares analysis as sensitive as the one provided by Daniel and Wood. A reasonable strategy would be to use the Wilcoxon or Winsorized Wilcoxon score function, for example, the 33% Winsorized Wilcoxon suggested by Example 2.9.5. If the results are in conflict with a standard least-squares analysis, then a deeper inspection of the experiment is in order to try to uncover the sources of the differences.

The estimates of the standard deviations for the R estimates are based on the length of confidence interval method. This requires that we assume symmetry of the error distribution. In the case of Wilcoxon scores, for example, the confidence interval method yields $\hat{\tau} = 3.06$. The density estimation approach yields $\hat{\tau} = 2.68$. The estimates $\hat{\beta}_1$, $\hat{\beta}_2$, and $\hat{\beta}_3$, and their distribution theory are not affected by the symmetry assumption. The intercept estimate $\hat{\beta}_0$ depends on whether we assume symmetry or not. Under the symmetry assumption, for the Wilcoxon scores, $\hat{\beta}_0$ is the median of the Walsh averages of the full-model residuals; $\hat{\beta}_0 = -40.10$. Without the symmetry assumptions, $\hat{\beta}_0$ is simply the median of the residuals; $\hat{\beta}_0 = -40.14$. Recall the discussion surrounding (5.2.31) and (5.2.32).

To test for the significance of $\hat{\beta}_3$, which was eliminated in the least-squares analysis, we could compare the square of the ratio of the estimate and its standard deviation to the F critical value $F_\alpha(1, 17)$. This is precisely the test based on B^*, (5.3.23). For Wilcoxon scores, from Table 5.3, $B^* = (-.12/.15)^2 = .64$ which is not significant for any reasonable significance level α. Again using Wilcoxon scores, but with the density estimate of τ and the test based on $D(\mathbf{Y} - \mathbf{X}\boldsymbol{\beta})$, we find from the RANK REGRESSION command in Minitab that D^*, (5.3.6), is .81. In fact, a quick glance at Table 5.3 reveals that all of the rank score tests based on B^* will fail to reject $H_0: \beta_3 = 0$ at any reasonable significance level.

In summary, we would recommend an equation based on the Wilcoxon or 33% Winsorized Wilcoxon score function. The respective equations are

$$y = -40.1 + 0.82x_1 + 0.89x_2 - 0.12x_3$$

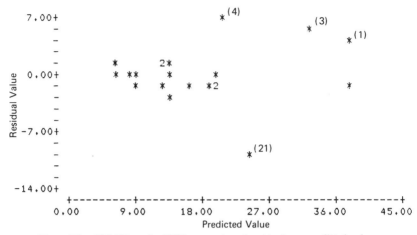

Figure 5.2. 33% Winsorized Wilcoxon plot of residuals versus fitted values.

or

$$y = -38.0 + 0.85x_1 + 0.72x_2 - 0.13x_3.$$

We have not removed the outliers nor dropped the x_3 term. Figure 5.2 provides a plot of residuals versus fitted values for the 33% Winsorized Wilcoxon case. Note the four outliers are more prominent than the least-squares plot (Fig. 5.1) and the remaining pattern is improved. An equation based on sign scores is also attractive and perhaps preferable, based on estimated efficiency. In any case, the fitted equation developed from $D(\mathbf{Y} - \mathbf{X}\boldsymbol{\beta})$ is better than the least-squares fit on the full data set. Cheng and Hettmansperger (1983) discuss an alternate computational approach based on iteratively reweighted least squares. The resulting estimates for this example are given there and are very close to the numbers reported in Table 5.3.

Example 5.4.2. In this example we illustrate the analysis of data arising from a two-way layout. The analysis parallels a two-way analysis of variance, and the least-squares analysis is provided for comparison. The data, from Box and Cox (1964) and given in Table 5.4, consists of the survival times of 48 animals exposed to three different poisons and four different treatments. A 3×4 factorial design was used with four observations per cell.

We wish to test for the presence of significant interaction and main effects. A plot of the cell medians, Fig. 5.3, indicates all effects may be present. The two most striking features of the plot are (1) the second

Table 5.4. Survival Times (unit = 10 hours)

Poison	Treatment A	B	C	D
I	0.31	0.82	0.43	0.45
	0.45	1.10	0.45	0.71
	0.46	0.88	0.63	0.66
	0.43	0.72	0.76	0.62
II	0.36	0.92	0.44	0.56
	0.29	0.61	0.35	1.02
	0.40	0.49	0.31	0.71
	0.23	1.24	0.40	0.38
III	0.22	0.30	0.23	0.30
	0.21	0.37	0.25	0.36
	0.18	0.38	0.24	0.31
	0.23	0.29	0.22	0.33

treatment seems to be the most efficacious and (2) the third poison is almost uniformly lethal.

Before describing the analysis, we must develop a linear model with a nonsingular design matrix for this example. We begin with $Y_{ijk} = \mu + P_i + T_j + I_{ij} + e_{ijk}$, $i = 1,2,3$; $j = 1,2,3,4$; $k = 1,2,3.4$. The conditions $\sum P_i = \sum T_j = \sum_i I_{ij} = \sum_j I_{ij} = 0$ are usually imposed so the parameters can be identified. We have used P_1, P_2, P_3 and T_1, T_2, T_3, T_4 to denote the poison and treatment effects, respectively. The term I_{ij} is an interaction term. We will outline the construction of the design matrix. This is illustrated for the

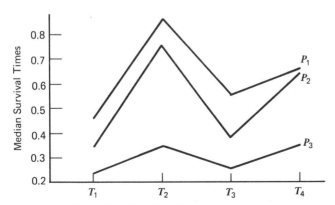

Figure 5.3. Plot of median survival times.

one-way layout in Example 5.3.1. The Y observations will be set in a 48×1 vector in the order $P_1T_1, P_1T_2, P_1T_3, P_1T_4, P_2T_1, \ldots$, where P_1T_1 means data corresponding to the first poison and first treatment, and so on. Using $\mathbf{1}$ to denote a 4×1 vector of ones and $\mathbf{D}_{ij}$ to denote the 4×1 vector of Y values in the (i, j) cell, we can represent the first 16 observations as follows:

	P_1	P_2	P_3	T_1	T_2	T_3	T_4
$\mathbf{D}_{11}$	1	0	0	1	0	0	0
$\mathbf{D}_{12}$	1	0	0	0	1	0	0
$\mathbf{D}_{13}$	1	0	0	0	0	1	0
$\mathbf{D}_{14}$	1	0	0	0	0	0	1

The pattern then continues for the remaining cells. In order to convert the matrix to a nonsingular one and also account for the interaction terms, we first subtract the P_3 column from the P_1 and P_2 columns and subtract the T_4 column from the T_1, T_2, and T_3 columns. We then have the starred columns in the following array.

	P_1^*	P_2^*	T_1^*	T_2^*	T_3^*	I_{11}	I_{12}	I_{13}	I_{21}	I_{22}	I_{23}
$\mathbf{D}_{11}$	1	0	1	0	0	1	0	0	0	0	0
$\mathbf{D}_{12}$	1	0	0	1	0	0	1	0	0	0	0
$\mathbf{D}_{13}$	1	0	0	0	1	0	0	0	1	0	0
$\mathbf{D}_{14}$	1	0	-1	-1	-1	-1	-1	-1	0	0	0

The I_{ij} column is found by multiplying ith P^*-column by the jth T^*-column to form the interaction dummy variables. The starred parameters are the simple contrasts, for example, $P_1^* = P_1 - P_3$, used to formulate hypotheses.

The entire data set consisting of a column of Y values and 11 columns of ones, negative ones, and zeros is imputed into the computer. We can now specify various hypotheses of interest. Let $\boldsymbol{\beta}' = (P_1^*, P_2^*, T_1^*, T_2^*, T_3^*, I_{12}, \ldots, I_{23})$ with 11 components. Let the 6×11 matrix $\mathbf{H}_I$ be given by $\mathbf{H}_I = [\mathbf{0}, \mathbf{I}]$ where $\mathbf{I}$ is a 6×6 identity matrix. Then $H_0: \mathbf{H}_I\boldsymbol{\beta} = \mathbf{0}$ specifies the hypothesis of no interaction. Let $\mathbf{H}_P = [\mathbf{I}, \mathbf{0}]$, where $\mathbf{I}$ is the 2×2 identity matrix. Then $H_0: \mathbf{H}_P\boldsymbol{\beta} = \mathbf{0}$ specifies the hypothesis of no poison effect. Note $\mathbf{H}_P\boldsymbol{\beta} = \mathbf{0}$ means $P_1^* = 0$ and $P_2^* = 0$ so that $P_1^* = P_1 - P_3 = 0$ and $P_2^* = P_2 - P_3 = 0$, and $P_1 = P_2 = P_3$. The matrix $\mathbf{H}_T$ needed to specify the hypothesis of no treatment effect is left for the reader.

Minitab was used to make the computations. Since survival time may be asymmetrically distributed, the density estimate of τ should be used. Wilcoxon scores and the test statistic D^*, (5.3.6), were applied. In Table 5.5

Table 5.5. Analysis of the Survival Data

Source	df	D^*/df	$F_{.05}(df, 36)$	F
P	2	42.53	3.26	23.22
T	3	22.85	2.87	13.81
$P \times T$	6	3.35	2.36	1.87
$\tau^* = 0.084, \hat{\sigma} = .149$				

we compare D^*/df to $F_\alpha(df, N - p - 1)$ with $N = 48$ and $p = 11$, as suggested at the beginning of this section. The column labeled F provides the classical F-test values.

From Table 5.5, based on the rank test D^*, we see a significant interaction term and highly significant main effects. On the other hand, the F test fails to reject, at the 5% level, the hypothesis of no interaction. In Fig. 5.4 a plot of residuals versus fitted values reveals a problem in the data.

The plot shows highly heterogeneous cell variances. Huber, in his 1972 Wald lecture (see Huber, 1973), points out that "uncontrollable inhomogeneity of variances among the [variables] and genuinely long-tailed error distributions have almost indistinguishable effects, both impairing the efficiency of the [least-squares] estimates."

The analysis based on the ranks of the residuals also depends upon the assumption of homogeneity of variances. Various authors have suggested transformations to correct the problem; see Box and Cox (1964) and Brown (1975). The reciprocal has been advocated and Table 5.6 provides the analysis for the transformed data. It should be noted that these rank test

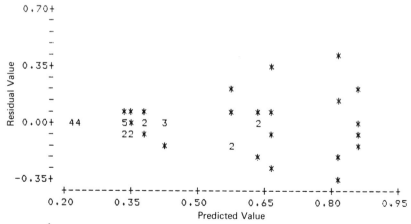

Figure 5.4. Plot of residuals versus fitted values with Wilcoxon scores.

Table 5.6. Analysis of Reciprocals of Data

Source	df	D^*/df	$F_{.05}(df, 36)$	F
P	2	56.52	3.26	72.56
T	3	25.26	2.87	28.36
$P \times T$	6	1.45	2.36	1.09
$\tau^* = 0.435, \hat{\sigma} = .490$				

statistics in the linear model are not generally invariant under monotone transformations, as is the case in the simple location model. In the location model, the data rather than the residuals is ranked. Hence the rank-based tests can be responsive to transformations of the data.

The situation has improved. The rank and least-squares analyses are compatible and Fig. 5.5 shows the cell variances to be less heterogeneous. Our recommended analysis would use D^* based on the Wilcoxon score function with the density estimate, and performed on the reciprocals of the data.

We have described the use of D^*, the model-fitting criterion, in testing hypotheses. The approach to data analysis, suggested by this example, is to compare the rank and least-squares analyses and look deeper into the data when they disagree. In this case a reciprocal transformation of the data is suggested.

The aligned rank test S^* could also be used. Provided a least-squares regression algorithm is available, no special programs, beyond the ability to rank a vector of numbers, are necessary. To test $H_0 : \mathbf{H}_I \boldsymbol{\beta} = \mathbf{0}$, no interac-

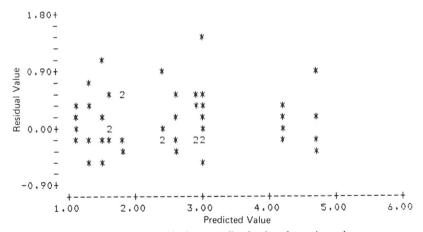

Figure 5.5. Plot of residuals versus fitted values for reciprocals.

tion, we first find the reduced-model least-squares residuals, say $\hat{r}_{01}, \ldots,$ $\hat{r}_{0N}$. Then compute $a(R_1), \ldots, a(R_N)$, where

$$a(R_i) = \sqrt{12}\left[\frac{R_i}{N+1} - \frac{1}{2}\right]$$

and R_i is the rank of $\hat{r}_{0i}$ among $\hat{r}_{01}, \ldots, \hat{r}_{0N}$. The values $a(R_1), \ldots, a(R_N)$ are then fed into a least-squares program and S^* is the numerator sum of squares in the F-test statistic for no interaction. When this is done on the raw data we find $S^* = 14.77$. The chi-squre critical value for 6 degrees of freedom is $\chi^2_{.05}(6) = 12.59$ and we would reject the null hypothesis of no interaction. For the reciprocal data, $S^* = 7.39$ and fails to reject at any reasonable level. Hence the analysis based on S^* is similar to that based on D^* displayed in Tables 5.5 and 5.6. If aligned rank tests for the main effects are to be computed, then new reduced-model residuals must be found in each case.

If we are willing to assume an additive model (no interaction), then we have a two-way layout with four observations per cell. A test for poison or treatment effects can be carried out using an extension of Friedman's test outlined in Exercise 4.5.10. The reader is asked to make the computations in Exercise 5.5.8.

Other more complex designs can be analyzed in a similar way. It is only necessary to set up the dummy variables to produce a nonsingular design matrix for a multiway layout or input the appropriate design matrix in the case of regression. In a like manner, by combining dummy variables and regression variables, an analysis of covariance can be carried out. Schrader and McKean (1977) and McKean and Schrader (1982) discuss further aspects of robust analysis of variance. For a discussion of robust analysis of variance based on an M-estimate approach, see Schrader and Hettmansperger (1980).

5.5. EXERCISES

5.5.1. Consider the simple linear regression model introduced in Section 5.1, $Y_i = \beta_1 + \beta_2 x_i + e_i$, $i = 1, \ldots, N$. Assume that $e_1, \ldots, e_N$ are i.i.d. $F \in \Omega_0$. Rather than use U, (5.1.4), to test $H_0: \beta_2 = 0$ versus $H_A: \beta_2 \neq 0$, we could use Kendall's τ defined by (4.4.6). Equivalently, we may use the numerator S of τ, defined in (4.4.5). If $x_1 < \cdots < x_N$, then $S = \sum\sum_{i<j} s(Y_j - Y_i)$ is the test statistic and

the mean, variance, and limiting distribution under $H_0 : \beta_2 = 0$ are given in Theorem 4.4.2. To derive the Hodges–Lehmann estimate, define $S(\beta_2) = \sum\sum_{i<j} s(Y_j - \beta_2 x_j - Y_i + \beta_2 x_i)$ and show that

$$S(\beta_2) = 2\{ \#[(Y_j - Y_i)/(x_j - x_i) > \beta_2]\} - \binom{N}{2}$$

$$= 2S^*(\beta_2) - N(N-1)/2.$$

Now show that the estimate is

$$\hat{\beta}_2 = \underset{i<j}{\mathrm{med}} \left(\frac{Y_j - Y_i}{x_j - x_i} \right).$$

Further, a $(1 - \alpha)$ 100% confidence interval for β_2 is given by $[S_{(k+1)}, S_{(N^*-k)}]$ where $N^* = N(N-1)/2$, $S_{(1)} \leq \cdots \leq S_{(N^*)}$ are the ordered pairwise slopes and $P(S^* \leq k) = \alpha/2$. Finally, show that k can be approximated as

$$k \doteq \frac{N(N-1)}{4} - Z_{\alpha/2} \sqrt{\frac{N(N-1)(2N+5)}{72}}$$

where $1 - \Phi(Z_{\alpha/2}) = \alpha/2$.

5.5.2. When $\boldsymbol{\beta}$ is the true parameter value, show that $E\mathbf{S}(\mathbf{Y} - \mathbf{X}\boldsymbol{\beta}) = \mathbf{0}$, where $S_j(\mathbf{Y} - \mathbf{X}\boldsymbol{\beta})$ is given by (5.2.7) for $j = 1, \ldots, p$.

5.5.3. a. Suppose we have the linear model specified by (5.2.1). Suppose $\boldsymbol{\beta}_0$ denotes the true parameter value and $\mathbf{S}(\mathbf{Y} - \mathbf{X}\boldsymbol{\beta}_0)$ denotes the $p \times 1$ vector of rank statistics with jth component given by (5.2.7). If assumptions in (5.2.11) hold, show that

$$\frac{1}{\sqrt{N}} \mathbf{S}(\mathbf{Y} - \mathbf{X}\boldsymbol{\beta}_0) \overset{D}{\to} \mathbf{Z} \sim MVN(\mathbf{0}, \boldsymbol{\Sigma}).$$

   b. Under the conditions of (a), show that

$$\sqrt{N}(\tilde{\boldsymbol{\beta}} - \boldsymbol{\beta}_0) \overset{D}{\to} \mathbf{Z} \sim MVN\left[\mathbf{0}, \left(\frac{1}{12\left(\int f^2(x)\,dx\right)^2} \boldsymbol{\Sigma}^{-1} \right)\right]$$

where $\tilde{\boldsymbol{\beta}}$ is given by (5.2.14).

5.5.4. Suppose that $N^{1/2}(\hat{\boldsymbol{\beta}}_0 - \boldsymbol{\beta}_0)$ is bounded in probability. Then show that, under assumptions in (5.2.11), $N^{1/2}(\hat{\boldsymbol{\beta}}_1 - \boldsymbol{\beta}_0)$, the one-step estimate defined in (5.2.19), is asymptotically $MVN(\mathbf{0}, \tau^2 \boldsymbol{\Sigma}^{-1})$ where $\tau = 1/(12^{1/2} \int f^2(x)\, dx)$. Hint: Write

$$\sqrt{N}(\hat{\boldsymbol{\beta}}_1 - \boldsymbol{\beta}_0) - \tau \frac{1}{\sqrt{N}} \boldsymbol{\Sigma}^{-1} \mathbf{S}(\boldsymbol{\beta}_0)$$

$$= \sqrt{N}(\hat{\boldsymbol{\beta}}_0 - \boldsymbol{\beta}_0) + \tau \frac{1}{\sqrt{N}} \boldsymbol{\Sigma}^{-1} [\mathbf{S}(\hat{\boldsymbol{\beta}}_0) - \mathbf{S}(\boldsymbol{\beta}_0)].$$

Use the asymptotic linearity, (5.2.10), and the boundedness in probability of $N^{1/2}(\hat{\boldsymbol{\beta}}_0 - \boldsymbol{\beta}_0)$ to show the right side is $o_p(1)$. Then argue the asymptotic distribution of the second term on the right side.

5.5.5. Consider the density estimator, (5.2.21). Show that for each fixed y

$$Ef_N(y) = \frac{1}{h_N}\left[F\left(y + \frac{h_N}{2}\right) - F\left(y - \frac{h_N}{2}\right)\right].$$

Derive a similar formula for $\text{Var } f_N(y)$. Then show $Ef_N(y) \to f(y)$ and $\text{Var } f_N(y) \to 0$. Hence $f_N(y)$ is a consistent estimator of $f(y)$.

5.5.6. Suppose $f_N(x)$ is given by (5.2.21). Show that $\int f_N^2(x)\, dx = \int f_N^*(x)\, dF_N(x)$ where $f_N^*(x)$ satisfies Definition 5.2.3 with window $w^*(x) = w * w(x)$, the convolution density; see Definition A3 in the Appendix.

5.5.7. Suppose $\hat{\boldsymbol{\beta}}$ is the full-model rank estimate, Definition 5.2.2. Under assumptions in (5.2.11), argue that $N^{1/2}(\mathbf{H}\hat{\boldsymbol{\beta}} - \mathbf{H}\boldsymbol{\beta})$ has an asymptotic distribution which is multivariate normal. Find the covariance matrix of the limiting distribution.

5.5.8. Using the data in Table 5.4 and assuming no interaction (additive model) use the extended Friedman test in Exercise 4.5.10 to test the null hypothesis of no treatment effect. Then test the null hypothesis of no poison effect.

5.5.9. Gini's mean difference measure of variability in $\mathbf{X}' = (X_1, \ldots, X_n)$ is defined to be

$$g(\mathbf{X}) = \frac{\sqrt{\pi}}{2n(n-1)} \sum_{i=1}^{n} \sum_{j=1}^{n} |X_i - X_j|.$$

a. If $\mathbf{X}' = (X_1, \ldots, X_n)$ is a random sample from a normal distribution with variance σ^2, then show that $Eg(\mathbf{X}) = \sigma$, so $g(\mathbf{X})$ is an unbiased estimate of the standard deviation.

b. Show that if R_i is the rank of X_i, then

$$g(\mathbf{X}) = \frac{2\sqrt{\pi}}{n(n-1)} \sum_{i=1}^{n} \left(R_i - \frac{n+1}{2} \right) X_i .$$

Hence, argue that the R estimate of $\boldsymbol{\beta}$ in the linear model minimizes Gini's measure of variability.

5.5.10. Consider $D(\mathbf{Z})$ defined by (5.2.3). Prove

a. For every $\mathbf{Z}$ in R^N, $D(\mathbf{Z}) \geqslant 0$.

b. For every $\mathbf{Z}, \mathbf{W}$ in R^N, $D(\mathbf{Z} + \mathbf{W}) \leqslant D(\mathbf{Z}) + D(\mathbf{W})$.

c. For every scalar b, $D(b\mathbf{Z}) = |b| D(\mathbf{Z})$

d. $D(\mathbf{Z}) = 0$ if and only if $Z_1 = \cdots = Z_N$.

The exercise shows that $D(\mathbf{Z})$, with scores satisfying the conditions given above, is a pseudo- (or semi-) norm. Hence the rank estimate $\hat{\boldsymbol{\beta}}$, Definition 5.2.2, is defined by $\mathbf{X}_c\hat{\boldsymbol{\beta}}$, the closest vector to the observation vector $\mathbf{Y}$, in the subspace defined by the model $\mathbf{X}\boldsymbol{\beta}$ where distance is defined by the pseudonorm $D(\mathbf{Y} - \mathbf{X}\boldsymbol{\beta})$.

5.5.11. Exercise 5.5.10 shows that D^*, (5.3.6), is a natural test statistic since it compares the distance from $\mathbf{Y}$ to the subspace specified by the null hypothesis and the distance from $\mathbf{Y}$ to the subspace defined by the full model. Hence D^* is analogous to the least-squares F statistic which can be interpreted in terms of the reduction in sums of squares due to fitting the reduced and full models. Analogy with the F statistic also suggests the test statistic:

$$D^o = \frac{D(\mathbf{Y} - \mathbf{X}\hat{\boldsymbol{\beta}}_0) - D(\mathbf{Y} - \mathbf{X}\hat{\boldsymbol{\beta}})}{N^{-1}D(\mathbf{Y} - \mathbf{X}\hat{\boldsymbol{\beta}})} .$$

Give a heuristic argument to show $N^{-1}D(\mathbf{Y} - \mathbf{X}\hat{\boldsymbol{\beta}})$ converges in probability to $\int_{-\infty}^{\infty} x\phi(F(x))\,dF(x) = \delta$, where $\phi(u)$ is a general score function such that $\int\phi(u)\,du = 0$, $\int\phi^2(u)\,du = 1$. Show that in the case of the Wilcoxon score function, (5.2.9), δ is not equal to $\tau/2$. Hence, the test based on D^o is not an asymptotically size α test and D^* is preferable.

5.5.12. In the proof of Theorem 5.2.3, show $T = \sigma^* \epsilon^2 3^{1/2} \int f^2(x)\,dx$, where σ^x is the minimum eigenvalue of Σ.

CHAPTER 6

The Multivariate Location Model

6.1. INTRODUCTION

In this brief chapter we indicate how one- and two-sample univariate tests and estimates can be extended to the multivariate case. If the multivariate problem has p components, then we will be interested in the vector of univariate statistics computed on each of the p components. For example, we will consider the vector of sign statistics in the multivariate one-sample model presented in the next section.

The p-component vector of statistics is denoted by $\mathbf{S}' = (S_1, \ldots, S_p)$ and the statistic S_i is centered so that $ES_i = 0$ under the null hypothesis. Even though the statistics are distribution free in the univariate model, the vector is no longer distribution free in the multivariate model. The difficulty occurs in the $\text{Cov}(S_i, S_j)$ which depends on the joint distribution of the observations. The solution is to estimate the covariances and construct asymptotically distribution-free tests. Hence, the emphasis is on the asymptotic distributions.

It is possible to construct conditional permutation or randomization tests which are distribution free for finite samples. These tests are only practical for small samples and we do not discuss them in this text. For a discussion of conditional permutation distribution-free tests see Maritz (1981) for applied issues and Puri and Sen (1971) for theoretical issues.

Our strategy for the construction of tests is as follows:

1. Establish, under the null hypothesis, that $n^{-1/2}\mathbf{S}$ is asymptotically $MVN(\mathbf{0}, \mathbf{V})$;
2. find a consistent estimate $\hat{\mathbf{V}}$ of $\mathbf{V}$;
3. then $S^* = n^{-1}\mathbf{S}'(\hat{\mathbf{V}})^{-1}\mathbf{S} = \mathbf{S}'(n\hat{\mathbf{V}})^{-1}\mathbf{S}$ will be asymptotically $\chi^2(p)$. The test rejects the null hypothesis at approximate level α if

$S^* \geqslant \chi_\alpha^2(p)$, the upper α critical value from a chi-square distribution with p degrees of freedom.

Estimates of location in the multivariate model are defined by the vector of univariate Hodges–Lehmann estimates corresponding to the univariate test statistics in **S**. Confidence regions with approximate confidence coefficients can also be constructed by inverting the test vectors. Another approach to confidence regions is to find the Hodges–Lehmann estimate $\hat{\boldsymbol{\theta}}$, estimate the covariance matrix **W** of the asymptotic multivariate normal distribution of $n^{1/2}(\hat{\boldsymbol{\theta}} - \boldsymbol{\theta})$, then define an approximate $(1 - \alpha)$ 100% confidence ellipsoid by

$$\left\{ \boldsymbol{\theta} : n(\hat{\boldsymbol{\theta}} - \boldsymbol{\theta})'\hat{\mathbf{W}}^{-1}(\hat{\boldsymbol{\theta}} - \boldsymbol{\theta}) \leqslant \chi_{1-\alpha}^2(p) \right\}$$

where $\chi_{1-\alpha}^2(p)$ is the upper $1 - \alpha$ percentile of the chi-square distribution with p degrees of freedom.

It should be noted that the rank procedures are not invariant under rotations of the axes. They are, however, scale and location invariant. This is in contrast to the methods based on the vector of component means which are invarient under nonsingular transformations.

6.2. ONE-SAMPLE MULTIVARIATE LOCATION MODEL

The sampling model is specified by n independent, identically distributed p-variate random vectors $\mathbf{X}_1, \ldots, \mathbf{X}_n$ each with the p-variate cdf $F(t_1 - \theta_1, \ldots, t_p - \theta_p)$. We assume the p-variate cdf F is absolutely continuous with absolutely continuous marginal cdfs $F_1(t_1 - \theta_1), \ldots, F_p(t_p - \theta_p)$ and absolutely continuous bivariate marginal cdfs $F_{ij}(t_i - \theta_i, t_j - \theta_j)$, $i, j = 1, \ldots, p$. The location parameters $\theta_1, \ldots, \theta_p$ are the marginal medians and $F_1, \ldots, F_p \in \Omega_0$. The data can be displayed in a $p \times n$ array:

$$
\begin{matrix}
X_{11} & X_{12} & \cdots & X_{1n} \\
\vdots & \vdots & & \vdots & . \\
X_{p1} & X_{p2} & \cdots & X_{pn}
\end{matrix}
$$

The ith column is the p-vector $\mathbf{X}_i$ and the jth row is a random sample of size n taken on the jth component. This model arises when p measurements are taken on each of n subjects.

We wish to test $H_0: \boldsymbol{\theta} = \mathbf{0}$ versus $H_A: \boldsymbol{\theta} \neq \mathbf{0}$ where $\boldsymbol{\theta}' = (\theta_1, \ldots, \theta_p)$ and $\mathbf{0}' = (0, \ldots, 0)$. The hypothesis $H_0: \boldsymbol{\theta} = \boldsymbol{\theta}^*$ can be transformed into $H_0: \boldsymbol{\theta} = \mathbf{0}$ by subtracting the given $\boldsymbol{\theta}^*$ from the observation vectors. Since we make no further structural assumptions on $F(t_1, \ldots, t_p)$, the vector of sign statistics is appropriate for the construction of a test. We use a form of the sign statistic that has zero expectation under the null hypothesis. Hence, for $i = 1, \ldots, p$, let

$$S_i = \sum_{t=1}^{n} \text{sgn } X_{it} = \#(X_{it} > 0) - \#(X_{it} < 0), \qquad t = 1, \ldots, n \quad (6.2.1)$$

where $\text{sgn}(x) = -1$ if $x < 0$, 0 if $x = 0$, and $+1$ if $x > 0$. We shall suppose that no observations equal 0 (true with probability one). The sign statistic, (1.2.1), is related to S_i since $\#(X_{it} > 0) + \#(X_{it} < 0) = n$, $t = 1, \ldots, n$, and $S_i = 2\#(X_{it} > 0) - n$; see Exercise 1.8.12. The symmetric form is more convenient.

Let $\mathbf{S}' = (S_1, \ldots, S_p)$, then the necessary asymptotic distribution theory is given in the following theorem. First, we need to introduce some additional notation. Let $\mathbf{X}' = (X_1, \ldots, X_p)$ be a random vector with joint cdf $F(\cdot)$. Let $p_{ij}(0,0) = P(X_i \leq 0, X_j \leq 0)$ and $p_{ij}(0,1) = P(X_i \leq 0, X_j > 0)$, $p_{ij}(1,0) = P(X_i > 0, X_j \leq 0)$ and $p_{ij}(1,1) = P(X_i > 0, X_j > 0)$.

Theorem 6.2.1. Under the null hypothesis $H_0: \boldsymbol{\theta} = \mathbf{0}$,

$$\frac{1}{\sqrt{n}}\mathbf{S} \xrightarrow{D} \mathbf{Z} \sim MVN_p(\mathbf{0}, \mathbf{V}),$$

where $\mathbf{V} = ((v_{ij}))$, $i, j = 1, \ldots, p$ and $v_{ii} = 1$

$$v_{ij} = p_{ij}(0,0) + p_{ij}(1,1) - p_{ij}(0,1) - p_{ij}(1,0).$$

Proof. First, $ES_i = 0$ and $\text{Var } S_i = n$ so that $v_{ii} = 1$. Next consider

$$\text{Cov}(S_i, S_j) = \sum_{t=1}^{n} E\left[\text{sgn } X_{it} \text{sgn } X_{jt}\right]$$

$$= n\left[p_{ij}(0,0) + p_{ij}(1,1) - p_{ij}(0,1) - p_{ij}(1,0)\right], \quad (6.2.2)$$

hence v_{ij} is as given in the statement of the theorem. Theorem A12 in the Appendix implies that the limiting distribution for $n^{-1/2}\mathbf{S}$ is $MVN(\mathbf{0}, \mathbf{V})$.

To construct the test, we need a consistent estimate of $\mathbf{V}$. From (6.2.2), the law of large numbers implies that

$$\hat{v}_{ij} = \frac{1}{n} \sum_{t=1}^{n} \text{sgn}\, X_{it} \text{sgn}\, X_{jt} \xrightarrow{P} v_{ij}. \tag{6.2.3}$$

Hence, define $\hat{\mathbf{V}}$ by $\hat{v}_{ii} = 1$ and $\hat{v}_{ij}$ given by (6.2.3). Then, under the null hypothesis, $\hat{\mathbf{V}}$ is a consistent estimate of $\mathbf{V}$ and

$$S^* = \mathbf{S}'(n\hat{\mathbf{V}})^{-1}\mathbf{S} \tag{6.2.4}$$

is asymptotically distributed as $\chi^2(p)$.

When $p = 2$, a particularly simple form for S^* is available. You are asked to show in Exercise 6.4.1 that S^* can be written as

$$S^* = \frac{(N_{00} - N_{11})^2}{N_{00} + N_{11}} + \frac{(N_{01} - N_{10})^2}{N_{01} + N_{10}} \tag{6.2.5}$$

where $N_{00} = \#(X_{1i} \leqslant 0, X_{2i} \leqslant 0)$, $N_{11} = \#(X_{1i} > 0, X_{2i} > 0)$; $N_{01} = \#(X_{1i} \leqslant 0, X_{2i} > 0)$ and $N_{10} = \#(X_{1i} > 0, X_{2i} \leqslant 0)$, $i = 1, \ldots, n$. These quantities are conveniently displayed in a two way table:

		X_{2i}
X_{1i}	$\leqslant 0$	> 0
$\leqslant 0$	N_{00}	N_{01}
> 0	N_{10}	N_{11}

The asymptotic distribution of the vector of sample medians is developed in Exercise 6.4.4.

It may be appropriate to impose a symmetry assumption on the joint cdf $F(t_1, \ldots, t_p)$. If the joint pdf $f(t_1, \ldots, t_p)$ satisfies

$$f(t_1, \ldots, t_p) = f(-t_1, \ldots, -t_p) \tag{6.2.6}$$

we say f is diagonally symmetric about $\mathbf{0}$. This results in symmetric marginal densities $f_1, \ldots, f_p$ and the location vector $\boldsymbol{\theta}' = (\theta_1, \ldots, \theta_p)$ represents the centers of symmetry of the marginal distributions.

Under this assumption, we will consider the vector of Wilcoxon signed rank statistics computed on the p components of the observation vectors. For $i = 1, \ldots, p$, let

$$T_i = \sum_{t=1}^{n} \frac{R_{it}}{n+1} \operatorname{sgn}(X_{it}) \tag{6.2.7}$$

where, as in (2.2.2), R_{it} is the rank of $|X_{it}|$ among $|X_{i1}|, \ldots, |X_{in}|$, and let $\mathbf{T}' = (T_1, \ldots, T_p)$. The following theorem provides the limiting distribution needed to develop a test of $H_0 : \boldsymbol{\theta} = \mathbf{0}$ versus $H_A : \boldsymbol{\theta} \neq \mathbf{0}$.

Theorem 6.2.2. Suppose $F(t_1, \ldots, t_p)$ has a density $f(t_1, \ldots, t_p)$, diagonally symmetric about $\mathbf{0}$. Then

$$\frac{1}{\sqrt{n}} \mathbf{T} \xrightarrow{D} \mathbf{Z} \sim MVN_p(\mathbf{0}, \mathbf{V})$$

where

$$v_{ii} = \tfrac{1}{3}$$

$$v_{ij} = 4 \int_{-\infty}^{\infty} F_i(u) F_j(v) \, dF_{ij}(u, v) - 1.$$

Proof. Write T_i as

$$T_i = \frac{2}{n+1} \left\{ \sum_{t=1}^{n} R_{it} s(X_{it}) - \frac{n(n+1)}{4} \right\} \tag{6.2.8}$$

where $T_i^+ = \Sigma R_{it} s(X_{it})$ is the form studied in Example 2.5.5. Using the projection result in Example 2.5.5, define

$$U_i = \frac{2}{\sqrt{n}} \sum_{t=1}^{n} \left[F_i(X_{it}) - \frac{1}{2} \right].$$

We have used the fact that $F_i \in \Omega_s$ so that $1 - F_i(-x) = F_i(x)$. Let $\mathbf{U}' = (U_1, \ldots, U_p)$ then $n^{-1/2}\mathbf{T} - \mathbf{U}$ converges in probability to $\mathbf{O}$. Hence the limiting distribution of $n^{-1/2}\mathbf{T}$ is the same as that of $\mathbf{U}$.

First note that $\operatorname{Var} U_i = 4(1/12) = 1/3$, since $F_i(X_{it})$ is uniformly distributed on $(0, 1)$. Next

$$\operatorname{Cov}(U_i, U_j) = 4E\left[F_i(X_{i1})F_j(X_{j1})\right] - 1$$

$$= 4\int_{-\infty}^{\infty}\int_{-\infty}^{\infty} F_i(u)F_j(v)\,dF_{ij}(u, v) - 1.$$

Hence the formulas for v_{ii} and v_{ij} are established. The limiting multivariate normality of $\mathbf{U}$ follows from Theorem A12 in the Appendix.

To construct the test we need a consistent estimate of $\mathbf{V}$ under the null hypothesis. Note that from (6.2.8) and (2.2.4)

$$\operatorname{Var} T_i = \frac{4}{(n+1)^2}\,\frac{n(n+1)(2n+1)}{24}$$

$$= \frac{n(2n+1)}{6(n+1)} \sim \frac{n}{3}.$$

Hence we take

$$\hat{v}_{ii} = \frac{1}{n}\operatorname{Var} T_i = \frac{2n+1}{6(n+1)}. \tag{6.2.9}$$

We next give a heuristic argument to show that $\hat{v}_{ij}$, defined by

$$\hat{v}_{ij} = \frac{1}{n}\sum_{t=1}^{n} \frac{R_{it}}{n+1}\,\frac{R_{jt}}{n+1}\,\operatorname{sgn} X_{it}\operatorname{sgn} X_{jt}, \tag{6.2.10}$$

is a consistent estimate of v_{ij}. Let $F_{ni}^{+}(x)$ be the empirical cdf computed on the absolute values of the ith component. Hence we can write (6.2.10) as

$$\hat{v}_{ij} = \frac{n^2}{n(n+1)^2}\sum_{t=1}^{n} F_{ni}^{+}(|X_{it}|)F_{nj}^{+}(|X_{jt}|)\operatorname{sgn}(X_{it})\operatorname{sgn}(X_{jt})$$

$$= \frac{n^2}{(n+1)^2}\int_{-\infty}^{\infty}\int_{-\infty}^{\infty} F_{ni}^{+}(|s|)F_{nj}^{+}(|t|)\operatorname{sgn}(s)\operatorname{sgn}(t)\,dF_{nij}(s, t)$$

where $F_{nij}(s, t)$ is the bivariate empirical cdf that assigns mass n^{-1} to each pair (x_{it}, x_{jt}), $t = 1, \ldots, n$. From the convergence in probability of the empirical cdfs, it can be shown that

$$\hat{v}_{ij} \xrightarrow{P} \int_{-\infty}^{\infty} \int_{-\infty}^{\infty} F_i^+(|s|) F_j^+(|t|) \operatorname{sgn}(s) \operatorname{sgn}(t) \, dF_{ij}(s,t).$$

The right side can be written

$$\int_0^\infty \int_0^\infty F_i^+(s) F_j^+(t) \, dF_{ij}(s,t)$$

$$- \int_{-\infty}^0 \int_0^\infty F_i^+(s) F_j^+(-t) \, dF_{ij}(s,t)$$

$$- \int_0^\infty \int_{-\infty}^0 F_i^+(-s) F_j^+(t) \, dF_{ij}(s,t)$$

$$+ \int_{-\infty}^0 \int_{-\infty}^0 F_i^+(-s) F_j^+(-t) \, dF_{ij}(s,t). \qquad (6.2.11)$$

Now recall $F_i^+(x) = P(|X_i| \leqslant x) = F_i(x) - F_i(-x) = 2F_i(x) - 1$ when $x > 0$, and $F_i^+(x) = 0$ when $x \leqslant 0$. Also $F^+(-x) = 1 - 2F(x)$ when $x < 0$ and $F^+(-x) = 0$ when $x \geqslant 0$. Substituting into (6.2.11), multiplying the factors in each integrand, and recombining the integrals yields

$$\int_{-\infty}^{\infty} \int_{-\infty}^{\infty} \left[4F_i(s) F_j(t) - 2F_i(s) - 2F_j(t) + 1 \right] dF_{ij}(s,t). \qquad (6.2.12)$$

Using $\int\int F_i(s) \, dF_{ij}(s,t) = \int F_i(s) \int dF_{ij}(s,t) = \int F_i(s) \, dF_i(s) = 1/2$, (6.2.12) reduces to v_{ij} given in Theorem 6.2.2.

Hence defining $\hat{\mathbf{V}} = ((\hat{v}_{ij}))$, $i, j = 1, \ldots, p$, with $\hat{v}_{ii}$ in (6.2.9) and $\hat{v}_{ij}$ in (6.2.10), we have, under the null hypothesis, that

$$T^* = \mathbf{T}'(n\hat{\mathbf{V}})^{-1}\mathbf{T} \qquad (6.2.13)$$

is asymptotically distributed as $\chi^2(p)$.

Both S^* and T^* are easy to compute and T^* is illustrated in the following example. Exercise 6.4.2 shows that the multivariate test may provide a distinct advantage over the naive strategy of carrying out univariate tests separately on each component.

If we assume that $f(t_1 - \theta_1, \ldots, t_p - \theta_p)$ is the multivariate normal density, centered at $\boldsymbol{\theta}' = (\theta_1, \ldots, \theta_p)$, with covariance matrix $\boldsymbol{\Sigma}$, then

Hotelling's statistic is the basis for testing $H_0: \boldsymbol{\theta} = \mathbf{0}$ versus $H_A: \boldsymbol{\theta} \neq \mathbf{0}$. The statistic is given by

$$H^2 = n\overline{\mathbf{X}}'\hat{\boldsymbol{\Sigma}}^{-1}\overline{\mathbf{X}} \qquad (6.2.14)$$

where $\overline{\mathbf{X}}' = (\overline{X}_1, \ldots, \overline{X}_p)$, $\hat{\boldsymbol{\Sigma}} = ((\hat{\sigma}_{ij}))$, $i, j = 1, \ldots, p$, with $\hat{\sigma}_{ij} = (n - 1)^{-1}\sum_{t=1}^{n}(X_{it} - \overline{X}_i)(X_{jt} - \overline{X}_j)$. Hotelling's test rejects $H_0: \boldsymbol{\theta} = \mathbf{0}$ at level α if

$$H^2 \geqslant \left[(n - 1)p/(n - p)\right]F_\alpha(p, n - p)$$

where $F_\alpha(p, n - p)$ is the F critical point with p and $n - p$ degrees of freedom.

Example 6.2.1. Ryan et al. (1976, p. 276) provide a table of measurements on male Peruvian Indians over 21 years old who were born at a high altitude and whose parents were also born at a high altitude. Table 6.1 gives systolic and diastolic blood pressures for 15 individuals. We will suppose that the bivariate distribution is diagonally symmetric about $\boldsymbol{\theta}' = (\theta_1, \theta_2)$, θ_1 is the center of the marginal distribution of systolic blood pressures and likewise for θ_2. We wish to test $H_0: \boldsymbol{\theta}' = (120, 80)$ versus $H_A: \boldsymbol{\theta}' \neq (120, 80)$, the standard values for healthy males over 21 in the United States. Hence we suspect that the Indian blood pressure differs from its American counterpart.

Table 6.1. Blood Pressure

X_1	$X_1 - 120$	$R(\|X_1 - 120\|) \times$ sgn$(X_1 - 120)$	X_2	$X_2 - 80$	$R(\|X_2 - 80\|) \times$ sgn$(X_2 - 80)$
170	50	15	76	− 4	− 3.5
125	5	5	75	− 5	− 5
148	28	14	120	40	15
140	20	13	78	− 2	− 1.5
106	− 14	− 10	72	− 8	− 8
108	− 12	− 8	62	− 12	− 10.5
124	4	3	70	− 10	− 9
134	14	10	64	− 16	13.5
116	− 4	− 3	76	− 4	− 3.5
114	− 6	− 6.5	74	− 6	− 6.5
118	− 2	− 1	68	− 12	− 10.5
138	18	12	78	− 2	− 1.5
134	14	10	86	6	6.5
124	4	3	64	− 16	− 13.5
114	− 6	− 6.5	66	− 14	− 12

Let $\mathbf{X}' = (X_1, X_2)$ where $X_1 =$ systolic and $X_2 =$ diastolic. From (6.2.9), $\hat{v}_{ii} = .323$ and using the third and sixth columns in Table 6.1 with (6.2.10), $\hat{v}_{ij} = .0684$. Now

$$(n\hat{\mathbf{V}})^{-1} = \begin{pmatrix} .216 & -.046 \\ -.046 & .216 \end{pmatrix}$$

and from (6.2.13), $T^* = 8.49$. When this is compared to $\chi^2_{.025}(2) = 7.38$, we are able to reject $H_0: \boldsymbol{\theta}' = (120, 80)$ at approximate level $\alpha = .025$. We conclude that, at this significance level, there is evidence that the Peruvian blood pressure differs from the American blood pressure. The point esti- mate of $\boldsymbol{\theta}' = (\theta_1, \theta_2)$ is the pair of medians of the Walsh averages. We find $\hat{\boldsymbol{\theta}}' = (126, 73)$.

Further, from Table 6.1, we have $\overline{X}_1 = 127.533$, $\overline{X}_2 = 75.267$, and

$$\hat{\Sigma} = \begin{pmatrix} 288.410 & 114.133 \\ 114.133 & 195.781 \end{pmatrix}.$$

If we assume the underlying bivariate distribution is bivariate normal, then Hotelling's test is appropriate. We find, for $\boldsymbol{\theta}_0' = (120, 80)$,

$$H^2 = n(\overline{\mathbf{X}} - \boldsymbol{\theta}_0)'\hat{\Sigma}^{-1}(\overline{\mathbf{X}} - \boldsymbol{\theta}_0) = 8.88.$$

The .025 critical point for this test is $[(n-1)p/(n-p)]F_\alpha(p, n-p) = (2.15)(4.965) = 10.694$. Hotelling's test fails to reject the null hypothesis at $\alpha = .025$. It should be noted that the significant T^* was refered to an approximate critical value based on the asymptotic distribution and not on the exact distribution. Further study of the approximating distribution of T^* for small to moderate samples is needed.

Since the covariances among the components of $\mathbf{S}$ and $\mathbf{T}$ depend on the underlying univariate and bivariate distributions, the multivariate exten- sions of the sign and Wilcoxon signed rank statistics are not distribution free under the null hypothesis. The tests just proposed use estimates of the needed marginal distributions, namely the empirical cdfs. We then have asymptotically distribution-free tests. This approach is quite general and can be applied to other score functions. Multivariate tests based on Winsorized Wilcoxon signed rank statistics are described and illustrated on a data set by Utts and Hettmansperger (1980). General scores tests are developed by Puri and Sen (1971, Chapter 4).

Asymptotic efficiency can be developed in a manner similar to the univariate case. It is necessary to establish the limiting distribution of the vector $n^{-1/2}\mathbf{S}$ along a sequence of alternatives $\boldsymbol{\theta}_n = n^{-1/2}\boldsymbol{\theta}$ converging to $\mathbf{0}$.

The result of the calculation is that the limiting distribution is altered only in the mean of the asymptotic distribution. Hence, referring to the notation in the introduction, $n^{-1/2}\mathbf{S}$ has an asymptotic $MVN(\boldsymbol{\mu}, \mathbf{V})$ distribution along $\boldsymbol{\theta}_n$. For example, given the vector of sign statistics $\mathbf{S}$, $E_{\boldsymbol{\theta}_n}(n^{-1/2}\mathbf{S})$ converges to $\boldsymbol{\mu} = 2(\theta_1 f_1(0), \ldots, \theta_p f_p(0))'$. The limiting distribution of the test statistic S^*, along $\boldsymbol{\theta}_n$, will be noncentral chi-square with p degrees of freedom, and noncentrality parameter $\boldsymbol{\mu}'\mathbf{V}^{-1}\boldsymbol{\mu}$. Since the asymptotic local power is an increasing function of the noncentrality parameter, the efficiency of two such test statistics is taken to be the ratio of their noncentrality parameters. Bickel (1965) provides an extensive analysis of the efficiency properties of the multivariate sign, Wilcoxon, and Hotelling tests. He concludes that the former tests are better than Hotellings test in the presence of gross errors, but they should be used with caution in situations where considerable degeneracy is present.

The vector of Hodges–Lehmann estimates derived from the vector of rank tests generally has a distribution that is asymptotically multivariate normal. Estimation efficiency is defined in terms of the generalized variance which is the determinant of the asymptotic covariance matrix. Hence, unlike the univariate case, testing and estimation efficiency have different formulas. The efficiency results are quite similar, however. See Bickel (1964) for a discussion of the estimation case. Exercises 6.4.4 and 6.4.5 develop the asymptotic distributions and provide some efficiency results for the estimators.

6.3. TWO-SAMPLE MULTIVARIATE LOCATION MODEL

We will discuss the comparison of two multivariate samples, size m and n, respectively. Let $\mathbf{X}_1, \ldots, \mathbf{X}_m$ and $\mathbf{Y}_1, \ldots, \mathbf{Y}_n$ be samples from p-variate distributions with cdfs $F(u_1, \ldots, u_p)$ and $F(v_1 - \Delta_1, \ldots, v_p - \Delta_p)$, respectively. The data can be displayed as

$$
\begin{array}{cccccc}
X_{11} & \cdots & X_{1m} & Y_{11} & \cdots & Y_{1n} \\
\vdots & & \vdots & \vdots & & \vdots \\
X_{p1} & \cdots & X_{pm} & Y_{p1} & \cdots & Y_{pn}
\end{array}
$$

We will suppose that F is absolutely continuous with absolutely continuous marginal distributions. No symmetry assumption is needed. The parameter $\boldsymbol{\Delta}' = (\Delta_1, \ldots, \Delta_p)$ represents the amount of shift in each component in passing from the $\mathbf{X}$ distribution to the $\mathbf{Y}$ distribution. We wish to construct a test of $H_0 : \boldsymbol{\Delta} = \mathbf{0}$ versus $H_A : \boldsymbol{\Delta} \neq \mathbf{0}$.

The test that we will consider is a multivariate version of the Mann–Whitney–Wilcoxon test developed in Section 3.2. Let $\mathbf{U}' = (U_1, \ldots, U_p)$, where

$$U_i = \sum_{t=1}^{n} \left[\frac{R_{it}}{N+1} - \frac{1}{2} \right], \tag{6.3.1}$$

$N = m + n$. Hence U_i is the centered rank sum statistic where R_{it} is the rank of Y_{it} in the combined samples of the ith component. Under $H_0: \boldsymbol{\Delta} = \mathbf{0}$, $E\mathbf{U} = \mathbf{0}$. The following theorem provides the limiting distribution of $N^{-1/2}\mathbf{U}$ under the null hypothesis.

Theorem 6.3.1.   Suppose the null hypothesis $H_0: \boldsymbol{\Delta} = \mathbf{0}$ is true. Suppose $m, n \to \infty$ and $m/N \to \lambda$, $0 < \lambda < 1$. Then

$$\frac{1}{\sqrt{N}} \mathbf{U} \overset{D}{\to} \mathbf{Z} \sim MVN(\mathbf{0}, \mathbf{V})$$

where $\mathbf{V} = ((v_{ij}))$, $i, j = 1, \ldots, p$, and

$$v_{ii} = \lambda(1 - \lambda)/12$$

$$v_{ij} = \lambda(1 - \lambda) \left\{ \int_{-\infty}^{\infty} \int_{-\infty}^{\infty} F_i(x) F_j(y) \, dF_{ij}(x, y) - 1/4 \right\}, \qquad i \neq j.$$

Proof.   The argument is sketched because it is similar to that of Theorem 6.2.2. We first note that

$$\frac{1}{\sqrt{N}} U_i = \frac{1}{\sqrt{N}} \sum_{t=1}^{n} \left[\frac{R_{it}}{N+1} - \frac{1}{2} \right] = \frac{1}{\sqrt{N}(N+1)} W^*$$

where W^* is defined in the proof of Theorem 3.2.4. Hence the projection is given by

$$V_p \left(\frac{1}{\sqrt{N}} U_i \right) = \frac{1}{\sqrt{N}(N+1)} \sum_{t=1}^{N} b_t \left[F_i(Z_{it}) - \tfrac{1}{2} \right]$$

where Z_{it}, $t = 1, \ldots, N$, has cdf $F_i(u)$ and

$$b_i = \begin{cases} -n & t = 1, \ldots, m \\ m & t = m + 1, \ldots, N. \end{cases}$$

Since $F_i(Z_{it}) - \frac{1}{2}$ is uniformly distributed on $(-1/2, 1/2)$ we have

$$\text{Var}\left\{ V_p\left(\frac{1}{\sqrt{N}} U_i \right) \right\} = \frac{1}{12N(N+1)^2} \sum_{t=1}^{N} b_t^2 = \frac{mn^2 + nm^2}{12N(N+1)^2} \rightarrow \frac{\lambda(1-\lambda)}{12}.$$

Further,

$$\text{Cov}\left\{ V_p\left[\frac{1}{\sqrt{N}} U_i \right], V_p\left[\frac{1}{\sqrt{N}} U_j \right] \right\}$$

$$= \frac{1}{N(N+1)^2} \left\{ \int_{-\infty}^{\infty}\int_{-\infty}^{\infty} F_i(x)F_j(y)\,dF_{ij}(x, y) - 1/4 \right\} \sum_{t=1}^{N} b_t^2$$

$$\rightarrow \lambda(1-\lambda)\left\{ \int_{-\infty}^{\infty}\int_{-\infty}^{\infty} F_i(x)F_j(y)\,dF_{ij}(x, y) - 1/4 \right\}. \tag{6.3.2}$$

The limiting multivariate normality of the projection follows from Theorem A12 or Theorem A13 with $a_t = b_t/(N+1)$.

By the Projection Theorem 2.5.2, $N^{-1/2}\mathbf{U}$, has the same multivariate normal limiting distribution and the proof is complete.

In order to construct a test of $H_0 : \boldsymbol{\Delta} = \mathbf{0}$ versus $H_A : \boldsymbol{\Delta} \neq \mathbf{0}$ we first need a consistent estimate of $\mathbf{V}$ the asymptotic covariance matrix. For v_{ii} we will take

$$\hat{v}_{ii} = \text{Var}\left(\frac{1}{\sqrt{N}} U_i \right) = \frac{mn}{12N(N+1)}. \tag{6.3.3}$$

If we substitute the single and bivariate marginal empirical distribution functions into (6.3.2), we will have a consistent estimate of v_{ij}. The result, slightly modified, is

$$\hat{v}_{ij} = \frac{mn}{N^2(N-1)} \left\{ \sum_{t=1}^{N} \frac{R_{it}}{N+1} \frac{R_{jt}}{N+1} - \frac{1}{4} \right\}$$

$$= \frac{mn}{N^2(N-1)(N+1)^2} \left\{ \sum_{t=1}^{N} R_{it}R_{jt} - \frac{N(N+1)^2}{4} \right\}. \tag{6.3.4}$$

Straight substitution in (6.3.2) results in N^2 rather than $N(N-1)$. Equation 6.3.4 is the appropriate covariance equation for the conditional joint distribution of U_i and U_j, given the combined sample observations. Equa-

Table 6.2. Levels of Biochemical Compound

	Control Group									
1:	1.21	0.92	0.80	0.85	0.98	1.15	1.10	1.02	1.18	1.09
2:	0.61	0.43	0.35	0.48	0.42	0.52	0.50	0.53	0.45	0.40

	Treatment Group										
1:	1.40	1.17	1.23	1.19	1.38	1.17	1.31	1.30	1.22	1.00	1.12	1.09
2:	0.50	0.39	0.44	0.37	0.42	0.45	0.41	0.47	0.29	0.30	0.27	0.35

tion (6.3.4) is the appropriate equation for the finite sample permutation test. See Maritz (1981, Chapter 7).

Now, using (6.3.3) and (6.3.4), the test statistic is

$$U^* = N^{-1}\mathbf{U}'(\hat{\mathbf{V}})^{-1}\mathbf{U} = \mathbf{U}'(N\hat{\mathbf{V}})^{-1}\mathbf{U}. \tag{6.3.5}$$

Under $H_0 : \mathbf{\Delta} = \mathbf{0}$, U^* is asymptotically chi-square with p degrees of freedom. Hence we reject $H_0 : \mathbf{\Delta} = \mathbf{0}$, at approximate level α, if $U^* \geq \chi_\alpha^2(p)$.

Example 6.3.1. This example is taken from Morrison (1976, p. 167). The data consists in the levels of two biochemical components found in the brains of mice. Twenty-four mice of the same strain were divided randomly into a treatment group (which received a drug) and a control group. It is hypothesized that the biochemical levels will be different for the two groups. Let $\mathbf{\Delta}' = (\Delta_1, \Delta_2)$ be the difference in medians in the two bivariate populations. Two mice in the control group died. Hence we have a control sample $\mathbf{X}_1, \ldots, \mathbf{X}_{10}$ from $F(t_1, t_2)$ and a treatment sample $\mathbf{Y}_1, \ldots, \mathbf{Y}_{12}$ from $F(t_1 - \Delta_1, t_2 - \Delta_2)$, and we wish to test $H_0 : \mathbf{\Delta} = \mathbf{0}$ versus $H_A : \mathbf{\Delta} \neq \mathbf{0}$. Table 6.2 provides the amounts of the components in micrograms per gram of brain tissue.

The data analysis proceeds by combining the first component of the control and treatment groups and recording the ranks. The same is done for the second component. Ties are assigned the average rank. Ranks are shown in Table 6.3.

Table 6.3. Ranks of the Data

	Control Group									
1:	16	3	1	2	4	11	9	6	14	7.5
2:	22	12	4.5	17	10.5	20	18.5	21	14.5	8

	Treatment Group										
1:	22	12.5	18	15	21	12.5	20	19	17	5	10	7.5
2:	18.5	7	13	6	10.5	14.5	9	16	2	3	1	4.5

We find from (6.3.1) that $U_1 = 1.80$ and $U_2 = -1.43$. Further $\hat{v}_{11} = \hat{v}_{22}$ $= mn/[12N(N + 1)] = .01976$ and $\hat{v}_{12} = .0029$. Hence, with $N = 22$,

$$(N\hat{\mathbf{V}})^{-1} = \begin{pmatrix} 2.3523 & -3.496 \\ -3.496 & 2.3523 \end{pmatrix}$$

and $U^* = \mathbf{U}'(N\hat{\mathbf{V}})^{-1}\mathbf{U} = 14.22$. A test at approximate level .05 compares U^* to $\chi^2_{.05}(2) = 5.99$; hence U^* easily rejects $H_0: \Delta = \mathbf{0}$. The Hodges–Lehmann estimate of Δ' is $(.19, -.08)$; see (3.2.5).

Maritz (1981, p. 243) carries out a test using normal scores. He finds a value of 15.00 which is also compared to $\chi^2_{.05}(2) = 5.99$. Hence the conclusion is the same. For an analysis based on the multivariate extension of Mood's test see Exercise 6.4.8. The normal theory test would be based on the two sample Hotelling statistic; see Morrison (1967, Chapter 4).

See Puri and Sen (1971, Chapter 5) for a developement of the theory for general score functions. They also extend the results from two-sample to multisample multivariate location models. Sen and Puri (1977) develop aligned rank tests, with general score functions, in the multivariate linear model.

6.4. EXERCISES

6.4.1. Verify that (6.2.5) can be derived from (6.2.4).

6.4.2. Suppose that a large high school is concerned that emphasis is placed on the science program at the expense of the language program. To test this hypothesis, 100 juniors were given the SAT (Scholastic Aptitude Test). We have a pair of measurements (V, Q), the verbal and quantitative parts of the SAT test. The following table provides the numbers below and above the national average on the two components.

	Q	
V	Below	Above
Below	34	22
Above	8	36

Let S_V = number of observations above the national verbal average (θ_V^o) and let S_Q = number of observations above the national quantitative average (θ_Q^o). Use S^* to test $H_0 : \boldsymbol{\theta}' = (\theta_V^o, \theta_Q^o)$ versus $H_A : \boldsymbol{\theta}'$ $\neq (\theta_V^o, \theta_Q^o)$ at approximate significance level $\alpha = .05$. Then test each component separately at $\alpha = .05$ using the sign test in Section 1.2.

6.4.3. Show that, in (6.2.2), $p_{ij}(0,0) + p_{ij}(1,1) - p_{ij}(0,1) - p_{ij}(1,0) = 4p_{ij}(0, 0) - 1$. Hint: Use the fact that the marginal medians are 0.

6.4.4. Let $\hat{\boldsymbol{\theta}}$ be p-component vector of sample medians. Then $\mathbf{S}(\hat{\boldsymbol{\theta}}) \doteq \mathbf{0}$ where $S_i(\boldsymbol{\theta}) = \sum_{t=1}^n \operatorname{sgn}(X_{it} - \theta_i)$; see (6.2.1).

a. Using Exercise 2.10.18, show that

$$\sqrt{n}\,\hat{\boldsymbol{\theta}} = \operatorname{diag}\!\left(\frac{1}{2f_1(0)}, \ldots, \frac{1}{2f_p(0)} \right) \frac{1}{\sqrt{n}}\, \mathbf{S}(\mathbf{0}) + \mathbf{o}_p(1)$$

where we let the true value of $\boldsymbol{\theta}$ be $\mathbf{0}$ without loss of generality, $\operatorname{diag}(a_1, a_2, \ldots, a_p)$ denotes a $p \times p$ diagonal matrix with ith diagonal element a_i, and $\mathbf{o}_p(1)$ is a $p \times 1$ vector that converges to $\mathbf{0}$ in probability.

b. Let $\mathbf{D}$ denote the diagonal matrix in (a) and argue that

$$\sqrt{n}\,\hat{\boldsymbol{\theta}} \xrightarrow{D} \mathbf{Z} \sim MVN(\mathbf{0}, \mathbf{DVD}')$$

with $\mathbf{V}$ given by Theorem 6.2.1.

c. For $p = 2$ show

$$\mathbf{DVD}' = \frac{1}{4} \begin{vmatrix} \dfrac{1}{f_1^2(0)} & \dfrac{\gamma}{f_1(0)f_2(0)} \\[2ex] \dfrac{\gamma}{f_1(0)f_2(0)} & \dfrac{1}{f_2^2(0)} \end{vmatrix}$$

where $\gamma = 4P(X_1 < 0, X_2 < 0) - 1$.

d. Suppose the bivariate central limit theorem applies and

$$\sqrt{n}\,\overline{\mathbf{X}} \xrightarrow{D} \mathbf{Z} \sim MVN(\mathbf{0}, \boldsymbol{\Sigma})$$

with variances σ_1^2, σ_2^2 and correlation ρ. Then the efficiency of $\hat{\boldsymbol{\theta}}$ relative to $\overline{\mathbf{X}}$ is the $(1/p)$th root of the ratio of the asymptotic generalized variances. Recall the generalized variance is the determinant of the covariance matrix. Hence show that

$$e(\hat{\boldsymbol{\theta}}, \overline{\mathbf{X}}) = 4f_1(0)\,f_2(0)\sigma_1\sigma_2 \left[\frac{1 - \rho^2}{1 - \gamma^2} \right]^{1/2}.$$

e. Suppose the underlying distribution is bivariate normal. Then show that the efficiency in (d) reduces to

$$e(\hat{\boldsymbol{\theta}}, \overline{\mathbf{X}}) = \frac{2}{\pi} \left\{ \frac{1 - \rho^2}{1 - (4/\pi^2)[\sin^{-1}(\rho)]^2} \right\}^{1/2}.$$

Plot $e(\hat{\boldsymbol{\theta}}, \overline{\mathbf{X}})$ as a function of ρ and discuss the efficiency. Hint: Use Exercise 1.8.14.

6.4.5. Let $\hat{\boldsymbol{\theta}}$ be the p component vector in which $\hat{\theta}_i$ is the median of the Walsh averages of the n observations in the ith component. Then $T(\hat{\boldsymbol{\theta}}) \doteq \mathbf{0}$ where

$$T_i(\boldsymbol{\theta}) = \sum_{t=1}^{n} \frac{R_{it}(\boldsymbol{\theta})}{n+1} \operatorname{sgn}(X_{it} - \theta_i)$$

and $R_{it}(\boldsymbol{\theta})$ is the rank of $|X_{it} - \theta_i|$ among $|X_{i1} - \theta_i|, \ldots, |X_{in} - \theta_i|$.

a. Using Theorem 2.7.2 and assuming the true value of $\boldsymbol{\theta}$ is $\mathbf{0}$, without loss of generality, show that

$$\frac{1}{2\sqrt{n}} T_i\left(\frac{b}{\sqrt{n}}\right) = \frac{1}{2\sqrt{n}} T_i(0) - b \int f_i^2(x) \, dx + o_p(1).$$

Hence, show

$$n^{1/2}\hat{\boldsymbol{\theta}} = \operatorname{diag}\left((2f_1^*)^{-1}, \ldots, (2f_p^*)^{-1}\right) \frac{1}{\sqrt{n}} \mathbf{T}(0) + \mathbf{o}_p(1),$$

where $f_i^* = \int_{-\infty}^{\infty} f_i^2(x) \, dx$ and $\operatorname{diag}(a_1, \ldots, a_p)$ is a $p \times p$ diagonal matrix with ith diagonal element a_i.

b. Let $\mathbf{D}$ be the diagonal matrix in (a). Show that

$$\sqrt{n}\,\hat{\boldsymbol{\theta}} \xrightarrow{D} \mathbf{Z} \sim MVN(\mathbf{0}, \mathbf{DVD'})$$

where $\mathbf{V}$ is given in Theorem 6.2.2.

c. For $p = 2$ show if $\mathbf{DVD'} = ((d_{ij}))$, $i, j = 1, 2$, then

$$d_{ii} = \frac{1}{12\left[\int_{-\infty}^{\infty} f_i^2(x) \, dx\right]^2} \qquad i = 1, 2$$

$$d_{ij} = \frac{4\int_{-\infty}^{\infty}\int_{-\infty}^{\infty} F_1(x) F_2(y) \, dF_{12}(x, y) - 1}{4\int_{-\infty}^{\infty} f_1^2(x) \, dx \int_{-\infty}^{\infty} f_2^2(x) \, dx} \qquad i \neq j.$$

d. Show that the efficiency of $\hat{\boldsymbol{\theta}}$ relative to $\overline{\mathbf{X}}$ for $p = 2$ is

$$e(\hat{\boldsymbol{\theta}}, \overline{\mathbf{X}}) = 12\sigma_1\sigma_2 \int_{-\infty}^{\infty} f_1^2(x)\,dx \int_{-\infty}^{\infty} f_2^2(x)\,dx \left\{ \frac{1 - \rho^2}{1 - 9\delta^2} \right\}^{1/2}$$

where $\delta = 4\int_{-\infty}^{\infty}\int_{-\infty}^{\infty} F_1(x)F_2(y)\,dF_{12}(x, y) - 1$.

e. Show that, if F_{12} is the bivariate normal distribution,

$$e(\hat{\boldsymbol{\theta}}, \overline{\mathbf{X}}) = \frac{3}{\pi} \left\{ \frac{1 - \rho^2}{1 - (36/\pi^2)\left[\sin^{-1}(\rho/2)\right]^2} \right\}^{1/2}.$$

Plot $e(\hat{\boldsymbol{\theta}}, \overline{\mathbf{X}})$ as a function of ρ and discuss the relative behavior of $\hat{\boldsymbol{\theta}}$ and $\overline{\mathbf{X}}$. Hint: Use (2.7.9).

6.4.6. For the data in Example 6.2.1, use sign scores to find the point estimate of $\hat{\boldsymbol{\theta}}$ and test the hypothesis $H_0: \boldsymbol{\theta}' = (120, 80)$ versus $H_A: \boldsymbol{\theta}' \neq (120, 80)$ where $\boldsymbol{\theta}$ is the vector of marginal medians.

6.4.7. The exercise extends Mood's test, Example 3.4.2, to the multivariate two-sample location model described at the beginning of Section 6.3. We consider testing $H_0: \boldsymbol{\Delta} = \mathbf{0}$ versus $H_A: \boldsymbol{\Delta} \neq \mathbf{0}$. Let, for $i = 1, \ldots, p$, $M_i = \#(Y_{it} > \hat{C}_i) - n/2$, $t = 1, \ldots, n$, where $\hat{C}_i$ is the median of the combined data in the ith coordinate. Hence M_i is the centered Mood's statistic. Let $\mathbf{M}' = (M_1, \ldots, M_p)$ and show that, if $m, n \to \infty$ so that $m/N \to \lambda$, $0 < \lambda < 1$, $N = m + n$, then

$$\frac{1}{\sqrt{N}} \mathbf{M}' \xrightarrow{D} \mathbf{Z} \sim MVN(\mathbf{0}, \mathbf{V})$$

where $\mathbf{V} = ((v_{ij}))$, $i, j = 1, \ldots, p$, and

$$v_{ii} = \lambda(1 - \lambda)/4$$

$$v_{ij} = \lambda(1 - \lambda)\left\{ P(Y_i > 0, Y_j > 0) - \tfrac{1}{4} \right\}.$$

Further, argue that $\hat{v}_{ii} = mn/[4n(N - 1)]$ and

$$\hat{v}_{ij} = \frac{mn}{N(N - 1)} \left\{ \frac{N_{11}}{N - 1} - \frac{1}{4} \right\}$$

(where N_{11} is the number of pairs, in the ith and jth components, in

the combined sample that are both positive), are consistent esti-
mates of v_{ii} and v_{ij}, respectively. Finally, argue that $M^* = M'(N\hat{V})^{-1}M$ is asymptotically chi-square with p degrees of freedom.

6.4.8. Based on the results of Exercise 6.4.7 carry out the test of hypothesis in Example 6.3.1. Find the estimate of Δ based on M.

6.4.9. Let $\hat{\Delta}$ be the Hodges–Lehmann estimate of Δ based on U, (6.3.1).
Suppose $m, n \to \infty$ in such a way that $m/N \to \lambda$, $0 < \lambda < 1$, $N = m + n$. Without loss of generality let $\Delta = 0$ and show that

$$\sqrt{N}\,\hat{\Delta} \xrightarrow{D} Z \sim MVN\left(0, [\lambda(1-\lambda)]^{-1}DVD'\right)$$

where DVD' is described in Exercise 6.4.5(b).

APPENDIX

Some Results from Mathematics and Mathematical Statistics

In this appendix we list many of the results from asymptotic or limiting distribution theory that are used throughout the book. We either provide an argument for the results or give a reference where a discussion can be found. The asymptotic theory is important for two reasons. First, the distribution of test statistics and estimates can be approximated for practical purposes and second, the asymptotic distributions of test statistics and estimates provide the basis for large sample power and efficiency comparisons. Other mathematical results, such as a short discussion of Stieltjes integrals, are included.

We first define the two main types of limiting behavior for sequences of random variables. We will denote a sequence of random variables Z_1, $Z_2, \ldots$ by $\{Z_n\}$.

Definition A1. A sequence of random variables $\{Z_n\}$ converges in probability to a constant c if for every $\epsilon > 0$, $\lim P(|Z_n - c| \geqslant \epsilon) = 0$. This is denoted by $Z_n \overset{P}{\to} c$ and we say Z_n converges in probability to c. Note that c may be replaced by a random variable Z in the definition. In this case we say Z_n converges in probability to Z.

Definition A2. A sequence of random variables $\{Z_n\}$ converges in distribution to the random variable Z if $\lim F_n(t) = F(t)$, at continuity points of F, where F_n and F are the cumulative distribution functions (cdfs) of Z_n and Z respectively. We write $Z_n \overset{D}{\to} Z$.

A random variable Z is degenerate at the constant c if $P(Z = c) = 1$. Definition A2 contains Definition A1; since $Z_n \overset{D}{\to} Z$, Z degenerate at c if

297

and only if $Z_n \overset{P}{\to} c$. This is discussed by Hogg and Craig (1978, p. 186) in their treatment of stochastic convergence. However, since convergence in probability and convergence in distribution are very different types of behavior, it is convenient to keep the definitions separate.

The standard normal distribution has mean 0 and variance 1, denoted by $n(0, 1)$. The cdf will be denoted by $\Phi(t)$. Often, in Definition A2, Z will have a $n(0, 1)$ distribution, denoted $Z \sim n(0, 1)$ and we say Z_n is asymptotically (or limiting) $n(0, 1)$.

The first theorem strengthens convergence in distribution for cases such as normal limiting distributions.

Theorem A1. (Polya). Suppose in Definition A2 that $Z_n \overset{D}{\to} Z$ and Z has a continuous cdf, then $F_n(t) \to F(t)$ uniformly in t. Hence $\sup_t |F_n(t) - F(t)| \to 0$ as $n \to \infty$. This is proved by Parzen (1960, p. 438).

Theorem A2. Suppose $g(x)$ is a continuous function.

    a. If $Z_n \overset{P}{\to} c$ then $g(Z_n) \overset{P}{\to} g(c)$.

    b. If $Z_n \overset{D}{\to} Z$ then $g(Z_n) \overset{D}{\to} g(Z)$.

For a proof that includes the case of random vectors as well as random variables, see Serfling (1980, p. 24).

The next result, due to Slutsky, combines convergence in probability and distribution. It will be used extensively throughout the test.

Theorem A3. (Slutsky). Suppose $Z_n \overset{D}{\to} Z$ and $Y_n \overset{P}{\to} c$, a finite constant. Then

    a. $Z_n + Y_n \overset{D}{\to} Z + c$.

    b. $Z_n Y_n \overset{D}{\to} cZ$, and if $c = 0$ then $Z_n Y_n \overset{P}{\to} 0$.

    c. $Z_n / Y_n \overset{D}{\to} Z/c$ provided $c \neq 0$.

For a proof (which covers the case of random vectors also), see Serfling (1980, p. 19). Note that Y_n may be degenerate at c_n, so $Y_n \to c$ may be replaced by $c_n \to c$ in the theorem.

We now turn to the problem of establishing criteria for determining convergence in probability and convergence in distribution.

Theorem A4. (Chebyshev). For any random variable Z and any positive constant c,

$$P(|Z| \geqslant c) \leqslant \frac{EZ^2}{c^2}.$$

This is usually proved in introductory texts in mathematical statistics, for example, Hogg and Craig (1978, p. 58).

Theorem A5. For the sequence of random variables $\{Z_n\}$, suppose $EZ_n \to c$ and $\operatorname{Var} Z_n \to 0$. Then $Z_n \overset{P}{\to} c$.

Proof. From Theorem A4, $P(|Z_n - c| \geqslant \epsilon) \leqslant E(Z_n - c)^2/\epsilon^2$. But $E(Z_n - c)^2 = E(Z_n - EZ_n + EZ_n - c)^2 = \operatorname{Var} Z_n + [EZ_n - c]^2 \to 0$. Hence $Z_n \overset{P}{\to} c$ by Definition A1. Note this theorem covers the case $EZ_n = \mu$ and $\operatorname{Var} Z_n \to 0$.

Suppose $X_1, X_2, \ldots$ are independent and identically distributed (i.i.d.) random variables with $EX_i = \mu$ and $\operatorname{Var} X_i = \sigma^2 < \infty$. Then $\overline{X} \overset{P}{\to} \mu$. This follows at once from Theorem A5 with $Z_n = \overline{X}$ and is known as the Weak Law of Large Numbers. Much stronger results are possible. For example the Strong Law of Large Numbers asserts that $\overline{X}$ converges to μ with probability 1 provided only $EX_i = \mu < \infty$. See Serfling (1980, p. 27) for a discussion.

We now turn to convergence in distribution. The major tool is the Central Limit Theorem. We give the most general form of this theorem and then specialize it to various types of situations that are encountered in the text. We use $I(\cdot)$ to denote the indicator function $I(A) = 1$ if A occurs and 0 otherwise.

Theorem A6. (Lindeberg Central Limit Theorem). For each n let $Z_{1n}, \ldots, Z_{nn}$ be independent random variables with $EZ_{in} = 0$ and $\operatorname{Var} Z_{in} = \sigma_{in}^2 < \infty$ for $i = 1, \ldots, n$. Let $B_n^2 = \sum_{i=1}^n \sigma_{in}^2$. If for each $\epsilon > 0$,

$$\lim_{n \to \infty} \frac{1}{B_n^2} \sum_{i=1}^n E\left[Z_{in}^2 I(|Z_{in}| > \epsilon B_n) \right] = 0,$$

then

$$\frac{\sum\limits_{i=1}^{n} Z_{in}}{B_n} \xrightarrow{D} Z \sim n(0,1).$$

This theorem is proved by Chung (1968, p. 187). The result is quite general and in fact the condition is necessary as well as sufficient. The double subscripts generate a triangular array of random variables and the independence is only assumed within the rows of the array, that is, for $Z_{1n}, \ldots, Z_{nn}$. Also note that the random variables do not need to be identically distributed.

We state next a very important theorem from real analysis that provides conditions under which limits and derivatives can be passed through integrals. In its statistical form it allows interchange of differentiation and expectation. We illustrate the theorem by deriving the usual central limit theorem from Theorem A6.

Theorem A7. (Lebesque Dominated Convergence Theorem). Suppose $h(x)$ is an integrable function and $\{g_n(x)\}$ is a sequence of (measurable) functions such that $|g_n(x)| \le h(x)$ and $\lim g_n(x) = g(x)$. Then $\lim \int g_n(x) \, dx = \int [\lim g_n(x)] \, dx = \int g(x) \, dx$. For a proof see Royden (1968, p. 229). In its statistical version we have a random variable X and a sequence of functions $\{g_n(x)\}$ such that $g_n(x) \to g(x)$. We also have a function $h(x)$ such that $|g_n(x)| \le h(x)$ for all n and $E|h(X)| < \infty$. Then $\lim Eg_n(X) = Eg(X)$.

Theorem A8. (Central Limit Theorem)

   a. Suppose $X_1, X_2, \ldots$ are i.i.d. random variables with $EX_i = \mu$ and $\operatorname{Var} X_i = \sigma^2$, $0 < \sigma^2 < \infty$. Then $\Sigma(X_i - \mu)/\sigma n^{1/2} \xrightarrow{D} Z \sim n(0,1)$.

   b. Suppose $\mathbf{X}_1, \mathbf{X}_2, \ldots$ are i.i.d. p-component random vectors with $E\mathbf{X}_i = \boldsymbol{\mu}$ and covariance matrix $\boldsymbol{\Sigma}$, positive definite. Then

$$\left[\sqrt{n} \left(\overline{\mathbf{X}}_{.1} - \mu_1 \right), \ldots, \sqrt{n} \left(\overline{\mathbf{X}}_{.p} - \mu_p \right) \right]' \xrightarrow{D} \mathbf{Z} \sim MVN(\mathbf{0}, \boldsymbol{\Sigma}),$$

where $\overline{\mathbf{X}}_j$ is the mean of the jth component. We will derive (a) from Theorem A6; see the remarks following A11 for (b).

Proof. a. In Theorem A6, let $Z_{in} = (X_i - \mu)$ $i = 1, \ldots, n$, then $EZ_{in} = 0$, $\text{Var } Z_{in} = \sigma^2 < \infty$ and $B_n^2 = n\sigma^2$. Without loss of generality let $\mu = 0$. Now

$$\lim_{n \to \infty} \frac{1}{B_n^2} \sum_{i=1}^{n} E\left[Z_{in}^2 I(|Z_{in}| > \epsilon B_n) \right]$$

$$= \lim_{n \to \infty} \frac{1}{\sigma^2} EX_1^2 I(|X_1| > \epsilon\sigma\sqrt{n}).$$

In Theorem A7, let $g_n(x) = x^2 I(|x| > \epsilon\sigma(n)^{1/2})$ and let $h(x) = x^2$. Then $E|h(X)| = EX^2 < \infty$ and $g_n(x) \to 0$; hence $EX_1^2 I(|X_1| > \epsilon\sigma(n)^{1/2}) \to E(0) = 0$ and the Lindeberg condition is satisfied.

There are various ways to write the conclusion of the Central Limit Theorem. Often in statistics we deal with averages, $\overline{X}$. When $X_1, \ldots, X_n$ are i.i.d. with mean μ and finite variance σ^2, we conclude from Theorem A8 that $n^{1/2}(\overline{X} - \mu) \xrightarrow{D} Z \sim n(0, \sigma^2)$, and we may say $n^{1/2}(\overline{X} - \mu)$ is asymptotically (or limiting) $n(0, \sigma^2)$. This means for large n we can make the following approximation:

$$P\left(\frac{\sqrt{n}\,(\overline{X} - \mu)}{\sigma} \leqslant t \right) \doteq \Phi(t)$$

or

$$P(\overline{X} \leqslant x) \doteq \Phi\left(\frac{\sqrt{n}\,(x - \mu)}{\sigma} \right).$$

These approximations can then be used to approximate critical values of tests based on $\overline{X}$.

We next discuss some additional central limit type results that are useful in nonparametrics.

Theorem A9. Suppose $W_1 W_2, \ldots$ are i.i.d. with $EW_i = 0$ and $\text{Var } W_i = \sigma^2, 0 < \sigma^2 < \infty$. Define $S = \sum_{i=1}^{n} a_i W_i / n^{1/2}$. If

$$\frac{\max|a_i|}{\sqrt{\sum_{i=1}^{n} a_i^2}} \to 0,$$

then $S/(\text{Var } S)^{1/2} \xrightarrow{D} Z \sim n(0, 1)$ with $\text{Var } S = \sigma^2 \sum_1^n a_i^2 / n$.

Proof. In Theorem A6 define $Z_{in} = a_i W_i / n^{1/2}$ then $EZ_{in} = 0$, $\operatorname{Var} Z_{in} = a_i^2 \sigma^2 / n < \infty$, and $B_N^2 = \sigma^2 \sum a_i^2 / n$. The Lindeberg condition reduces to consideration of:

$$\frac{1}{\sigma^2 \sum a_i^2} \sum_{i=1}^{n} E\left[a_i^2 W_i^2 I\left[|W_i| > \epsilon\sigma \frac{\sqrt{\sum a_i^2}}{|a_i|} \right] \right]$$

$$\leqslant \frac{1}{\sigma^2 \sum a_i^2} \sum_{i=1}^{n} E\left[a_i^2 W_1^2 I\left[|W_1| > \epsilon\sigma \frac{\sqrt{\sum a_i^2}}{\max|a_i|} \right] \right]$$

$$= \frac{1}{\sigma^2} E\left[W_1^2 I\left[|W_1| > \epsilon\sigma \frac{\sqrt{\sum a_i^2}}{\max|a_i|} \right] \right].$$

From the hypothesis of the theorem, $(\sum a_i^2)^{1/2}/\max|a_i| \to \infty$ and now the argument is identical to that given in the proof of Theorem A8. Hence the Lindeberg condition holds and the proof is complete.

A simple variation on this theorem is quite useful in identifying the asymptotic parameters. The proof is adapted from the proof of a lemma by Arnold (1980).

Theorem A10. Suppose $W_1, W_2, \ldots$ are i.i.d. with $EW_i = 0$ and $\operatorname{Var} W_i = \sigma^2$, $0 < \sigma^2 < \infty$. Define $S = \sum_{i=1}^{n} a_i W_i / n^{1/2}$. If

$$\frac{1}{n} \sum_{i=1}^{n} a_i^2 \to \tau^2, \qquad 0 < \tau^2 < \infty$$

then $S \xrightarrow{D} Z \sim n(0, \sigma^2\tau^2)$ and $S/(\operatorname{Var} S)^{1/2} \xrightarrow{D} Z \sim n(0, 1)$.

Proof. First note that, as $n \to \infty$,

$$\frac{a_n^2}{n} = \frac{1}{n} \sum_{i=1}^{n} a_i^2 - \frac{(n-1)}{n} \frac{1}{(n-1)} \sum_{i=1}^{n-1} a_i^2 \to 0.$$

Hence $n^{-1/2}|a_n| \to 0$ as $n \to \infty$.

The proof follows from Theorem A9, provided $n^{-1/2}\max|a_i| \to 0$. Suppose $n^{-1/2}\max|a_i|$ does not tend to zero. Then there exists an $\epsilon > 0$ such that for any integer N there exists an integer $n_N \geqslant N$ such that $n_N^{-1/2}\max|a_i| \geqslant \epsilon$ for $1 \leqslant i \leqslant n_N$. By letting $N = 1, 2, \ldots$ it is possible to construct a sequence $n_1 < n_2 < \cdots$ of such values. For n_i let $m_i \leqslant n_i$ be such that $|a_{m_i}| = \max|a_i|, 1 \leqslant i \leqslant n_i$. Now

$$\epsilon \leqslant n_i^{-1/2} \max_{1 \leqslant i \leqslant n_i} |a_i| = n_i^{-1/2}|a_{m_i}| \leqslant m_i^{-1/2}|a_{m_i}|$$

which contradicts the first paragraph. Hence $n^{-1/2}\max|a_i| \to 0$ and $S/(\text{Var } S)^{1/2} \xrightarrow{D} Z \sim n(0,1)$. Further, $\text{Var } S = \sigma^2 \sum a_i^2 / n \to \sigma^2 \tau^2$ so $S \xrightarrow{D} Z \sim n(0, \sigma^2\tau^2)$ by Slutsky's Theorem A3 and the proof is complete.

We will need multivariate versions of some of the above limit theorems. These multivariate extensions are often proved using the "Cramér–Wold device."

Theorem A11. A sequence of random vectors $\{\mathbf{Z}_n\}$ converges in distribution to $\mathbf{Z}$ if and only if the sequence of (univariate) random variables $\{\boldsymbol{\lambda}'\mathbf{Z}_n\}$ converges in distribution to $\boldsymbol{\lambda}'\mathbf{Z}$ for every vector of constants $\boldsymbol{\lambda}$ such that $\sum\lambda_i^2 = 1$. See Serfling (1980, p. 18) for a proof.

The multivariate central limit theorem, Theorem A8(b), follows from the univariate case with an application of Theorem A11. We now consider multivariate versions of Theorem A9.

Theorem A12. Suppose $W_1, W_2, \ldots$ are i.i.d. with $EW_i = 0$ and $\text{Var } W_i = \sigma^2, 0 < \sigma^2 < \infty$. Define $S_k = \sum_{i=1}^n a_{ik} W_i / n^{1/2}$ for $k = 1, 2, \ldots, p$. Let

$$\mathbf{S} = \begin{bmatrix} S_1 \\ \vdots \\ S_p \end{bmatrix} = \frac{1}{\sqrt{n}} \mathbf{A}'\mathbf{W}$$

where

$$\mathbf{W} = \begin{bmatrix} W_1 \\ \vdots \\ W_n \end{bmatrix} \quad \text{and} \quad \mathbf{A} = \begin{bmatrix} a_{11} & \cdots & a_{1p} \\ \vdots & & \vdots \\ a_{n1} & & a_{np} \end{bmatrix}.$$

If

$$\frac{1}{n} \mathbf{A}'\mathbf{A} \to \mathbf{\Lambda}, \quad \text{positive definite}$$

then $\mathbf{S} \xrightarrow{D} \mathbf{Z} \sim MVN(\mathbf{0}, \sigma^2\mathbf{\Lambda})$.

Proof. Let λ be a p-vector of length 1 and consider $T = \lambda'\mathbf{S}$ $= \lambda'\mathbf{A}'\mathbf{W}/n^{1/2}$. Let $\mathbf{C} = \mathbf{A}\lambda$ so $T = \sum C_i W_i/n^{1/2}$ and we need to show that $\mathbf{C}$ satisfies the condition in Theorem A10.

Note that, since Λ is positive definite, $n^{-1}\sum C_i^2 = n^{-1}\lambda'\mathbf{A}'\mathbf{A}\lambda \to \lambda'\Lambda\lambda$ > 0, so we can take $\tau^2 = \lambda'\Lambda\lambda > 0$. Hence $\lambda'\mathbf{S} \xrightarrow{D} Z \sim n(0, \sigma^2\lambda'\Lambda\lambda)$ and the theorem follows from Theorem A11.

The next theorem is a slight variation on Theorem A12 that is sometimes useful.

Theorem A13. Suppose the same setup as in Theorem A12. If the covariance matrix of $\mathbf{S}$ converges to a positive definite matrix, then $\mathbf{S}$ has an asymptotic multivariate normal distribution.

The proof is immediate upon writing out the $\text{Cov}(\mathbf{S})$.

As an example of the application of Theorem A12 we consider the asymptotic distribution of the vector of estimates of the regression coefficients in the linear model. Let $\mathbf{Y} = \mathbf{X}\beta + \mathbf{e}$ where $\mathbf{Y}$ is an $n \times 1$ vector of observations, $\mathbf{X}$ an $n \times p$ design matrix of full rank p, β a $p \times 1$ vector of regression coefficients and $\mathbf{e}$ an $n \times 1$ vector of i.i.d. random variables. We suppose $Ee_i = 0$, $\text{Var}\, e_i = \sigma^2$, $0 < \sigma^2 < \infty$. Finally, suppose $n^{-1}\mathbf{X}'\mathbf{X} \to \Sigma$, positive definite.

The least-squares (maximum likelihood under normality) estimate of β is $\hat{\beta} = (\mathbf{X}'\mathbf{X})^{-1}\mathbf{X}'\mathbf{Y}$. This can be written as follows

$$\hat{\beta} = (\mathbf{X}'\mathbf{X})^{-1}\mathbf{X}'(\mathbf{X}\beta + \mathbf{e}) = \beta + (\mathbf{X}'\mathbf{X})^{-1}\mathbf{X}'\mathbf{e}.$$

Hence, identifying with the notation of Theorem A12, we have

$$\sqrt{n}\,(\hat{\beta} - \beta) = \sqrt{n}\,(\mathbf{X}'\mathbf{X})^{-1}\mathbf{X}'\mathbf{e} = \frac{1}{\sqrt{n}}\,\mathbf{A}'\mathbf{e},$$

where $\mathbf{A}' = (n^{-1}\mathbf{X}'\mathbf{X})^{-1}\mathbf{X}'$. Then it follows that $n^{-1}\mathbf{A}'\mathbf{A} = n(\mathbf{X}'\mathbf{X})^{-1} \to \Sigma^{-1}$ and the condition of Theorem A12 is satisfied.

Hence Theorem A12 applies and $n^{1/2}(\hat{\beta} - \beta)$ is asymptotically $MVN(\mathbf{0}, \sigma^2\Sigma^{-1})$ or $\hat{\beta}$ is approximately $MVN(\beta, \sigma^2(\mathbf{X}'\mathbf{X})^{-1})$. See Huber (1981, Section 7.2) for an alternative approach.

The Central Limit Theorem A8 for i.i.d. random variables with finite variance is the most common version discussed in statistics tests and courses. It says that $P(n^{1/2}(\overline{X} - \mu)/\sigma \leqslant t) \to \Phi(t)$, where $EX_i = \mu$ and $\text{Var}\, X_i = \sigma^2$, $0 < \sigma^2 < \infty$. We also know from Polya's Theorem A1 that the convergence is uniform in t. We cannot tell at this point, however, how fast the convergence is taking place. With information on the third absolute moment, the rate can be established.

Theorem A14. (Berry–Esseen). Suppose $X_1, X_2, \ldots$ are i.i.d. with $EX_i = \mu$, $\operatorname{Var} X_i = \sigma^2$, $0 < \sigma^2 < \infty$ and $E|X_i - \mu|^3 = \rho^3$, $0 < \rho^3 < \infty$. Then for all n

$$\left| P\left(\sqrt{n}\, \frac{(\overline{X} - \mu)}{\sigma} \leqslant t \right) - \Phi(t) \right| \leqslant \frac{c\rho^3}{\sqrt{n}\, \sigma^3}$$

where c is a numerical constant independent of n. This is proved by Chung (1968, p. 206), who gives a numerical value for c of approximately 28. Serfling (1980, p. 33) points out that the value of c has been sharpened to .7975.

This theorem is useful for establishing uniform convergence over a class of underlying distributions. As an application of Theorems A1 and A14 we have:

Theorem A15. Suppose $X_1, X_2, \ldots$ are i.i.d. with cdf $F_\theta(x)$, $\theta \in \Omega$. Suppose there exist constants K and M such that $0 < K \leqslant \sigma_\theta^2 < \infty$ and $\rho_\theta^3 \leqslant M < \infty$ for all $\theta \in \Omega$. Then

$$P_\theta\left(\frac{\sqrt{n}\,(\overline{X} - \mu)}{\sigma} \leqslant t \right)$$

converges to $\Phi(t)$ uniformly in $\theta \in \Omega$ as well as in t.

The next theorem describes conditions under which the conclusion of the Central Limit Theorem continues to hold when $X_1, X_2, \ldots$ are no longer independent. The sequence $X_1, X_2, \ldots$ is called stationary if the joint distribution of $(X_i, X_{i+1}, \ldots, X_{i+r})$ is independent of i for all r. The sequence is said to be m-dependent if $(X_1, \ldots, X_r)$ and $(X_s, X_{s+1}, \ldots)$ are independent provided $s - r > m$. The following theorem is a special case of a more general theorem due to Hoeffding and Robbins (1948). For further discussion see Fraser (1957, Section 6.4) and Serfling (1968a).

Theorem A16. Suppose $X_1, X_2, \ldots$ is a stationary, m-dependent sequence with $EX_1 = \mu$ and $E|X_1|^3 < \infty$. Then $n^{1/2}(\overline{X} - \mu) \xrightarrow{D} Z \sim n(0, \sigma^2)$ with $\sigma^2 = \operatorname{Var} X_1^2 + 2\sum_{i=1}^{m} \operatorname{Cov}(X_1, X_{i+1})$.

Throughout the book the concept of an absolutely continuous function is important. A function $F(x)$ is absolutely continuous on the real line, provided there is an integrable function $f(x)$ such that for every $x < y$ $F(y) - F(x) = \int_x^y f(t)\,dt$. It then follows that $F(x)$ has a derivative equal to $f(x)$ (almost everywhere). Hence $F(x)$ is the indefinite integral of its derivative. When $F(x)$ is an absolutely continuous cumulative distribution function, the function $f(x)$ is the probability density function (pdf).

Definition A3. The convolution of any two cdfs, $F_1(x)$ and $F_2(x)$, is defined by $F_1 * F_2(x) = \int_{-\infty}^{\infty} F_1(x - y)\,dF_2(y)$ and is the cdf of $X_1 + X_2$, where X_1, X_2 are independent random variables with cdfs $F_1(x)$ and $F_2(x)$, respectively. The following theorem is important in the analysis of properties of the Wilcoxon rank procedures. See Definition A4 for the Stieltjes integral.

Theorem A17. Suppose $F_1(x)$ and $F_2(x)$ are absolutely continuous cdfs with pdfs $f_1(x)$ and $f_2(x)$, respectively. Then the convolution $F_1 * F_2(x)$ is also absolutely continuous with density $f_1 * f_2(x) = \int_{-\infty}^{\infty} f_1(x - y)f_2(y)\,dy$. A proof is given by Chung (1968, p. 135). The theorem allows differentiation under the integral that defines the convolution. Hence, if $F^*(x) = P(X_1 + X_2 \leqslant x)$ where X_1, X_2 are i.i.d. with absolutely continuous $F(x)$, then

$$f^*(x) = \frac{d}{dx} F^*(x) = \frac{d}{dx} \int_{-\infty}^{\infty} F(x - y)f(y)\,dy$$

$$= \int_{-\infty}^{\infty} f(x - y)f(y)\,dy.$$

For further discussion of this integral and the Wilcoxon rank procedures see Mehra and Sarangi (1967) or Olshen (1967). It is not necessary that the random variables be identically distributed to define the convolution. The proof of Theorem 4.4.2 contains an application of convolutions for discrete, not identically distributed random variables.

The Lebesque Dominated Convergence Theorem A7 has been used to exchange limits and integrals in Theorems A8 and A9. Another important application of Theorem A7 is the exchange of differentiation and integration. Statistical applications involve exchange of differentiation and expectation.

Theorem A18. Suppose the function $h(x, \theta)$ is absolutely continuous as a function of $\theta \in \Omega$, an open interval. Suppose $(\partial/\partial\theta)h(x, \theta)$ exists and is bounded by $q(x)$ for all $\theta \in \Omega$. Suppose $\int(\partial/\partial\theta)h(x, \theta)\,dx$ and $\int q(x)\,dx$ exist. Then $(d/d\theta)\int h(x, \theta)\,dx = \int(\partial/\partial\theta)h(x, \theta)\,dx$.

In the statistical version of this theorem we suppose $E(\partial/\partial\theta)h(X,\theta)$ and $Eq(X)$ both exist. Then Theorem 18 implies that $(d/d\theta)Eh(X,\theta)$ $= E(\partial/\partial\theta)h(X,\theta)$.

Two other important inequalities from mathematics are important in statistics.

Theorem A19. (Cauchy–Schwarz Inequality). Suppose $g(x)$ and $h(x)$ are square integrable functions. Then $(\int g(x)h(x)\,dx)^2 \leqslant \int g^2(x)\,dx \int h^2(x)\,dx$. Equality holds if and only if there exist constants a and b such that $ag(x) + bh(x) = 0$. In the statistical version we have two random variables X and Y such that $EX^2 < \infty$ and $EY^2 < \infty$. Then $(EXY)^2 \leqslant EX^2EY^2$. Equality holds if and only if $aX + bY = 0$ with probability one. The statistical version is sometimes called the correlation inequality, since if X and Y have been standardized to have means 0 and variances 1, the Cauchy–Schwarz inequality is the same as $\rho^2 \leqslant 1$, where ρ is the correlation between X and Y.

Next, recall that a function $q(x)$ is convex if, for each $\lambda \in (0,1)$, $g(\lambda x + (1 - \lambda)y) \leqslant \lambda g(x) + (1 - \lambda)g(y)$.

Theorem A20. (Jensen's Inequality). Suppose X and $g(X)$ have finite expectations. Suppose $g(x)$ is convex. Then $Eg(X) \geqslant g(EX)$. If $g(x)$ is strictly convex, then $Eg(X) > g(EX)$, unless X is degenerate. See Fraser (1957, Section 2.2.). The most common example is $g(x) = x^2$, so that $EX^2 \geqslant (EX)^2$ and $\operatorname{Var} X \geqslant 0$.

We now discuss the Stieltjes integral which is convenient for defining the expectation of a random variable.

Definition A4. Let $g(x)$ and $h(x)$ be given with $g(x)$ nonnegative, and let $a = x_0 < x_1 < \cdots < x_n = b$ be a partition of $[a,b]$. The Stieltjes integral of $g(x)$ with respect to $h(x)$ is defined by

$$\int_a^b g(x)\,dh(x) = \lim \sum_{i=1}^n g(x_i)\big[h(x_i) - h(x_{i-1})\big]$$

where the limit is taken over partitions with $\max(x_i - x_{i-1}) \to 0$. Stieltjes integrals over the real line are defined by letting $a \to -\infty$ and/or $b \to +\infty$. We will say the Stieltjes integral for an arbitrary function $f(x)$ relative to $h(x)$ exists if $\int |f(x)|\,dh(x) < \infty$.

A simple example is provided by the distribution function that assigns mass 1 to the point y. Let $\delta_y(x) = 1$ if $x \geqslant y$ and 0 otherwise. Then, when $g(x)$ is continuous, $\int g(x)\,d\delta_y(x) = g(y)$. This follows immediately from the definition because $\delta_y(x_i) - \delta y(x_{i-1}) = 1$ only when $y \in (x_{i-1}, x_i)$ and $g(x_i) \rightarrow g(y)$ as $x_i \rightarrow y$. This is simply a formal way of showing that if X is degenerate at y and if $g(x)$ is continuous, then $Eg(X) = g(y)$.

Suppose $F(x)$ is a discrete distribution function with jumps at x_1, $x_2, \ldots$, then, when the expectation exists, we have $Eg(X) = \Sigma g(x_i)$ $[F(x_i) - F(x_{i-1})] = \int g(x) dF(x)$. For example, let $F_n(x)$ denote the empirical distribution function $[F_n(x)$ puts mass $1/n$ at each of the observed values $x_1, x_2, \ldots, x_n]$, then $\int x \, dF_n(x) = \Sigma x_i [F_n(x_i) - F_n(x_{i-1})]$, $\Sigma x_i(1/n)$ $= \bar{x}$, the sample mean. When the distribution function $F(x)$ has a density $f(x) = F'(x)$, we have $Eg(X) = \int g(x)f(x) dx = \int g(x) dF(x)$, when it exists. Hence, using the Stieltjes integral, we do not need to separate the discrete and continuous cases to define expectation. In general, if X has cdf $F(x)$, discrete or continuous, then $Eg(X) = \int g(x) dF(x)$, when it exists. As pointed out previously, this reduces to the correct equations for the individual cases.

Most of the properties of the Riemann integral extend to the Stieltjes integral. Integration by parts is often useful and the formula is given by:

$$\int_a^b g(x) \, dh(x) = g(x)h(x)\big|_a^b - \int_a^b h(x) \, dg(x).$$

Cramér (1946, Chapter 7) contains a development of Stieltjes integrals.

Summation formulas for integers appear frequently in the computations of moments for rank sums. We list several of the most common equations.

Theorem A21.

$$\sum_{i=1}^{n} i = n(n+1)/2$$

$$\sum_{i=1}^{n} i^2 = n(n+1)(2n+1)/6$$

$$\sum_{i=1}^{n} (i - \bar{i})^2 = n(n^2 - 1)/12, \qquad \bar{i} = \sum_{i=1}^{n} i/n$$

$$\sum_{i=1}^{n} i^3 = n^2(n+1)^2/4$$

$$\sum_{i=1}^{n} i^4 = n(n+1)(2n+1)(3n^2 + 3n - 1)/30$$

$$\sum_{i=1}^{n} i^k = O(n^{k+1}).$$

Proof. We will describe a device that can be used to develop the formulas for any k. Note that

$$\sum_{i=2}^{n+1} i^k - \sum_{i=1}^{n} i^k = (n+1)^k - 1.$$

Make the change of variable $j = i - 1$ in the first sum on the left. Then $\sum_{i=2}^{n+1} i^k = \sum_{j=1}^{n} (j+1)^k$. Combining this with the second sum we have

$$\sum_{j=1}^{n} (j+1)^k - \sum_{j=1}^{n} j^k = \sum_{j=1}^{n} \left[(j+1)^k - j^k \right]$$

$$= (n+1)^k - 1.$$

Now $[(j+1)^k - j^k]$ is a polynomial in j of degree $k - 1$ and which, when summed on j, allows us to solve for $\sum_{j=1}^{n} j^{k-1}$ in terms of prior sums.

For example, let $k = 2$, then $\sum_{j=1}^{n} [(j+1)^2 - j^2] = \sum_{j=1}^{n} (2j+1) = 2\sum_{j=1}^{n} j + n = (n+1)^2 - 1$. This yields $\sum_{j=1}^{n} j = n(n+1)/2$. If we let $k = 3$, then $3\sum_{j=1}^{n} j^2 + 3\sum_{j=1}^{n} j + n = n^3 + 3n^2 + 3n$ can be solved for $\sum_{j=1}^{n} j^2$.

References

Adichie, J. N. (1967a). Asymptotic effiency of a class of nonparametric tests for regression parameters. *Ann. Math. Stat.* **38**:884–893.

Adichie, J. N. (1967b). Estimates of regression parameters based on rank tests. *Ann. Math. Stat.* **38**:894–904.

Adichie, J. N. (1978). Rank tests of sub-hypotheses in the general linear regression. *Ann. Stat.* **6**:1012–1026.

Alling, D. W. (1963). Early decision in the Wilcoxon two-sample test. *J. Stat. Assoc.* **58**:713–720.

Andrews, D. F. (1974). A robust method for multiple linear regression. *Technometrics* **16**:523–531.

Andrews, D. F., Bickel, P. J., Hampel, F. R., Huber, P. J., Rogers, W. H., and J. W. Tukey (1972). *Robust estimates of location: Survey and advances.* Priceton University Press, Princeton, New Jersey.

Andrews, F. C. (1954) Asymptotic behavior of some rank tests for analysis of variance. *Ann. Math. Stat.* **25**:724–736.

Arbuthnott, J. (1710). An argument for devine providence taken from the constant regularity observed in the birth of both sexes. *Philos. Trans.* **27**:186–190.

Arnold, H. (1965). Small sample power for the one-sample Wilcoxon test for nonnormal shift alternatives. *Ann. Math. Stat.* **36**:1767–1778.

Arnold, S. F. (1980). Asymptotic validity of F tests for the ordinary linear model and the multiple correlation model. *J. Am. Stat. Assoc.* **75**:890–894.

Arnold, S. F. (1981). *The theory of linear models and multivariate analysis.* Wiley, New York.

Aubuchon, J. C. (1982). *Rank tests in the linear model: Asymmetric errors.* Unpublished PhD Thesis, Department of Statistics, The Pennsylvania State University.

Aubuchon, J. C. and T. P. Hettmansperger (1982a). On the use of rank tests and estimates in the linear model. Tech. Rpt. #41, Department of Statistics, The Pennsylvania State University. (To appear in *Handbook of Statistics. V4: Nonparametric Methods*, eds., Krishnaiah and Sen).

Aubuchon, J. C. and T. P. Hettmansperger (1982b). Rank-based inference based on Wilcoxon scores without symmetric errors. Tech. Rpt. #42, Department of Statistics, The Pennsylvania State University. (To appear in the *J. Stat. Inference and Planning.*)

310

Bahadur, R. R. (1967). Rates of convergence of estimates and test statistics. *Ann. Math. Stat.* **31**:276–295.

Barlow, R. E., Bartholomew, D. J., Brenner, J. M., and H. D. Brunk (1972). *Statistical inference under order restrictions*. Wiley, New York.

Bauer, D. F. (1972). Constructing confidence sets using rank statistics. *J. Am. Stat. Assoc.* **67**:687–690.

Bean, S. J. and C. P. Tsokos (1980). Developments in nonparametric density estimation. *Int. Stat. Rev.* **43**:267–287.

Benard, A. and P. van Elteren (1953). A generalization of the method of *m* rankings. *Indagationes Math.* **15**:358–369.

Bhattacharyya, B. K. and G. G. Roussas (1969). Estimation of a certain functional of a probability density. *Skand. Aktuar. Tidskr.* **52**:201–206.

Bickel, P. J. (1964). On some alternative estimates for shift in the *p*-variate one sample problem. *Ann. Math. Stat.* **35**:1079–1090.

Bickel, P. J. (1965). On some asymptotically nonparametric competitors of Hotelling's T^2. *Ann. Math. Stat.* **36**:160–173.

Bickel, P. J. (1974). Edgeworth expansions in nonparametric statistics. *Ann. Stat.* **2**:1–20.

Bickel, P. J. and K. A. Doksum (1977). *Mathematical statistics.* Holden-Day, San Francisco.

Box, G. E. P. and D. R. Cox. (1964). An analysis of transformations. *J. R. Stat. Soc.* (B) **2**:211–252.

Brown, M. B. (1975). Exploring interaction effects in the ANOVA. *Appl. Stat.* **24**:288–298.

Brownlee, K. A. (1965). *Statistical theory and methodology*. 2nd ed. Wiley, New York.

Chacko, V. J. (1963). Testing homogeneity against ordered alternatives. *Ann. Math. Stat.* **34**:945–956.

Cheng, K. F. and R. J. Serfling (1981). On estimation of a class of efficiency-related parameters. *Scand. Actuarial J.*, 83–92.

Cheng, K. and T. P. Hettmansperger (1983). Weighted least squares rank estimates. *Commun. Stat.—Theor. Meth.* **12**(9), 1069–1086.

Chernoff, H. and I. R. Savage (1958). Asymptotic normality and efficiency of certain nonparametric test statistics. *Ann. Math. Stat.* **39**:972–994.

Chung, K. L. (1968). *A course in probability theory*. Harcourt, Brace and World, New York.

Clason, C. B. (1958). *Exploring the distant stars*. G. P. Putnam's Sons, New York.

Cochran, W. G. and G. M. Cox (1957). *Experimental designs*. 2nd ed. Wiley, New York.

Collins, J. R. (1983). On the minimax property for *R*-estimators of location. *Ann. Stat.* **11**:1190–1195.

Conover, W. J. (1973). On methods of handling ties in the Wilcoxon signed rank test. *J. Am. Stat. Assoc.* **68**:985–988.

Conover, W. J., Wehmanen, O., and F. L. Ramsey (1978). A note on the small-sample power functions for nonparametric tests of location in the double exponential family. *J. Am. Stat. Assoc.* **73**:188–190.

Cook, R. D. and S. Weisberg (1982). *Residuals and influence in regression*. Chapman and Hall, New York.

Cox, D. R. and A. Stuart (1955). Some quick tests for trend in location and dispersion. *Biometrika* **42**:80–95.

Cramér, H. (1946). *Mathematical methods of statistics.* Princeton University Press, Princeton. New Jersey.

Daniel, C. and F. S. Wood (1971). *Fitting equations to data.* Wiley, New York.

Denby, L. and C. L. Mallows (1977). Two diagnostic displays for robust regression analysis. *Technometrics* **19**:1–13.

Doksum, K. A. (1977). Some graphical methods in statistics. A review and some extensions. *Stat. Neerl.* **31**:53–68.

Draper, D. C. (1983). Rank-based robust analysis of linear models. II. Estimation of the asymptotic variance of rank-based estimators and associated statistics. Unpublished manuscript.

Draper, N. and H. Smith (1981). *Applied regression analysis.* 2nd ed. Wiley, New York.

Dunn, O. J. (1964). Multiple comparisons using rank sums. *Technometrics* **6**:241–252.

Durbin, J. (1951). Incomplete blocks in ranking experiments. *Br. J. Psychol.* **4**:85–90.

Fellingham, S. A. and D. J. Stoker (1964). An approximation for the exact distribution of the Wilcoxon test for symmetry. *J. Am. Stat. Assoc.* **59**:899–905.

Feustal, E. A. and L. D. Davisson (1967). The asymptotic relative efficiency of mixed statistical tests. *IEEE Trans. on Inf. Theory* IT-13, pp. 247–255.

Fisher, R. A. (1970). *Statistical methods for research workers.* 14th ed. Hafner, New York.

Fisher, R. A. and F. Yates (1938). *Statistical tables for biological, agricultural and medical research.* Oliver and Boyd, Edinburgh, London.

Fix, E. and J. L. Hodges, Jr. (1955). Significance probabilities of the Wilcoxon test. *Ann. Math. Stat.* **26**:301–312.

Fligner, M. A. and G. E. Policello, II. (1981). Robust rank procedures for the Behrens–Fisher problem. *J. Am. Stat. Assoc.* **76**:162–168.

Fligner, M. A. and S. W. Rust (1982). A modification of Mood's median test for the generalized Behrens–Fisher problem. *Biometrika* **69**:221–226.

Fraser, D. A. S. (1957). *Nonparametric methods in statistics.* Wiley, New York.

Friedman, M. (1937). The use of ranks to avoid the assumption of normality implicit in the analysis of variance. *J. Am. Stat. Assoc.* **32**:675–701.

Friedman, M. (1940). A comparison of alternative tests of significance for the problem of *m* rankings. *Ann. Math. Stat.* **11**:86–92.

Gastwirth, J. L. (1968). The first median test: A two-sided version of the control median test. *J. Am. Stat. Assoc.* **63**:692–706.

Gastwirth, J. L. (1970a). On asymptotic relative efficiencies of a class of rank tests. *J. R. Stat. Soc.* (B) **32**:227–232.

Gastwirth, J. L. (1970b). On robust rank tests. *Proc. First Int. Symp. on Nonparametric Techniques in Statistical Inference*, Cambridge University Press, Cambridge.

Gastwirth, J. L. (1971). On the sign test for symmetry. *J. Am. Stat. Assoc.* **66**:821–823.

Gastwirth, J. L. and H. Rubin (1971). Effect of dependence on the level of some one-sample tests. *J. Am. Stat. Assoc.* **66**:816–820.

Gastwirth, J. L. and H. Rubin (1975). The behavior of robust estimators on dependent data. *Ann. Stat.* **3**:1070–1100.

Gastwirth, J. L. and S. S. Wolff (1968). An elementary method for obtaining lower bounds on the asymptotic power of rank tests. *Ann. Math. Stat.* **39**:2128–2130.

Govindarajulu, Z. and S Eisenstat (1965). Best estimates of location and scale parameters of a chi distribution using ordered observations. *Nippon Kagaku Gijutus* **12**:149–164.

Graybill, F. A. (1969). *Introduction to matrices with applications in statistics.* Wadsworth, Belmont, CA.

Haigh, J. (1971). A neat way to prove asymptotic normality. *Biometrika* **58**:677–678.

Hajek, J. (1968). Asymptotic normality of simple linear rank statistics under alternatives. *Ann. Math. Stat.* **39**:325–346.

Hajek, J. and Z. Sidak (1967). *Theory of rank tests.* Academic, New York.

Hampel, F. R. (1974). The influence curve and its role in robust estimation. *J. Am. Stat. Assoc.* **69**:383–393.

Hannan, E. J. (1956). The asymptotic power of tests based upon multiple correlation. *J. R. Stat. Soc.* (B) **18**:227–233.

Hardy, G. H., Littlewood, J. E., and G. Polya (1952). *Inequalities.* 2nd ed. Cambridge University Press, Cambridge.

Hayman, G. E. and Z. Govindarajulu (1966). Exact power of the Mann–Whitney tests for exponential and rectangular alternatives. *Ann. Math. Stat.* **37**:945–953.

Hettmansperger, T. P. and M. A. Keenan (1975). Tail weight, statistical inference and families of distributions—a brief survey. In: Patil et al., eds., *Statistical distributions in scientific work*, VI, pp. 161–172. D. Reidel, Dordrecht, Holland.

Hettmansperger, T. P. and J. S. Malin (1975). A modified Mood's test for location with no shape assumptions on the underlying distributions. *Biometrika* **62**:527–529.

Hettmansperger, T. P. and J. W. McKean (1977). A robust alternative based on ranks to least squares in analyzing linear models. *Technometrics* **19**:275–284.

Hettmansperger, T. P. and J. W. McKean (1983). A geometric interpretation of inferences based on ranks in the linear model. *J. Am. Stat. Assoc.* **78**:885–893.

Hettmansperger, T. P. and J. M. Utts (1977). Robustness properties for a simple class of rank estimates. *Comm. Stat.* **A6**(9):855–868.

Hodges, J. L., Jr. (1967). Efficiency in normal samples and tolerance of extreme values for some estimates of location. *Proc. 5th Berkeley Symp. Math. Stat. Prob*, **1**:163–186. University of California Press, Berkeley.

Hodges, J. L., Jr. and E. L. Lehmann (1956). The efficiency of some nonparametric competitors of the *t*-test. *Ann. Math. Stat.* **27**:324–335.

Hodges, J. L., Jr. and E. L. Lehmann (1960). Comparison of the normal scores and Wilcoxon tests. *Proc. 4th Berkeley Symp.* **1**:307–317.

Hodges, J. L., Jr. and E. L. Lehmann (1962). Rank methods for combinations of independent experiments in analysis of variance. *Ann. Math. Stat.* **33**:482–497.

Hodges, J. L., Jr. and E. L. Lehmann (1963). Estimates of location based on rank tests. *Ann. Math. Stat.* **34**:598–611.

Hodges, J. L., Jr. and E. L. Lehmann (1973). Wilcoxon and *t*-tests for matched pairs of typed subjects. *J. Am. Stat. Assoc.* **68**:151–158.

Hoeffding, W. (1951). Optimum nonparametric tests. *Proc. 2nd Berkeley Symp. on Math. Stat. and Prob.*, pp. 83–92.

Hoeffding, W. and H. Robbins (1948). The central limit theorem for dependent random variables. *Duke Math. J.* **15**:773–780.

Hogg, R. V. and A. T. Craig (1978). *Introduction to mathematical statistics.* 4th ed. Macmillan, New York.

Hollander, M. (1967). Rank tests for randomized blocks when the alternatives have an a priori ordering. *Ann. Math. Stat.* **38**:867–877.

Hollander, M. and J. Sethuraman (1978). Testing for agreement between two groups of judges. *Biometrika* **65**:403–411.

Huber, P. J. (1964). Robust estimation of a location parameter. *Ann. Math. Stat.* **35**:73–101.

Huber, P. J. (1973). Robust regression: Asymptotics, conjectures, and Monte Carlo. *Ann. Stat.* **1**:799–821.

Huber, P. J. (1981). *Robust statistics.* Wiley, New York.

Huber, P. J. (1983). Minimax aspects of bounded-influence regression (with discussion). *J. Am. Stat. Assoc.* **78**:66–80.

Jaeckel, L. A. (1971). Robust estimates of location: Symmetry and asymmetric contamination. *Ann. Math. Stat.* **42**:1020–1034.

Jaeckel, L. A. (1972). Estimating regression coefficients by minimizing the dispersion of the residuals. *Ann. Math. Stat.* **43**:1449–1458.

Jirina, M. (1976). On the asymptotic normality of Kendall's rank correlation statistic. *Ann. Stat.* **4**:214–215.

Johnson, R. A. and K. G. Mehrotra (1971). Some *c*-sample nonparametric tests for ordered alternatives. *J. Indian Stat. Assoc.* **9**:8–23.

Joiner, B. L. and J. R. Rosenblatt (1971). Some properties of the range in samples from Tukey's symmetric lambda distribution. *J. Am. Stat. Assoc.* **66**:394–399.

Jonckheere, A. R. (1954). A distribution-free *k*-sample test against ordered alternatives. *Biometrika* **41**:133–145.

Jureckova, J. (1969). Asymptotic linearity of a rank statistic in regression parameter. *Ann. Math. Stat.* **40**:1889–1900.

Jureckova, J. (1971). Nonparametric estimate of regression coefficients. *Ann. Math. Stat.* **42**:1328–1338.

Kendall, M. G. (1938). A new measure of rank correlation. *Biometrika* **30**:81–93.

Kendall, M. G. (1970). *Rank correlation methods.* 4th ed. Hafner, New York.

Kendall, M. G. and A. Stuart (1973). *The advanced theory of statistics.* Vol. 2, 3rd ed. Hafner, New York.

Klotz, J. (1963). Small sample power and efficiency for the one-sample Wilcoxon and normal scores tests. *Ann. Math. Stat.* **34**:624–632.

Klotz, J. (1964). On the normal scores two-sample rank test. *J. Am. Stat. Assoc.* **59**:652–664.

Klotz, J. (1965). Alternative efficiencies for signed rank tests. *Ann. Math. Stat.* **36**:1759–1766.

Kruskal, W. H. (1952). A nonparametric test for the several sample problem. *Ann. Math. Stat.* **23**:525–540.

Kruskal, W. H. (1957). Historical notes on the Wilcoxon unpaired two-sample test. *J. Am. Stat. Assoc.* **52**:356–360.

Kruskal, W. H. (1958). Ordinal measures of association. *J. Am. Stat. Assoc.* **53**:814–861.

Kruskal, W. H. and W. A. Wallis (1952). Use of ranks in one criterion variance analysis. *J. Am. Stat. Assoc.* **57**:583–621.

Lambert, D. (1982). Qualitative robustness of tests. *J. Am. Stat. Assoc.* **77**:352–357.

Lehmann, E. L. (1951). Consistency and unbiasedness of certain nonparametric tests. *Ann. Math. Stat.* **22**:165–179.

Lehmann, E. L. (1953). The power of rank tests. *Ann. Math. Stat.* **24**:23–43.

Lehmann, E. L. (1959). *Testing statistical hypotheses.* Wiley, New York.

Lehmann, E. L. (1963). Nonparametric confidence intervals for a shift parameter. *Ann. Math. Stat.* **34**:1507–1512.

Lehmann, E. L. (1975). *Nonparametrics: Statistical methods based on ranks.* Hoden-Day, San Francisco.

Li, L. and W. R. Schucany (1975). Some properties of a test for concordance of two groups of rankings. *Biometrika* **62**:417–423.

Mann, H. B. (1945). Nonparametric tests against trend. *Econometrika* **13**:245–259.

Mann, H. B. and D. R. Whitney (1947). On a test of whether one of two random variables is stochastically larger than the other. *Ann. Math. Stat.* **18**:50–60.

Maritz, J. S. (1981). *Distribution-free statistical methods.* Chapman and Hall, London.

Markowski, E. P. and T. P. Hettmansperger (1982). Inference based on simple rank step score statistics for the location model. *J. Am. Stat. Assoc.* **77**: 901–907.

Mathisen, H. C. (1943). A method of testing the hypothesis that two samples are from the same population. *Ann. Math. Stat.* **14**:188–194.

Matthews, G. V. T. (1952). Sun navigation in homing pigeons. *J. Exp. Biol.* **30**:243–267.

Matthews, G. V. T. (1974). On bird navigation, with some statistical undertones (with discussion). *J. R. Stat. Soc.* (B) **36**:349–364.

McKean, J. W. and T. P. Hettmansperger (1976). Tests of hypotheses based on ranks in the general linear model. *Comm. Stat.—Theory and Methods* **A5**(8):693–709.

McKean, J. W. and T. P. Hettmansperger (1978). A robust analysis of the general linear model based on one-step *R*-estimates. *Biometrika* **65**:571–579.

McKean, J. W. and R. M. Schrader (1980). The geometry of robust procedures in linear models, *J. R. Stat. Soc.* (B) **42**:366–371.

McKean, J. W. and R. M. Schrader (1982). The use and interpretation of robust analysis of variance. In: R. L. Laurer and A. F. Siegel, eds., *Modern data analysis.* Academic, New York, pp. 171–188.

Mehra, K. L. and J. Sarangi (1967). Asymptotic efficiency of certain rank tests for comparative experiments. *Ann. Math. Stat.* **38**:90–107.

Miller, R. G., Jr. (1981). *Simultaneous statistical inference.* 2nd ed. Springer-Verlag, New York.

Mood, A. M. (1950). *Introduction to the theory of statistics.* McGraw-Hill, New York.

Morrison, D. F. (1967). *Multivariate statistical methods.* 2nd ed. McGraw-Hill, New York.

Mosteller, F. and J. W. Tukey (1977). *Data analysis and regression.* Addison-Wesley, Reading, MA.

Noether, G. E. (1955). On a theorem of Pitman. *Ann. Math. Stat.* **26**:64–68.

Noether, G. E. (1967). Wilcoxon confidence intervals for location parameters in the discrete case. *J. Am. Stat. Assoc.* **62**:184–188.

Noether, G. E. (1970). A central limit theorem with nonparametric applications. *Ann. Math. Stat.* **41**:1753–1755.

Noether, G. E. (1973). Some simple distribution-free confidence intervals for the center of a symmetric distribution. *J. Am. Stat. Assoc.* **68**:716–719.

Olshen, R. A. (1967). Sign and Wilcoxon tests for linearity. *Ann. Math. Stat.* **38**:1759–1769.

Orban, J. and D. A. Wolfe (1982). A class of distribution-free two-sample tests based on placements. *J. Am. Stat. Assoc.* **77**:666–671.

Page, E. B. (1963). Ordered hypotheses for multiple treatments: A significance test for linear ranks. *J. Am. Stat. Assoc.* **58**:216–230.

Parr, W. C. (1981). Minimum distance estimation: A bibliography. *Comm. Stat.—Theory Method* **A20**(12):1205–1224.

Parzen, E. (1960). *Modern probability theory and its applications.* Wiley, New York.

Parzen, E. (1962). On estimation of a probability density function and mode. *Ann. Math. Stat.* **33**:1065–1076.

Pirie, W. R. (1974). Comparing rank tests for ordered alternatives in randomized blocks. *Ann. Stat.* **2**:374–382.

Pitman, E. J. G. (1948). Notes on nonparametric statistical inference. Unpublished notes.

Policello, G. E., II and T. P. Hettmansperger (1976). Adaptive robust procedures for the one-sample location model. *J. Am. Stat. Assoc.* **71**:624–633.

Pratt, J. W. (1959). Remarks on zeros and ties in the Wilcoxon signed rank procedures. *J. Am. Stat. Assoc.* **54**:655–667.

Puri, M. L. and P. K. Sen (1968). On Chernoff–Savage tests for ordered alternatives in randomized blocks. *Ann. Math. Stat.* **39**:967–972.

Puri, M. L. and P. K. Sen (1971). *Nonparametric methods in multivariate analysis.* Wiley, New York.

Putter, J. (1955). The treatment of ties in some nonparametric tests. *Ann. Math. Stat.* **26**:368–386.

Randles, R. H. and R. V. Hogg (1973). Adaptive distribution free tests. *Comm. Stat.* **2**(4):337–356.

Randles, R. H. and D. A. Wolfe (1979). *Introduction to the theory of nonparametric statistics.* Wiley, New York.

Rieder, H. (1981). Robustness of one- and two-sample rank tests against gross errors. *Ann. Stat.* **9**:245–265.

Rieder H. (1982). Qualitative robustness of rank tests. *Ann. Stat.* **10**:205–211.

Rosenzweig, M., Bennett, E. L., and M. C. Diamond (1972). Brain changes in response to experience. *Sci. Am.*, **234**:22–24.

Royden, H. L. (1968). *Real analysis.* 2nd ed. Macmillan, New York.

Rosenblatt, M. (1956). Remarks on some nonparametric estimates of a density function. *Ann. Math. Stat.* **27**:832–837.

Ryan, T. A., Jr., Joiner, B. L., and B. F. Ryan (1976). *Minitab student handbook.* Minitab Project Inc., University Park, Pa.

Ryan, T. A., Jr., Joiner, B. L., and B. F. Ryan (1982). *Minitab reference manual.* Minitab Project Inc., University Park, Pa.

Sacks, J. and D. Ylvisaker (1982). L- and R-estimates and the minimax property. *Ann. Stat.* **10**:643–645.

Salk, L. (1973). The role of the heartbeat in the relations between mother and infant. *Sci. Am.* **235**:26–29.

Savage, I. R. (1956). Contributions to the theory of rank order statistics—the two sample case. *Ann. Math. Stat.* **27**:590–615.

Scheffé, H. (1959). *The analysis of variance.* Wiley, New York.

Schrader, R. M. and T. P. Hettmansperger (1980). Robust analysis of variance based upon a likelihood ratio criterion. *Biometrika* **67**:93–101.

Schrader, R. M. and J. W. McKean (1977). Robust analysis of variance. *Comm. Stat.—Theory Method* **46**(9):879–894.

Schucany, W. R. and W. H. Frawley (1973). A rank test for two group concordance. *Pschometrika* **38**:249–258.

Schuster, E. (1974). On the rate of convergence of an estimate of a functional of a probability density. *Scand. Actuarial J.* **1**:103–107.

Schweder, T. (1975). Window estimation of the asymptotic variance of rank estimators of location. *Scand. J. Stat.* **2**:113–126.

Sen, P. K. (1966). On a distribution-free method of estimating asymptotic efficiency of a class of nonparametric tests. *Ann. Math. Stat.* **37**:1759–1770.

Sen, P. K. (1968). Estimates of the regression coefficient based on Kendall's tau. *J. Am. Stat. Assoc.* **63**:1379–1389.

Sen, P. K. and M. L. Puri (1977). Asymptotically distribution-free aligned rank order tests for composite hypotheses for general multivariate linear models. *Z. Wahrscheinlichkeitstheorie und verw. Geb.* **39**:175–186.

Serfling, R. J. (1968a). Contributions to central limit theory for dependent variables. *Ann. Math. Stat.* **39**:1158–1175.

Serfling, R. J. (1968b). The Wilcoxon two-sample statistic on strongly mixing processes. *Ann. Math. Stat.* **39**:1202–1209.

Serfling, R. J. (1980). *Approximation theorems of mathematical statistics.* Wiley, New York.

Sievers, G. L. and J. W. McKean (1983). On the robust analysis of linear models with nonsymmetric error distributions. Unpublished manuscript, Dept. of Math., Western Michigan University, Kalamazoo, Mich.

Skillings, J. H. and D. A. Wolfe (1978). Distribution-free tests for orderd alternatives in a randomized block design. *J. Am. Stat. Assoc.* **73**:427–431.

Spearman, C. (1904). The proof and measurement of association between two things. *Am. J. Psychol.* **15**:72–101.

Terkel, J. and J. S. Rosenblatt (1968). Maternal behavior indced by maternal blood plasma injected into virgin rats. *J. Comp. and Physio. Psych.* **65**:479–482.

Terpstra, T. J. (1952). The asymptotic normality and consistency of Kendall's test against trend, when ties are present in one ranking. *Indag. Math.* **14**:327–333.

Terry, M. E. (1952). Some rank order tests which are most powerful against specific parametric alternatives. *Ann. Math. Stat.* **23**:364–366.

Todd, J. T., Mark, L. S., Shaw, R. E., and J. B. Pittenger (1980). The perception of growth. *Sci. Am.* **242**(Feb.), 132–145.

Tryon, P. V. and T. P. Hettmansperger (1974). A class of nonparametric tests for homogeneity against ordered alternatives. *Ann. Stat.* **1**:1061–1070.

Tukey, J. W. (1949). The simplest signed rank tests. Princeton University Stat. Res. Group, Memo Report No. 17.

Tukey, J. W. (1960). A survey of sampling from contaminated distributions, contributions to probability and statistics. I. Olkin, ed. Stanford University Press, Stanford, CA.

Utts, J. M. and T. P. Hettmansperger (1980). A robust class of tests and estimates for multivariate location. *J. Am. Stat. Assoc.* **75**:939–946.

van Eeden, C. (1972). An analogue, for signed rank statistics, of Jureckova's asymptotic linearity theorem for rank statistics. *Ann. Math. Stat.* **43**:791–802.

van Elteren, P. and G. E. Noether (1959). The asymptotic efficiency of the χ^2-test for a balanced incomplete block design. *Biometrika* **46**:475–477.

van der Waerden, B. L. (1952/1953). Order tests for the two sample problem and their power, I, II, III. *Indag. Math.* **14**:453–458; *Indag. Math.* **15**:303–310, 311–316; Correction: *Indag. Math.* **15**:80.

van Zwet, W. R. (1970). Convex tranformations of random variables. Mathematical Centre Tracts No. 7, Mathematesch Centrum Amsterdam.

Wegman, E. J. (1972). Nonparametric probability density estimation: I. A summary of available methods. *Technometrics* **14**:533–546.

Welch, B. L. (1937). The significance of the difference between two means when the population variances are unequal. *Biometrika* **29**:350–362.

Wilcoxon, F. (1945). Individual comparisons by ranking methods. *Biometrics* **1**:80–83.

Wilks, S. S. (1962). *Mathematical statistics*. Wiley, New York.

Witting, H. (1960). A generalized Pitman efficiency for nonparametric tests. *Ann. Math. Stat.* **31**:405–416.

Ylvisaker, D. (1977). Test resistance. *J. Am. Stat. Assoc.* **72**:551–556.

Index

319